Ship Construction

Ship Construction

Seventh edition

D. J. Eyres

M.Sc., F.R.I.N.A.

Formerly Lecturer in Naval Architecture, Department of Maritime Studies, Plymouth Polytechnic, (now University of Plymouth)

G. J. Bruce

M.B.A, F.R.I.N.A., MSNAME.

Formerly Professor of Shiprepair and Conversion, School of Marine Science and Technology, Newcastle University

AMSTERDAM • BOSTON • HEIDELBERG • LONDON • NEW YORK • OXFORD
PARIS • SAN DIEGO • SAN FRANCISCO • SINGAPORE • SYDNEY • TOKYO
Butterworth-Heinemann is an imprint of Elsevier

Butterworth-Heinemann is an imprint of Elsevier
The Boulevard, Langford Lane, Kidlington, Oxford, OX5 1GB
225 Wyman Street, Waltham, MA 02451, USA

First published 1971
Second edition 1978
Third edition 1988
Fourth edition 1994
Fifth edition 2001
Sixth edition 2007
Seventh Edition 2012

British Library Cataloguing in Publication Data
A catalogue record for this book is available from the British Library

Library of Congress Number: 2012936092

ISBN: 978-0-08-097239-8

Printed and bound in the United States

12 13 14 15 10 9 8 7 6 5 4 3 2 1

Contents

Preface vii
Acknowledgments ix

Part One Introduction to Shipbuilding 1

1 Basic design of the ship 3
2 Ship dimensions, form, size, or category 11
3 Development of ship types 17

Part Two Materials and Strength of Ships 35

4 Classification societies 37
5 Steels 45
6 Other shipbuilding materials 53
7 Testing of materials 61
8 Stresses to which a ship is subject 67

Part Three Welding and Cutting 79

9 Welding and cutting processes used in shipbuilding 81
10 Welding practice and testing welds 103

Part Four Shipyard Practice 117

11 Shipyard layout 119
12 Design information for production 125
13 Plate and section preparation and machining 135
14 Assembly of ship structure 147
15 Launching 161

Part Five Ship Structure 173

16 Bottom structure 175
17 Shell plating and framing 189

18 Bulkheads and pillars 207
19 Decks, hatches, and superstructures 225
20 Fore end structure 241
21 Aft end structure 249
22 Tanker construction 265
23 Liquefied gas carriers 279

Part Six Outfit 291

24 Cargo lifting arrangements 293
25 Cargo access, handling, and restraint 307
26 Pumping and piping arrangements 315
27 Corrosion control and antifouling systems 327
28 Heating, ventilation, air-conditioning, refrigeration, and insulation 345

Part Seven International Regulations 353

29 International Maritime Organization 355
30 Tonnage 359
31 Load Line Rules 363
32 Structural fire protection 371

Index 377

Preface

This text is designed as an introductory text for students of marine sciences and technology, including those following BTEC National and Higher National programs in preparation for careers at sea and in marine related industries. The subject matter is presented in sufficient depth to be of help to more advanced students on undergraduate programs in Marine Technology and Naval Architecture, as well as those preparing for the Extra Master examination. Students converting from other disciplines for higher degrees will also find the information useful. Other students following professional courses in shipbuilding will also find the book useful as background reading.

Many professionals from other disciplines, including law, insurance, accounting, and logistics joining the businesses will find the basic technical information on ship construction of value.

Considerable changes have occurred in ship design and shipbuilding practice with the introduction of new technology, and this book attempts to present current shipyard techniques without neglecting basic principles. Shipbuilding covers a wide field of crafts and, with new developments occurring regularly, it would be difficult to cover every aspect fully within the scope of a single textbook. For this reason further reading lists are given at the end of most chapters, these being selected from books, transactions, and periodicals that are likely to be found in the libraries of universities and other technical institutions.

In this edition the authors have also added a listing of some useful websites at the end of most chapters relating to the subject matter of the chapter. Those listed contain further information, drawings, and photographs that complement the text and/or add further knowledge to the subject. Some of the websites that are referenced also deal with regulations that apply to ships and their construction. The rapid development of available information makes it impossible to provide a completely up-to-date set of websites. Therefore, there is space for students to add further websites recommended by their tutors or that they may have found informative. However, it is important to consider the sources of information on any new sites to confirm their currency and validity.

Preface

Acknowledgments

The authors are grateful to the following firms and organizations who were kind enough to provide information and drawings from which material for the book was extracted:

Appledore Shipbuilders Ltd
Blohm and Voss, A.G.
British Maritime Technology
British Oxygen Co. Ltd
E.I. Du Pont De Nemours & Co. Ltd
ESAB AB
Irish Shipping Ltd
MacGregor-Navire International A.B.
Mitsubishi Heavy Industries Ltd
Ocean Steamship Co. Ltd
Shell Tankers (UK) Ltd
Shipping Research Services A/S
Hugh Smith (Glasgow) Ltd
Stone Manganese Marine Ltd
Wavemaster International

Lloyd's Register of Shipping also gave permission to refer to various requirements of their 'Rules and Regulations for the Classification of Ships'.

D.J.E. and G.J.B.

Part One

Introduction to Shipbuilding

1 Basic design of the ship

Chapter Outline

Preparation of the design 3
Information provided by design 4
Purchase of a new vessel 6
Ship contracts 7
Further reading 8
Some useful websites 9

The key requirement of a new ship is that it can trade profitably, so economics is of prime importance in designing a merchant ship. An owner requires a ship that will give the best possible returns for the owner's initial investment and running costs. The final design should be arrived at taking into account not only present economic considerations, but also those likely to develop within the life of the ship. This is especially the case for some trades, for example LNG, where the ship is expected to work the same route for its working life. Design for operation is the result. For other ships, including bulk carriers, the first cost of the ship is the major factor for the owner and the ship may be designed for ease of production. Resale value is also often a major consideration, leading to design for maintenance.

With the aid of computers it is possible to make a study of a large number of varying design parameters and to arrive at a ship design that is not only technically feasible but, more importantly, is the most economically efficient. Ideally the design will take into consideration first cost, operating cost, and future maintenance.

Preparation of the design

The initial design of a ship generally proceeds through three stages: concept; preliminary; and contract design. The process of initial design is often illustrated by the design spiral (Figure 1.1), which indicates that given the objectives of the design, the designer works towards the best solution adjusting and balancing the interrelated parameters as the designer goes.

A concept design should, from the objectives, provide sufficient information for a basic techno-economic assessment of the alternatives to be made. Economic criteria that may be derived for commercial ship designs and used to measure their profitability are net present value, discounted cash flow, or required freight rate.

Ship Construction. DOI: 10.1016/B978-0-08-097239-8.00001-5

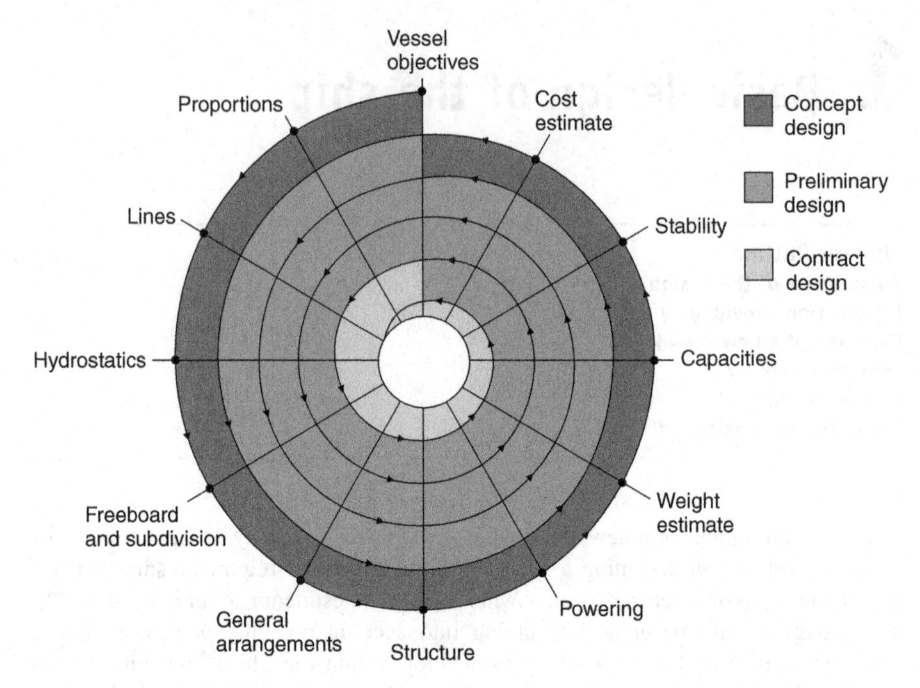

Figure 1.1 Design spiral.

Preliminary design refines and analyzes the agreed concept design, fills out the arrangements and structure, and aims to optimize service performance. At this stage the builder should have sufficient information to tender. Contract design details the final arrangements and systems agreed with the owner and satisfies the building contract conditions.

The design of the ship is not complete at this stage, rather for the major effort in resources it has only just started. Post-contract design requires confirmation that the ship will meet all operational requirements, including safety requirements from regulators. It also entails in particular design for production where the structure, outfit, and systems are planned in detail to achieve a cost- and time-effective building cycle. Production of the ship must also be given consideration in the earlier design stages, particularly where it places constraints on the design or can affect costs. The post-contract design will also ideally consider the future maintainability of the ship in the arrangement of equipment and services.

Information provided by design

When the preliminary design has been selected the following information is available:

- Dimensions
- Displacement
- Stability

- Propulsive characteristics and hull form
- Preliminary general arrangement
- Principal structural details.

Each item of information may be considered in more detail, together with any restraints placed on these items by the ship's service or other factors outside the designer's control.

1. The dimensions of most ships are primarily influenced by the cargo-carrying capacity of the vessel. In the case of the passenger vessel, dimensions are influenced by the height and length of superstructure containing the accommodation. Length, where not specified as a maximum, should be a minimum consistent with the required speed and hull form. Increase of length produces higher longitudinal bending stresses requiring additional strengthening and a greater displacement for the same cargo weight. Breadth may be such as to provide adequate transverse stability. A minimum depth is controlled by the draft plus statutory freeboard, but an increase in depth will result in a reduction of the longitudinal bending stresses, providing an increase in strength, or allowing a reduction in scantlings (i.e. plate thickness/size of stiffening members etc.). Increased depth is therefore preferred to increased length. Draft is often limited by area of operation, but if it can be increased to give a greater depth this can be an advantage.

Many vessels are required to make passages through various canals and straits and pass under bridges within enclosed waters, and this will place a limitation on their dimensions. For example, locks in the Panama Canal and St Lawrence Seaway limit length, breadth, and draft. At the time of writing, the Malacca Straits main shipping channel is about 25 meters deep and the Suez Canal could accommodate ships with a beam of up to 75 meters and maximum draft of 16 metres. A maximum air draft on container ships of around 40 meters is very close to clear the heights of the Gerard Desmond Bridge, Long Beach, California and Bayonne Bridge, New York. Newer bridges over the Suez Canal at 65 meters and over the Bosporus at 62 meters provide greater clearance.

2. Displacement is made up of lightweight plus deadweight. The lightweight is the weight of vessel as built and ready for sea. Deadweight is the difference between the lightweight and loaded displacement, i.e. it is the weight of cargo plus weights of fuel, stores, water ballast, fresh water, crew and passengers, and baggage. When carrying high-density cargoes (e.g. ore) it is desirable to keep the lightweight as small as possible, consistent with adequate strength. Since only cargo weight of the total deadweight is earning capital, other items should be kept to a minimum as long as the vessel fulfills its commitments.

3. In determining the dimensions, statical stability is kept in mind in order to ensure that this is sufficient in all possible conditions of loading. Beam and depth are the main influences. Statutory freeboard and sheer are important together with the weight distribution in arranging the vessel's layout.

4. Adequate propulsive performance will ensure that the vessel attains the required speeds. The hull form is such that economically it offers a minimum resistance to motion so that a minimum power with economically lightest machinery is installed without losing the specified cargo capacity.

A service speed is the average speed at sea with normal service power and loading under average weather conditions. A trial speed is the average speed obtained using the maximum power over a measured course in calm weather with a clean hull and specified load condition. This speed may be a knot or so more than the service speed.

Unless a hull form similar to that of a known performance vessel is used, a computer-generated hull form and its predicted propulsive performance can be determined. The propulsive performance can be confirmed by subsequent tank testing of a model hull, which may suggest further beneficial modifications.

The owner may specify the type and make of main propulsion machinery installation with which their operating personnel are familiar.

5. The *general arrangement* is prepared in cooperation with the owner, allowing for standards of accommodation particular to that company, also specific cargo and stowage requirements. Efficient working of the vessel must be kept in mind throughout and compliance with the regulations of the various authorities involved on trade routes must also be taken into account. Some consultation with shipboard employees' representative organizations may also be necessary in the final accommodation arrangements.

6. Almost all vessels will be built to the requirements of a classification society such as Lloyd's Register. The standard of classification specified will determine the structural scantlings and these will be taken out by the shipbuilder. The determination of the minimum hull structural scantlings can be carried out by means of computer programs made available to the shipyard by the classification society. Owners may specify thicknesses and material requirements in excess of those required by the classification societies and special structural features peculiar to the trade or owner's fleet may be asked for.

Purchase of a new vessel

In recent years the practice of owners commissioning 'one-off' designs for cargo ships from consultant naval architects, shipyards, or their own technical staff has increasingly given way to the selection of an appropriate 'stock design' to suit their particular needs. To determine which stock design, the shipowner must undertake a detailed project analysis involving consideration of the proposed market, route, port facilities, competition, political and labor factors, and cash flow projections. Also taken into account will be the choice of shipbuilder, where relevant factors such as the provision of government subsidies or grants or supplier credit can be important as well as the price, date of delivery, and the yard's reputation. Most stock designs offer some features that can be modified, such as outfit, cargo handling equipment, or alternate manufacture of main engine, for which the owner will have to pay extra.

Purchase of a passenger vessel will still follow earlier procedures for a 'one-off' design, but there are shipyards concentrating on this type of construction and the owner may be drawn to them for this reason. A nonstandard cargo ship of any form and a number of specialist ships will also require a 'one-off' design. Having decided on the basic ship requirements, based on the intended trade, after an appropriate project analysis the larger shipowners may employ their own technical staff to prepare the tender specification and submit this to shipbuilders who wish to tender for the building of the ship. The final building specification and design is prepared by the successful tendering shipbuilder in cooperation with the owner's technical staff. The latter may oversee construction of the vessel and approve the builder's drawings and calculations. Other shipowners may retain a firm of consultants or approach a firm who may assist with preliminary design studies and will prepare the tender specifications and in some cases call tenders on behalf of the owner. Often the consultants will also assist the owners in evaluating the tenders and oversee the construction on their behalf.

Ship contracts

The successful tendering shipbuilder will prepare a building specification for approval by the owner or the owner's representative that will form an integral part of the contract between the two parties and thus have legal status. This technical specification will normally include the following information:

- Brief description and essential qualities and characteristics of the ship
- Principal dimensions
- Deadweight, cargo and tank capacities, etc.
- Speed and power requirements
- Stability requirements
- Quality and standard of workmanship
- Survey and certificates
- Accommodation details
- Trial conditions
- Equipment and fittings
- Machinery details, including the electrical installation, will normally be produced as a separate section of the specification.

Most shipbuilding contracts are based on one of a number of standard forms of contract that have been established to obtain some uniformity in the contract relationship between builders and purchasers. There are a number of 'standard' contract forms, all very similar in structure and content. Four of the most common standard forms of contract have been established by:

1. CESA—Community of European Shipyards Associations
2. MARAD Maritime Administration, USA
3. SAJ—Shipbuilders Association of Japan
4. Norwegian Shipbuilding Contract—Norwegian Shipbuilders Association and Norwegian Shipowners Association.

The CESA standard form of contract was developed by the predecessor organization, the Association of Western European Shipyards (AWES).The contract form can be downloaded from the CESA website. The sections of the contract are:

1. Subject of contract (vessel details, etc.)
2. Inspection and approval

3. Modifications
4. Trials
5. Guarantee (speed, capacity, fuel consumption)
6. Delivery of vessel
7. Price
8. Property (rights to specifications, plans, etc. and to vessel during construction and on delivery)
9. Insurance
10. Default by the purchaser
11. Default by the contractor
12. Guarantee (after delivery)
13. Contract expenses
14. Patents
15. Interpretation, reference to expert and arbitration
16. Condition for the contract to become effective
17. Legal domicile (of purchaser and contractor)
18. Assignment (transfer of rights)
19. Limitation of liability
20. Addresses for correspondence.

Irrespective of the source of the owner's funds for purchasing the ship, payment to the shipbuilder is usually made as progress payments that are stipulated in the contract under item 7 above. A typical payment schedule may have been five equal payments spread over the contract period, but in recent years payment arrangements advantageous to the purchaser and intended to attract buyers to the shipyard have delayed a higher percentage of payment until delivery of the ship. The payment schedule may be as follows:

- 10% on signing contract
- 10% on arrival of materials on site
- 10% on keel laying
- 20% on launching
- 50% on delivery.

Because many cargo ships are of a standard design, and built in series, and modification can be very disruptive to the shipyard building program, item 3 in the standard form of contract where modifications are called for at a late date by the owner can have a dramatic effect on costs and delivery date given the detail now introduced at an early stage of the fabrication process. Many shipyards will refuse to accept modifications once a design is agreed and detailed work and purchasing commences. Item 3 also covers the costs and delays of compulsory modifications resulting from amendment of laws, rules, and regulations of the flag state and classification society.

Further reading

Rawson, Tupper: *Basic Ship Theory.* ed 5, vol 2. Chapter 15: Ship design, 2001, Butterworth Heinemann.

Watson DGM: *Practical Ship Design*, 2002, Elsevier.

Some useful websites

www.cesa.eu Community of European Shipyards Associations.
www.sajn.or.jp/e Shipbuilders Association of Japan; provides links to member shipyard sites.

2 Ship dimensions, form, size, or category

Chapter Outline

Oil tankers 13
Bulk carriers 13
Container ships 15
IMO oil tanker categories 15
Panama canal limits 15
Suez canal limits 16
Some useful websites 16

The hull form of a ship may be defined by a number of dimensions and terms that are often referred to during and after building the vessel. An explanation of the principal terms is given below:

After Perpendicular (AP): A perpendicular drawn to the waterline at the point where the after side of the rudder post meets the summer load line. Where no rudder post is fitted it is taken as the center line of the rudder stock.

Forward Perpendicular (FP): A perpendicular drawn to the waterline at the point where the fore-side of the stem meets the summer load line.

Length Between Perpendiculars (LBP): The length between the forward and aft perpendiculars measured along the summer load line.

Amidships: A point midway between the after and forward perpendiculars.

Length Overall (LOA): Length of vessel taken over all extremities.

Lloyd's Length: Used for obtaining scantlings if the vessel is classed with Lloyd's Register. It is the same as length between perpendiculars except that it must not be less than 96% and need not be more than 97% of the extreme length on the summer load line. If the ship has an unusual stem or stern arrangement the length is given special consideration.

Register Length: The length of ship measured from the fore-side of the head of the stem to the aft side of the head of the stern post or, in the case of a ship not having a stern post, to the fore-side of the rudder stock. If the ship does not have a stern post or a rudder stock, the after terminal is taken to the aftermost part of the transom or stern of the ship. This length is the official length in the register of ships maintained by the flag state and appears on official documents relating to ownership and other matters concerning the business of the ship. Another important length measurement is what might be referred to as the *IMO Length*. This length is found in various international conventions such as the Load Line, Tonnage, SOLAS and MARPOL conventions, and determines the application of requirements of those conventions to a ship. It is defined as 96% of the total length on a waterline at 85% of

Ship Construction. DOI: 10.1016/B978-0-08-097239-8.00002-7

the least molded depth measured from the top of keel, or the length from the fore-side of stem to the axis of rudder stock on that waterline, if that is greater. In ships designed with a rake of keel the waterline on which this length is measured is taken parallel to the design waterline.

Molded dimensions are often referred to; these are taken to the inside of plating on a metal ship.

Base Line: A horizontal line drawn at the top of the keel plate. All vertical molded dimensions are measured relative to this line.
Molded Beam: Measured at the midship section, this is the maximum molded breadth of the ship.
Molded Draft: Measured from the base line to the summer load line at the midship section.
Molded Depth: Measured from the base line to the heel of the upper deck beam at the ship's side amidships.
Extreme Beam: The maximum beam taken over all extremities.
Extreme Draft: Taken from the lowest point of keel to the summer load line. Draft marks represent extreme drafts.
Extreme Depth: Depth of vessel at ship's side from upper deck to lowest point of keel.
Half Breadth: Since a ship's hull is symmetrical about the longitudinal centre line, often only the half beam or half breadth at any section is given.
Freeboard: The vertical distance measured at the ship's side between the summer load line (or service draft) and the freeboard deck. The freeboard deck is normally the uppermost complete deck exposed to weather and sea that has permanent means of closing all openings, and below which all openings in the ship's side have watertight closings.
Sheer: A rise in the height of the deck (curvature or in a straight line) in the longitudinal direction. Measured as the height of deck at side at any point above the height of deck at side amidships.
Camber (or Round of Beam): Curvature of decks in the transverse direction. Measured as the height of deck at center above the height of deck at side. Straight line camber is used on many large ships to simplify construction.
Rise of Floor (or Deadrise): The rise of the bottom shell plating line above the base line. This rise is measured at the line of molded beam. Large cargo ships often have no rise of floor.
Half Siding of Keel: The horizontal flat portion of the bottom shell measured to port or starboard of the ship's longitudinal center line. This is a useful dimension to know when dry-docking.
Tumblehome: The inward curvature of the side shell above the summer load line. This is unusual on modern ships.
Flare: The outward curvature of the side shell above the waterline. It promotes dryness and is therefore associated with the fore end of ship.
Stem Rake: Inclination of the stem line from the vertical.
Keel Rake: Inclination of the keel line from the horizontal. Trawlers and tugs often have keels raked aft to give greater depth aft where the propeller diameter is proportionately larger in this type of vessel. Small craft occasionally have forward rake of keel to bring propellers above the line of keel.
Tween Deck Height: Vertical distance between adjacent decks measured from the tops of deck beams at ship's side.
Parallel Middle Body: The length over which the midship section remains constant in area and shape.

Entrance: The immersed body of the vessel forward of the parallel middle body.
Run: The immersed body of the vessel aft of the parallel middle body.
Tonnage: This is often referred to when the size of the vessel is discussed, and the gross tonnage is quoted from Lloyd's Register. Tonnage is a measure of the enclosed internal volume of the vessel (originally computed as 100 cubic feet per ton). This is dealt with in detail in Chapter 30.
Deadweight: This is defined in Chapter 1. It should be noted that for tankers deadweight is often quoted in 'long tons' rather than 'metric tons (tonnes)'; however, MARPOL regulations for oil tankers are in metric tons.

The principal dimensions of the ship are illustrated in Figure 2.1.

TEU and FEU: Indicate the cargo-carrying capacity of container ships. TEU (twenty-foot equivalent unit) indicates the number of standard shipping containers that may be carried on some shipping routes; container ships may carry standard containers that are 40 feet in length. FEU is forty-foot equivalent unit.

An indication of the size by capacity of oil tankers, bulk carriers, and container ships is often given by the following types:

Oil tankers

- ULCC (Ultra-Large Crude Carrier) is a tanker usually between 300,000 and 550,000 tonnes deadweight.
- VLCC (Very Large Crude Carrier) is a tanker usually between 200,000 tonnes and 300,000 tonnes deadweight.
- Suezmax indicates the largest oil tanker that can transit the current Suez Canal fully laden, being about 150,000 tonnes deadweight.
- Aframax the standard designation of smaller crude oil tankers, being the largest tanker size in the AFRA Freight Rate Assessment Scale Large One Category. AFRA stands for 'American Freight Rate Association'. Variously reported as being 80,000 to 115,000 tones deadweight.
- Panamax is the maximum size of oil tanker, with beam restriction of 32.2 meters and length restriction of 275 meters, that can transit the Panama Canal prior to completion of the planned new locks. Typical size is about 55,000–70,000 tonnes deadweight.
- Handysize/Handymax are typical product tankers of about 35,000–45,000 tonnes deadweight.

Bulk carriers

- Capesize ships that are too large to transit the current Panama Canal and therefore voyage around Cape Horn. All bulk carriers above 80,000 tonnes deadweight fall into this category. Most are up to 170,000 tonnes deadweight but a small number are larger for specific trade routes, the biggest being 365,000 tonnes deadweight.
- Panamax—As for oil tankers.
- Handymax ships are between around 35,000 and 60,000 tonnes deadweight.
- Ships between 10,000 and 35,000 tonnes deadweight have formed the majority of the fleet for many years and are designated 'Handysize'. In recent years the size of these ships has been increasing and the term 'Handymax' has been applied to designate the larger bulk carriers.

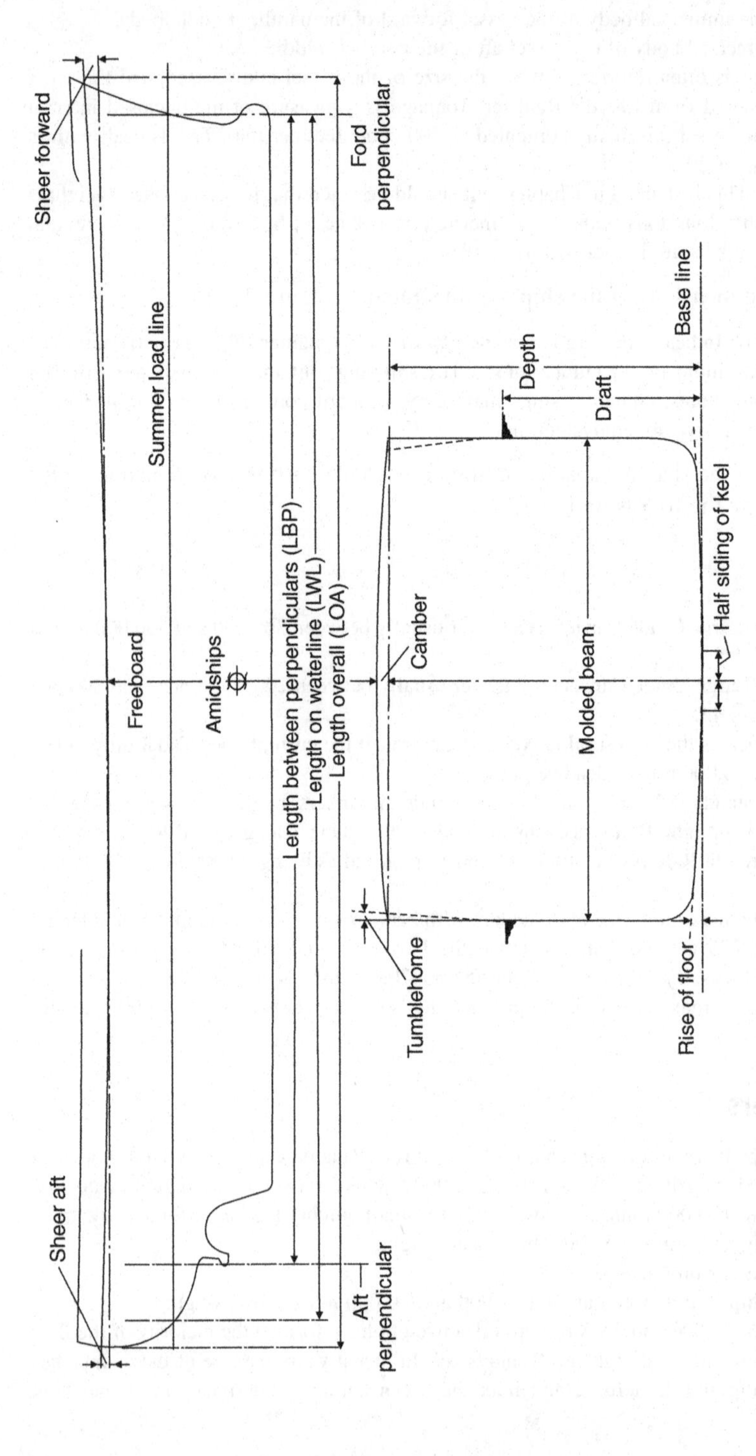

Figure 2.1 Principal ship dimensions.

Container ships

- Ultra-large container ships. Ships with a capacity of over 14,000 TEU. Few have been built to date. These ships are too large for any canals.
- Post-Panamax ships are too large to transit the current Panama Canal and undertake trans-ocean voyages. Their size is typically 5500–8000 TEU though larger ships with over 10,000 TEU capacity have been built.
- New Panamax ships (including most Post-Panamax ships) would be able to transit the expanded Panama Canal. They may carry up to around 12,000 TEU.
- Panamax ships that can transit the current Panama Canal carry between 3000 and 5000 TEU.
- Feeder ships are smaller vessels that do not undertake oceanic voyages but are generally engaged in shipping containers. The smallest of these may only carry several hundred TEU. There is no specific subclass below Panamax size.

IMO oil tanker categories

- Category 1 (commonly known as Pre-MARPOL tankers) includes oil tankers of 20,000 tonnes deadweight and above carrying crude oil, fuel oil, heavy diesel oil, or lubricating oil as cargo, and of 30,000 tonnes deadweight and above carrying other oils, which do not comply with the requirements for protectively located segregated ballast tanks. These ships have been phased out under IMO regulations.
- Category 2 (commonly known as MARPOL tankers) includes oil tankers of 20,000 tonnes deadweight and above carrying crude oil, fuel oil, or lubricating oil as cargo, and of 30,000 tonnes deadweight and above carrying other oils, which do comply with the protectively located segregated ballast tank requirements. These ships are due to be phased out.
- Category 3 includes oil tankers of 5000 tonnes deadweight and above but less than the tonnes deadweight specified for Categories 1 and 2. Also due to be phased out.

Note: For tankers carrying HGO (heavy gas oil) the lower limits for Categories 2 and 3 fall to 600 tonnes deadweight.

Panama canal limits

These are set by lock sizes. Current locks are 'Panamax'. New locks will be larger for 'New Panamax' ships (see Table 2.1).

Table 2.1 Panama Canal limits

	Panamax ships	New Panamax ships
Length (m)	294.13	366
Breadth (m)	32.81	49
Draft (m)	12.04	15.2

Suez canal limits

There are no locks and ship size is limited by the canal dimensions. There is a maximum breadth limit of 75 meters. With no locks the ship length is also unrestricted. The maximum draft is 20 meters.

The Saint Lawrence Seaway links the North American Great Lakes to the Atlantic. The limits for ships based on the locks are length 226 m, breadth 24 m, and draft 7.92 m.

Some useful websites

www.pancanal.com/eng/general For details of Panama Canal.
http://www.suezcanal.gov.eg
http://www.greatlakes-seaway.com

3 Development of ship types

Chapter Outline

Dry cargo ships 17
- Container ships 21
- Barge-carrying ships 21
- Ro-ro ships 21
- Hull form 23
- Cargo handling equipment 23

Bulk carriers 23
Car carriers 26
Oil tankers 26
Passenger ships 30
Further reading 33

A breakdown into broad working groups of the various types that the shipbuilder or ship designer might be concerned with are shown in Figure 3.1. This covers a wide range and reflects the adaptability of the shipbuilding industry. It is obviously not possible to cover the construction of all those types in a single volume. The development of the vessels with which the text is primarily concerned, namely dry cargo ships (including container ships and dry bulk carriers), tankers (oil, liquid gas and chemical) and passenger ships, follows.

Dry cargo ships

If the development of the dry cargo ship from the time of introduction of steam propulsion is considered, the pattern of change is similar to that shown in Figure 3.2. The first steam ships followed in most respects the design of the sailing ship, having a flush deck with the machinery openings protected only by low coamings and glass skylights. At quite an early stage it was decided to protect the machinery openings with an enclosed bridge structure. Erections forming a forecastle and poop were also introduced at the forward and after respectively for protection. This resulted in what is popularly known as the 'three island type'. A number of designs at that time also combined bridge and poop, and a few combined bridge and forecastle, so that a single well was formed.

Another form of erection introduced was the raised quarter deck. Raised quarter decks were often associated with smaller deadweight carrying vessels, e.g. colliers.

Ship Construction. DOI: 10.1016/B978-0-08-097239-8.00003-9

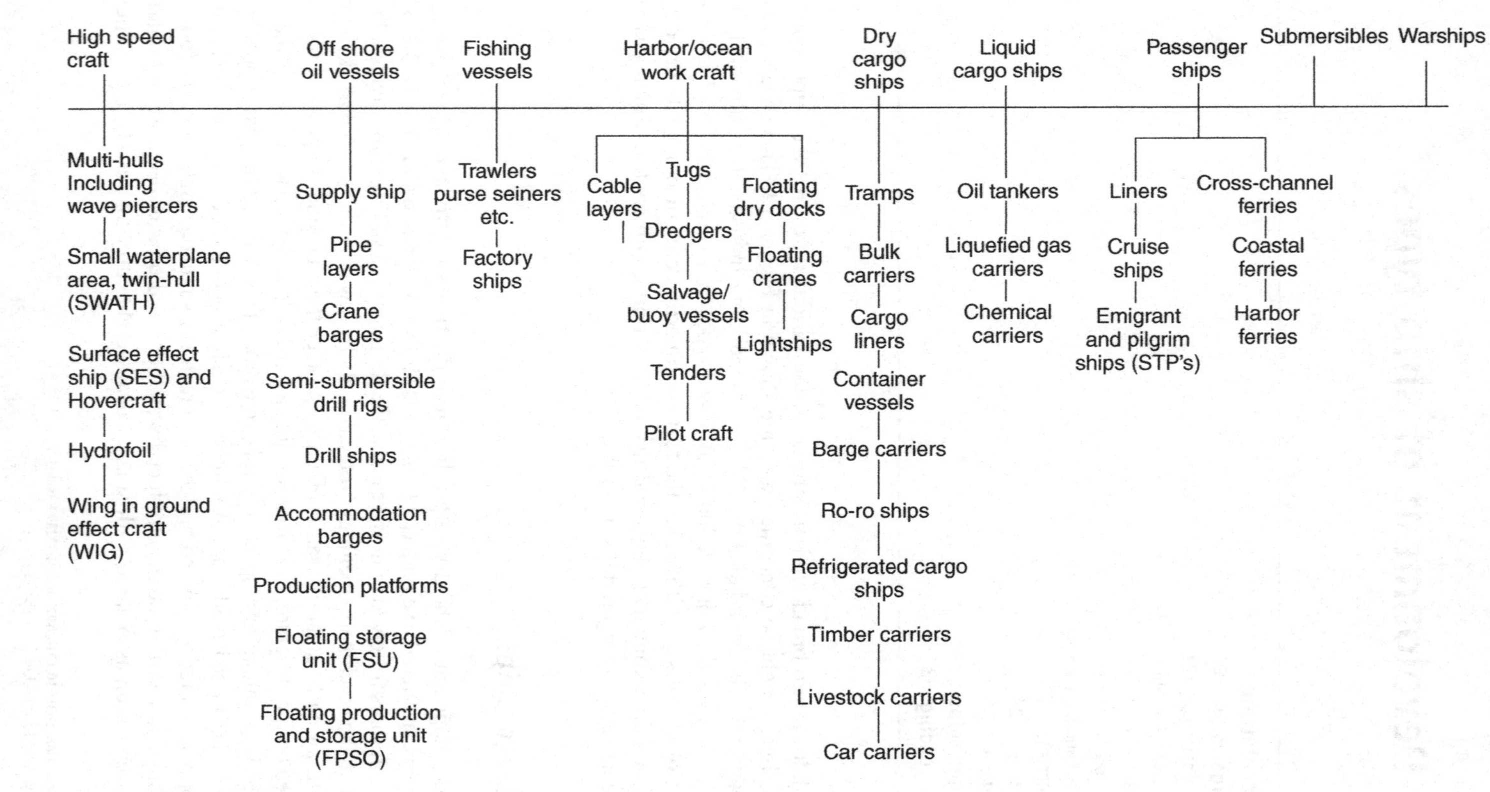

Figure 3.1 Ship types.

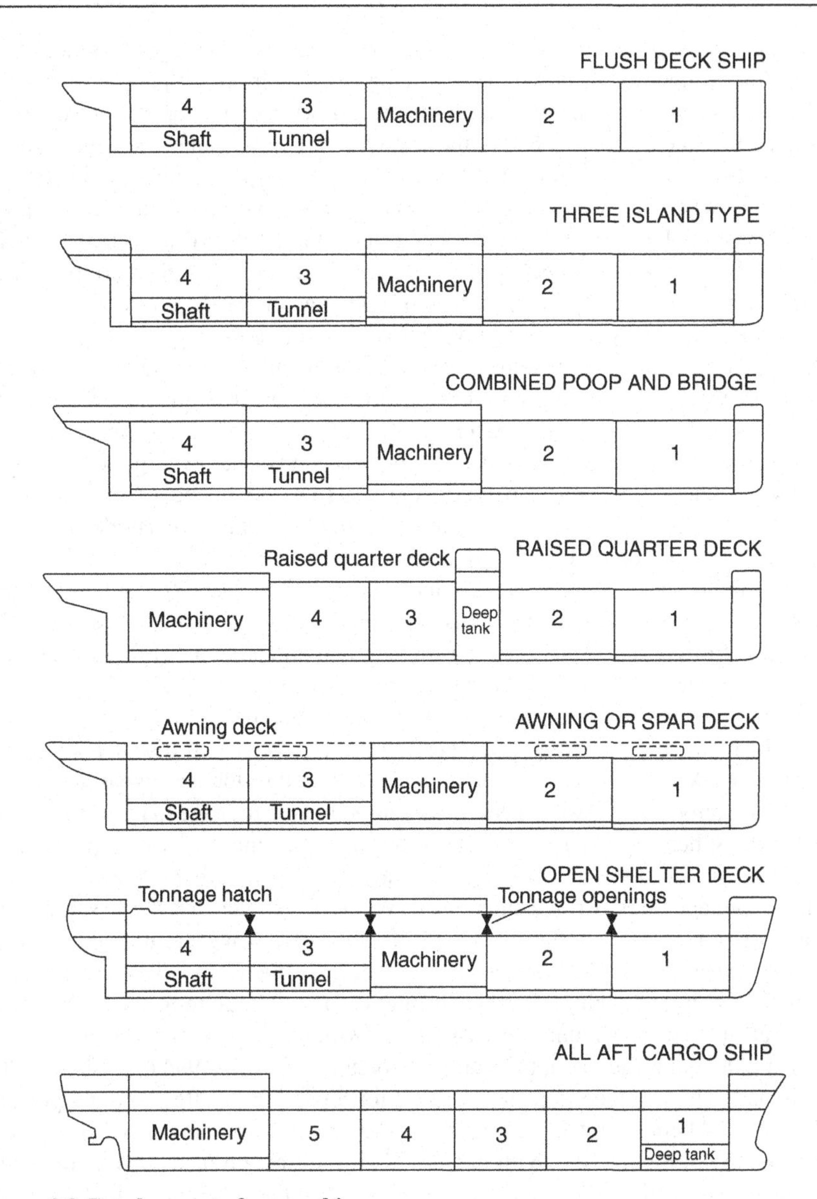

Figure 3.2 Development of cargo ship.

With the machinery space aft, which is proportionately large in a small vessel, there is a tendency for the vessel to trim by the bow when fully loaded. By fitting a raised quarter deck in way of the after holds this tendency was eliminated. A raised quarter deck does not have the full height of a tween deck, above the upper deck.

Further departures from the 'three island type' were brought about by the carriage of cargo and cattle on deck, and the designs included a light covering built over the wells for the protection of these cargoes. This resulted in the awning or spar deck type of ship, the temporary enclosed spaces being exempt from tonnage measurement since they were not permanently closed spaces. These awning or spar deck structures eventually became an integral part of the ship structure but retained a lighter structure than the upper deck structure of other two-deck ships, later referred to as 'full scantling' vessels. The 'shelter deck type', as this form of vessel became known, apart from having a lighter upper structure was to have the freeboard measured from the second deck, and the tween deck space was exempt from tonnage measurement. This exemption was obtained by the provision of openings in the shelter deck and tween deck bulkheads complying with certain statutory regulations.

At a later date, what were known as open/closed shelter deck ships were developed. These were full scantling ships having the prescribed openings so that the tween deck was exempt from tonnage measurement when the vessel was operating at a load draft where the freeboard was measured from the second deck. It was possible to close permanently these temporary openings and reassign the freeboard, it then being measured from the upper deck so that the vessel might load to a deeper draft, and the tween deck was no longer exempt from tonnage measurement.

Open shelter deck vessels were popular with shipowners for a long period. However, during that time much consideration was given to their safety and the undesirable form of temporary openings in the main hull structure. Eliminating these openings without substantially altering the tonnage values was the object of much discussion and deliberation. Finally, Tonnage Regulations introduced in 1966 provided for the assignment of a tonnage mark, at a stipulated distance below the second deck. A vessel having a 'modified tonnage' had tonnage measured to the second deck only, i.e. the tween deck was exempt, but the tonnage mark was not to be submerged. Where a vessel was assigned 'alternative tonnages' (the equivalent of previous open/closed shelter deck ship), tonnage was taken as that to the second deck when the tonnage mark was not submerged. When the tonnage mark was submerged, tonnage was taken as that to the upper deck, the freeboard being a minimum measured from the upper deck. The tonnage mark concept effectively dispensed with the undesirable tonnage openings. Further changes to tonnage requirements in 1969 led to the universal system of tonnage measurement without the need for tonnage marks, although older ships did retain their original tonnages up until 1994 (see Chapter 30).

Originally the machinery position was amidships with paddle wheel propulsion. Also, with coal being burnt as the propulsive fuel, bunkers were then favorably placed amidships for trim purposes. With the use of oil fuel this problem was more or less overcome, and with screw propulsion there are definite advantages in having the machinery aft. Taking the machinery right aft can produce an excessive trim by the stern in the light condition and the vessel is then provided with deep tanks forward. This may lead to a large bending moment in the ballast condition, and a compromise is often reached by placing the machinery three-quarters aft. That is, there are say three or four holds forward and one aft of the machinery space. In either arrangement the amidships portion with its better stowage shape is reserved for cargo, and shaft spaces lost to cargo are reduced. The all-aft cargo ship illustrating the final evolution of the dry cargo ship in Figure 3.2 could represent the sophisticated cargo liners of the mid 1960s. By the mid 1970s many of the cargo liner trades had been taken over by the container ship and much of the short haul trade undertaken by the conventional dry cargo ship had passed to the 'roll-on roll-off' (ro-ro) type of vessel.

Container ships

A feature of the container ship is the stowage of the rectangular container units within the fuller rectangular portion of the hull and their arrangement in tiers above the main deck level. In order to facilitate removal and placing of the container units of internationally agreed standard (ISO) dimensions hold and hatch widths are common. The narrow deck width outboard of the hatch opening forms the crown of a double shell space containing wing ballast tanks and passageways (see Figure 17.9). Later container ship designs feature hatchless vessels that provide a faster turnaround in port. These may have hatch covers on the forward holds only, or none at all, and are provided with substantial stripping pumps for removing rain and green water from the holds. In recent years the size of container ships making oceanic voyages has substantially increased. The largest ships are those operated by Maersk, which can carry a reported 13,500 TEU. These are unusual and most large ships are between with one classification society reporting more than 60 vessels of at least 8000 TEU classed (see Figure 3.3b).

Barge-carrying ships

Another development in the cargo liner trade was the introduction of the barge-carrying vessel. An early version of this type of ship had a particular advantage in maintaining a scheduled service between the ports at mouths of large river systems such as between the Mississippi river in the USA and the Rhine in Europe. Standard unit cargo barges (sometimes referred to as LASH—lighter aboard ship—barges) are carried on board ship and placed overboard or lifted onboard at terminal ports by large deck-mounted gantries or elevator platforms in association with traveling rails. Other designs make provision for floating the barges in and out of the carrying ship, which can be ballasted to accommodate them. This development appears not to have been as successful as was initially envisaged in the late 1970s, and whilst the merits of this type of craft are still often referred to, the type is now rarely seen.

Ro-ro ships

These ships are characterized by the stern and in some cases the bow or side doors giving access to a vehicle deck above the waterline but below the upper deck (see Figure 3.3a). Access within the ship may be provided in the form of ramps or lifts leading from this vehicle deck to upper decks or hold below. Ro-ro ships may be fitted with various patent ramps for loading through the shell doors when not trading to regular ports where link span and other shore-side facilities that are designed to suit are available. Cargo is carried in vehicles and trailers or in unitized form loaded by fork-lift and other trucks. In order to permit the drive-through vehicle deck

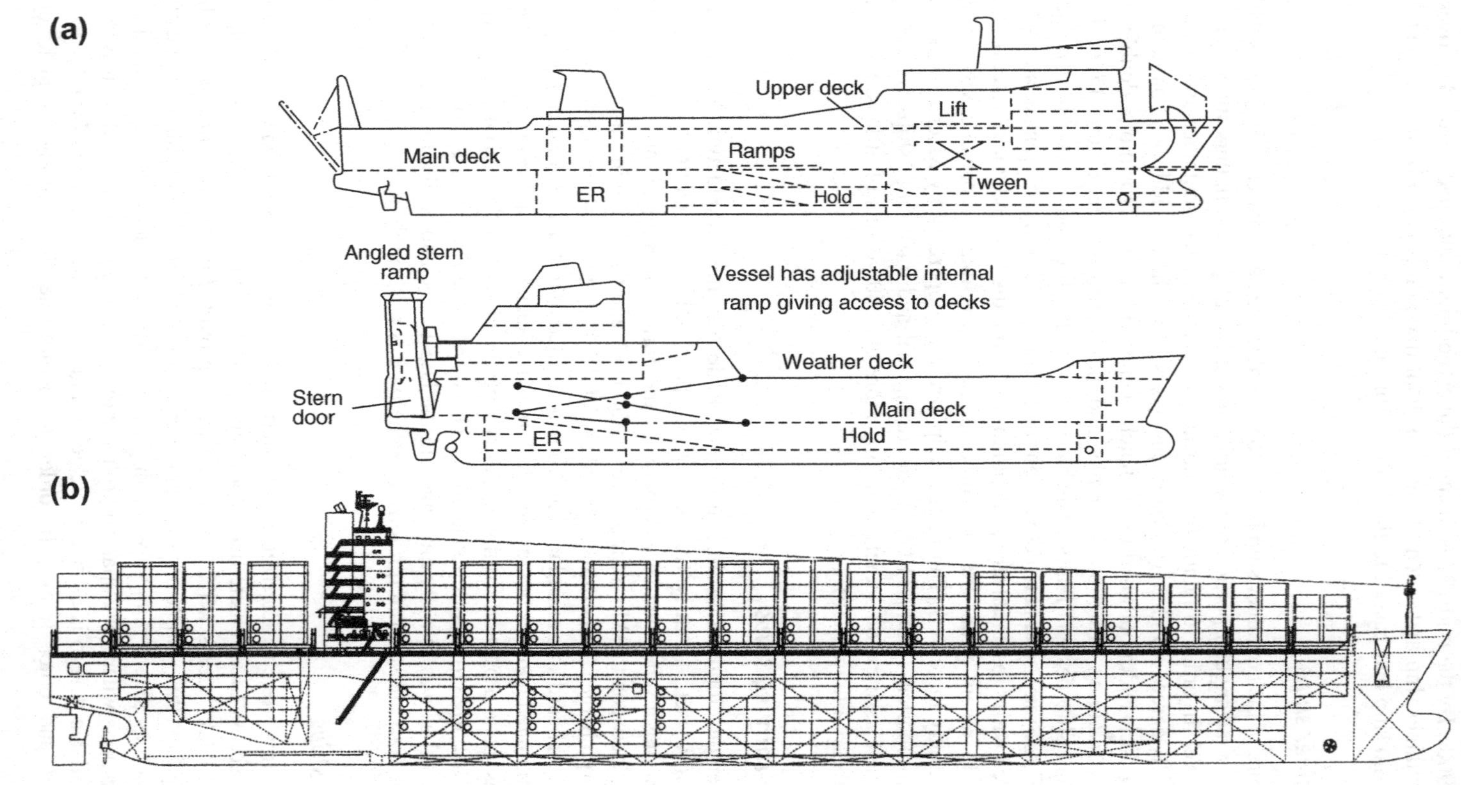

Figure 3.3 (a) Roll-on roll-off ships. (b) 7700 TEU container ship.

a restriction is placed on the height of the machinery space and the ro-ro ship was among the first to popularize the geared medium-speed diesel engine with a lesser height than its slow-speed counterpart. The dramatic loss of the ro-ro passenger ships *Herald of Free Enterprise* in 1987 and *Estonia* in 1994 saw much attention directed at the damage stability of this type of passenger ship when water entered the open unsubdivided deck space. This has resulted in international regulation requiring, amongst other things, strengthening and surveillance of bow doors, surveillance of internal watertight doors used at sea, enhanced damage stability criteria (SOLAS 90) and additional simplified stability information for the master. The *Estonia* loss led to further stringent damage stability requirements adopted on a regional basis by northern European countries (Stockholm Agreement 1997). A midship section of a ro-ro passenger/vehicle/train ferry complying with the requirements of the latter agreement is shown in Figure 17.10.

Hull form

Between the 1940s and 1970 there was a steady increase in the speed of the dry cargo ship and this was reflected in the hull form of the vessels. A much finer hull is apparent in modern vessels, particularly in those ships engaged in the longer cargo liner trades. Bulbous bow forms and open water sterns are used to advantage and considerable flare may be seen in the bows of container ships to reduce wetness on deck where containers are stowed. In some early container ships it is thought that this was probably overdone, leading to an undesirable tendency for the main hull to whip during periods when the bows pitched into head seas. Larger container ships may have the house three-quarters aft with the full beam maintained right to the stern to give the largest possible container capacity.

Cargo handling equipment

Cargo handling equipment, which remained relatively unchanged for a long period, has received considerable attention since the 1960s. This was primarily brought about by an awareness of the loss of revenue caused by the long periods of time the vessel may spend in port discharging and loading cargoes. Conventional cargo ships are now fitted with steel folding and/or rolling steel hatch covers of one patent type or another or liftable slab covers of steel, which reduce maintenance as well as speed cargo handling. Various new lifting devices, derrick forms, and winches have been designed and introduced with marine shipborne cranes now almost completely replacing rigged derrick installations on modern ships. These provide further increased rates of loading and discharge.

Bulk carriers

A wide range of bulk commodities are carried in bulk carriers, including coal, grain, ore, cement, alumina, bauxite, and mineral sand plus shipments of products such as packaged steel and timber.

The large bulk carrier originated as an ore carrier on the Great Lakes at the beginning of the twentieth century. For the period of the Second World War dedicated bulk carriers were only built spasmodically for ocean trading, since a large amount of these cargoes could be carried by general cargo tramps with the advantage of their being able to take return cargoes.

A series of turret-deck steamers were built for ore-carrying purposes between 1904 and 1910; a section through such a vessel is illustrated in Figure 3.4a. Since 1945 a substantial number of ocean-going ore carriers have been built of uniform design. This form of ore carrier with a double bottom and side ballast tanks first appeared in 1917, only at that time the side tanks did not extend to the full hold depth (see Figure 3.4b). To overcome the disadvantage that the ore carrier was only usefully employed on one leg of the voyage, the oil/ore carrier also evolved at that time. The latter ship type carried oil in the wing tanks, as shown in Figure 3.4c, and had a passageway for crew protection in order to obtain the deeper draft permitted tankers.

The common general bulk carrier that predominated in the latter half of the twentieth century took the form shown in Figure 3.4d with double bottom, hopper sides, and deck wing tanks. These latter tanks have been used for the carriage of light grain cargoes as well as water ballast. Specific variations of this type have been built; Figure 3.4e shows a 'universal bulk carrier' patented by the McGregor International Organization that offered a very flexible range of cargo stowage solutions. Another type, shown in Figure 3.4f, had alternate holds of short length. On single voyages the vessel could carry high-density cargoes only in the short holds to give an acceptable cargo distribution. Such stowage is not uncommon on general bulk carriers with uniform hold lengths where alternate hold loading or block hold loading may be utilized to stow high-density cargoes. With such loading arrangements high shear forces occur at the ends of the holds, requiring additional strengthening of the side shell in way of the bulkheads.

A general arrangement of a typical bulk carrier shows a clear deck with machinery aft. Large hatches with steel covers are designed to facilitate rapid loading and discharge of the cargo. Since the bulk carrier makes many voyages in ballast, a large ballast capacity is provided to give adequate immersion of the propeller. The size of this type of ship has also steadily increased and the largest bulk carriers have reached 365,000 tonnes deadweight.

Ships of the general bulk carrier form experienced a relatively high casualty rate during the late 1980s and early 1990s (between 1980 and 2000 some 170 bulk carriers were totally lost), giving rise to concern as to their design and construction. Throughout the late 1990s bulk carrier safety received considerable attention in the work of IMO, the classification societies and elsewhere. Based on experience of failures of lesser consequence, it was concluded that the casualties occurred through local structural failure leading to loss of watertight integrity of the side shell, followed by progressive flooding through damaged bulkheads. The flooding resulted either in excessive hull bending stresses or excessive trim, and loss of the ship. Much of this work concentrated on the structural hull details, stresses experienced as the result of loading and discharging cargoes (past experience showed that ships were often loaded in patterns not approved in the ship's loading manual), damage to structure and

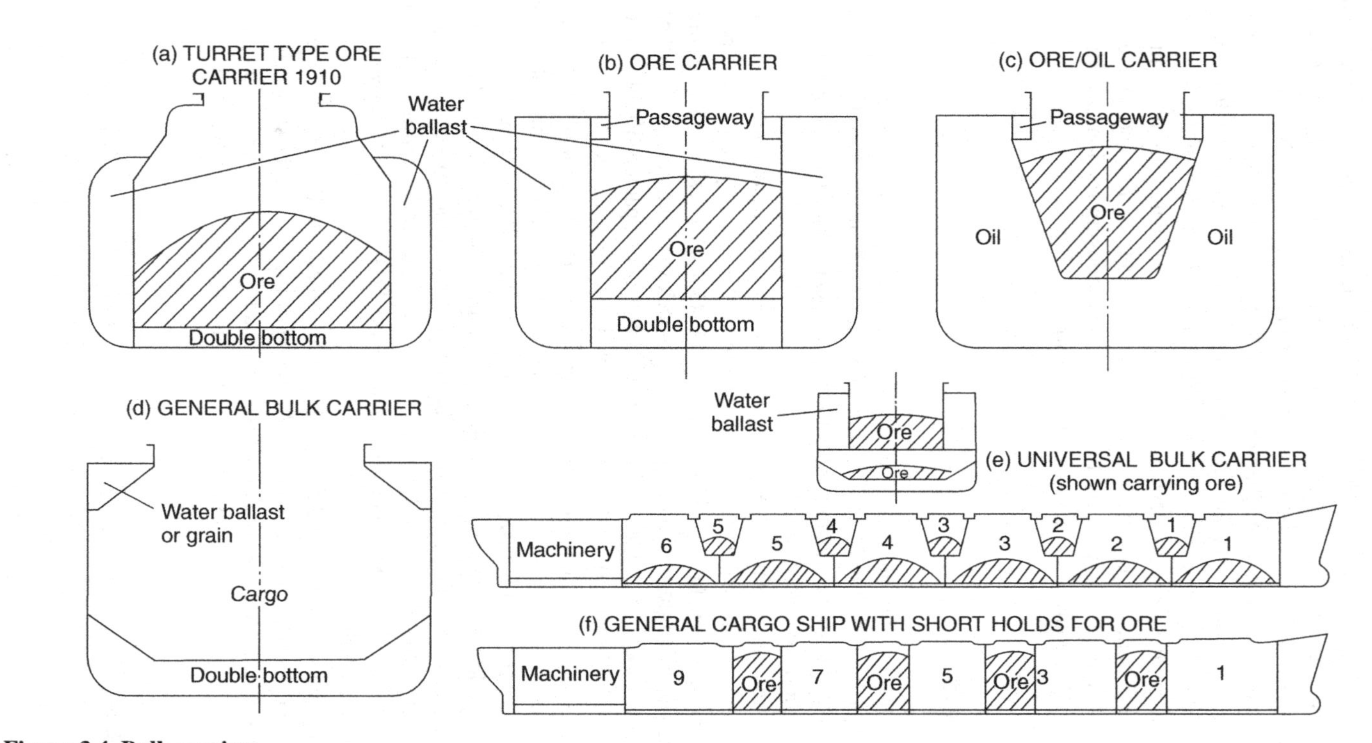

Figure 3.4 Bulk carriers.

protective coatings arising from discharging cargoes, poor maintenance, and subsequent inadequate inspection of the ship structure. The initial outcome of this work was the introduction of a new Chapter XII of SOLAS covering damage stability requirements, structural strength requirements, and enhanced survey procedures for bulk carriers. At its 79th session in December 2004, the Maritime Safety Committee of IMO adopted a new text of Chapter XII of SOLAS that included restrictions on sailing with any hold empty and requirements for double-skin construction as an optional alternative to single side-skin construction. The option of double side-skin construction applies only to new bulk carriers of 150 meters or more in length, carrying solid bulk cargoes having a density of 1000 kg/m^3 and above. These amendments entered into force on 1 July 2006. The midship section of a Handysize bulk carrier with double-skin construction is shown in Figure 17.8.

Car carriers

The increasing volume of car and truck production in the East (Japan, Korea, and China) and a large customer base in the West has seen the introduction and rapid increase in the number of ships specifically designed and built to facilitate the delivery of these vehicles globally.

Probably the ugliest ships afloat, car carriers are strictly functional, having a very high boxlike form above the waterline to accommodate as many vehicles as possible on, in some cases, as many as a dozen decks. Whilst most deck spacing is to suit cars, some tween deck heights may be greater and the deck strengthened to permit loading of higher and heavier vehicles. Within such greater deck spacing liftable car decks may be fitted for flexibility of stowage. The spacing of fixed car decks can vary from 1.85 to 2.3 meters to accommodate varying shapes and heights of cars. Transfer arrangements for vehicles from the main deck are by means of hoistable ramps, which can be lifted and lowered whilst bearing the vehicles. Loading and discharging vehicles onto and off the ship is via a large quarter ramp at the stern and a side shell or stern ramp. The crew accommodation and forward wheelhouse, providing an adequate view forward, sit atop the uppermost continuous weather deck. Propulsion machinery is situated aft with bow thruster(s) forward to aid mooring/maneuvering.

The ship shown in Figure 3.5 has an overall length of 148 meters, a beam of 25 meters, and a speed of 19 knots on a 7.2-meter draft. It can carry some 2140 units. A unit is an overall stowage area of 8.5 square meters per car and represents a vehicle 4.125 meters in length and 1.55 meters wide plus an all-round stowage margin.

Oil tankers

Until 1990 the form of vessels specifically designed for the carriage of oil cargoes had not undergone a great deal of change since 1880, when the vessel illustrated in

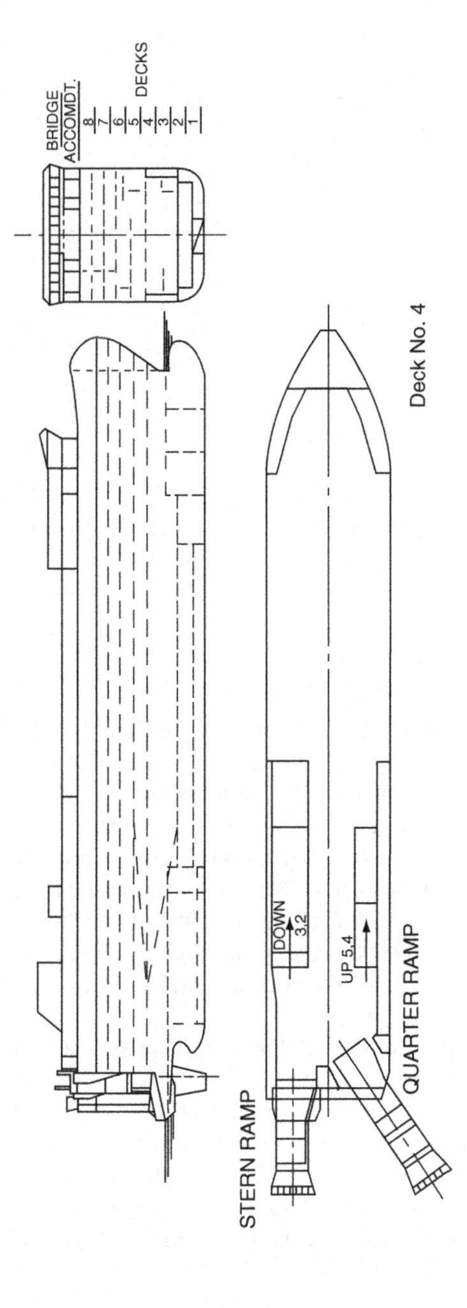

Figure 3.5 Car carrier.

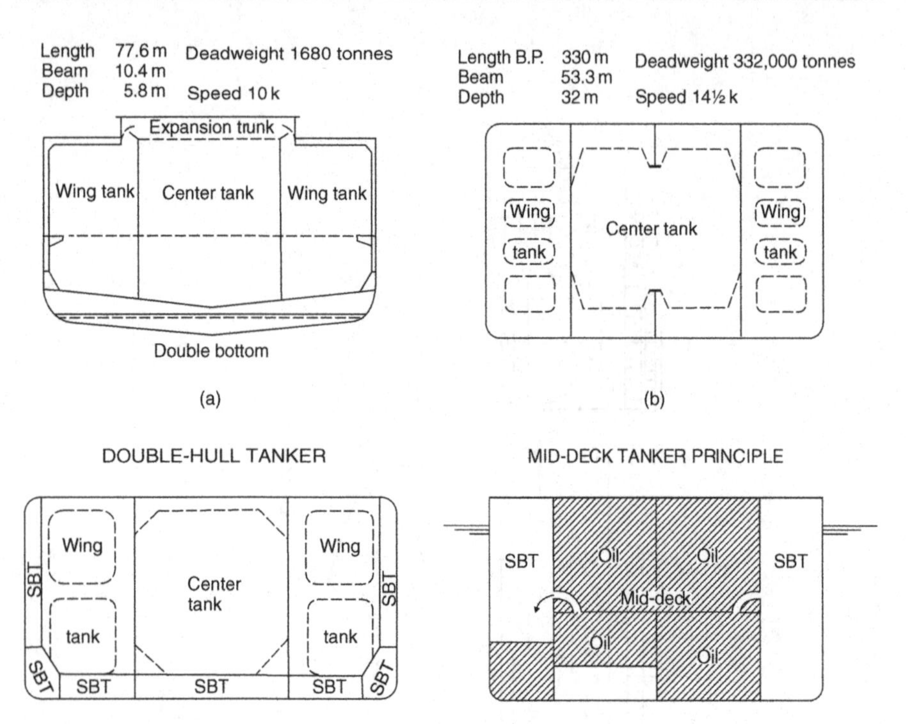

Figure 3.6 Oil tankers.

Figure 3.6a was constructed, the expansion trunk and double bottom within the cargo space having been eliminated much earlier. The greatest changes in that period were the growth in ship size and nature of the structure (see Figure 3.6b).

The growth in size of ocean-going vessels from 1880 to the end of the Second World War was gradual, the average deadweight rising from 1500 tonnes to about 12,000 tonnes. Since then the average deadweight increased rapidly to about 20,000 tonnes in 1953 and about 30,000 tonnes in 1959. Today there are afloat tankers ranging from 100,000 to 500,000 tonnes deadweight. It should be made clear that the larger size of vessel is the crude oil carrier, and fuel oil carriers tend to remain within the smaller deadweights.

Service speeds of oil tankers have shown an increase since the late 1940s, going from 12 to 17 knots. The service speed is related to the optimum economic operation of the tanker. Also, the optimum size of the tanker is very much related to current market economics. The tanker fleet growth increased enormously to meet the expanding demand for oil until 1973/1974, when the OPEC price increases slowed that expansion and led to a slump in the tanker market. It is unlikely that such a significant rise in tanker size and rise in speed will be experienced in the foreseeable future.

Structurally, one of the greatest developments has been in the use of welding, oil tankers being amongst the first vessels to utilize the application of welding. Little difficulty is experienced in making and maintaining oiltight joints: the same cannot be said of riveting. Welding has also allowed cheaper fabrication methods to be adopted. Longitudinal framing was adopted at an early date for the larger ships and revision of the construction rules in the late 1960s allowed the length of tank spaces to be increased, with a subsequent reduction in steel weight, making it easier to pump discharge cargoes.

As far as the general arrangement is concerned, there appears always to have been a trend towards placing the machinery aft. Moving all the accommodation and bridge aft was a later feature and is desirable from the fire protection point of view. Location of the accommodation in one area is more economic from a building point of view, since all services are only to be provided at a single location.

The requirements of the International Convention for the Prevention of Pollution from Ships 1973 (see Chapter 29) and particularly its Protocol of 1978 have greatly influenced the arrangement of the cargo spaces of oil tankers. A major feature of the MARPOL Convention and its Protocol has been the provision in larger tankers of clean water ballast capacity. Whilst primarily intended to reduce the pollution risk, the fitting of segregated water ballast tanks in the midship region aids the reduction of the still water bending moment when the tanker is fully loaded. It also reduces corrosion problems associated with tank spaces, which are subject to alternate oil and sea water ballast cargoes.

In March 1989 the tanker *Exxon Valdez*, which complied fully with the then current MARPOL requirements, ran aground and discharged 11 million gallons of crude oil into the pristine waters of Prince William Sound in Alaska. The subsequent public outcry led to the United States Congress passing the Oil Pollution Act 1990 (OPA 90). This unilateral action by the United States government made it a requirement that existing single-hull oil tankers operating in United States waters were to be phased out by an early date, after which all oil tankers were to have a double hull (see Figures 3.6 and 22.2).

In November 1990 the USA suggested that the MARPOL Convention should be amended to make double hulls compulsory for new tankers. A number of other IMO member states suggested that alternative designs offering equivalent protection against accidental oil spills should be accepted. In particular, Japan proposed an alternative, the mid-deck tanker. This design has side ballast tanks providing protection against collision but no double bottom. The cargo tank space (see Figure 3.6) had a structural deck running its full length at about 0.25–0.5 the depth from the bottom, which ensures that should the bottom be ruptured the upward pressure exerted by the sea would prevent most of the oil from escaping into the sea.

In 1992 IMO adopted amendments to MARPOL that required tankers of 5000 tonnes deadweight and above contracted for after July 1993, or which commenced construction after January 1994, to be of double-hulled or mid-deck construction, or of other design offering equivalent protection against oil pollution. Existing tankers with single hulls without segregated ballast tanks with protective location were to be phased out by June 2007. Those with segregated ballast tanks with protective location were to be phased out by July 2021.

Studies by IMO and the US National Academy of Sciences confirmed the effectiveness of the double hull in preventing oil spills caused by grounding and collision where the inner hull is not breached. The mid-deck tanker was shown to have more favorable outflow performance in extreme accidents where the inner hull is breached. The United States authorities considered grounding the most prevalent type of accident in their waters and believed only the double-hull type prevented spills from tanker groundings in all but the most severe incidents. Thus, whilst MARPOL provided for the acceptance of alternative tanker designs, the United States legislation did not, and no alternative designs were built.

As the result of the break-up of the tanker *Erika* and subsequent pollution of the French coastline in 1999, IMO members decided to accelerate the phase-out of single-hull tankers. As a result, in April 2001 a stricter timetable for the phasing out of single-hull tankers entered into force in September 2003. In December 2003 a decision to further accelerate the phase-out dates of single-hull tankers was agreed, Pre-MARPOL tankers being phased out in 2005 and MARPOL tankers and small tankers in 2010 (see Chapter 2 for definitions).

Oil tankers now generally have a single pump space aft, adjacent to the machinery, and specified slop tanks into which tank washings and oily residues are pumped. Tank cleaning may be accomplished by water-driven rotating machines on the smaller tankers but for new crude oil tankers of 20,000 tonnes deadweight and above the tank cleaning system uses crude oil washing.

Passenger ships

Early passenger ships did not have the tiers of superstructure associated with modern vessels, and they also had a narrower beam in relation to their length. The reason for the absence of superstructure decks was the Merchant Shipping Act 1894, which limited the number of passengers carried on the upper deck. An amendment to this Act in 1906 removed this restriction and vessels were then built with several tiers of superstructures. This produced problems of strength and stability, stability being improved by an increase in beam. The transmission of stresses to the superstructure from the main hull girder created much difference of opinion as to the means of overcoming the problem. Both light structures of a discontinuous nature, i.e. fitted with expansion joints, and superstructures with-heavier scantlings able to contribute to the strength of the main hull girder were introduced. Present practice, where the length of the superstructure is appreciable and has its sides at the ship side, does not require the fitting of expansion joints.

The introduction of aluminum alloy superstructures provided increased passenger accommodation on the same draft, and/or a lowering of the lightweight center of gravity with improved stability. This was brought about by the lighter weight of the aluminum alloy structure. Subsequent experience, however, has shown that for passenger liners, that are required to maintain a service speed in a seaway, the maintenance costs of aluminum alloy superstructures can be higher.

A feature of the general arrangement is the reduction in size of the machinery space in this time. It is easy to see the reason for this if the *Aquitania*, built in 1914 and having direct drive turbines with 21 double-ended scotch boilers, is compared with the *Queen Elizabeth 2*. The latter as originally built had geared drive turbines with three water tube boilers. Many modern passenger ships have had their machinery placed aft; this gives over the best part of the vessel amidships entirely to passenger accommodation. Against this advantage, however, allowance must be made for an increased bending moment if a suitable trim is to be obtained. The more recent provision of electric podded propulsors as fitted on the *Queen Mary 2* has, with the removal of shaft lines, permitted optimization of the internal arrangements of the passenger liner and cruise ship.

Passenger accommodation standards have increased substantially, the volume of space allotted per passenger rising steadily. Tween deck clearances are greater and public rooms extend through two or more decks, whilst enclosed promenade and atrium spaces are now common in these vessels. The provision of air-conditioning and stabilizing devices has also added to passenger comfort. Particular attention has been paid to fire safety in the modern passenger ship, structural materials of low fire risk being utilized in association with automatic extinguishing and detection systems.

There has been a demise of the larger passenger liner and larger passenger ships are now either cruise ships, short-haul ferries, or special trade passenger (STP) ships, the latter being unberthed immigrant or pilgrim passenger ships operating in the Middle East to South East Asian region.

Whilst the safety of passenger ships in general has been good in recent years, the growth in the size and number of cruise ships has led IMO to initiate a review of passenger ship safety. In particular, it is looking at placing greater emphasis on the prevention of a casualty from occurring in the first place. That is, future passenger ships should be designed for improved survivability so that in the event of a casualty passengers and crew can stay safely on board as the ship proceeds to port.

The development of high-speed passenger ferries of lightweight construction and often of radical hull form and/or nondisplacement modes of operation has been notable since the early 1980s. Initially relatively small, these craft may now be more than 100 meters in length and carry upwards of 500 persons plus 100 cars/30 trucks or more. The lightweight construction is usually of aluminum alloy but some have been constructed of lighter higher-tensile steels, and fiber-reinforced plastics may be used in the superstructure and accommodation areas. With speeds of up to 50 knots, many craft are of twin-hull form and include conventional catamarans, wave piercers with twin hulls and a faired buoyant bridging structure forward, and small waterplane twin-hulled (SWATH) ships. The latter have a high proportion of their twin-hull buoyancy below the waterline (see Figure 3.7). Other high-speed craft include hydrofoils and various surface effect ships (SESs) including hovercraft, which maintain a cushion of air, fully or partially, between the hull and the water to reduce drag. The increasing use of these vessels led in 1994 to the promulgation by IMO of specific international regulations concerning their design, safety, and operation. An updated version of this Code of Safety was adopted in December 2000. Figure 3.7

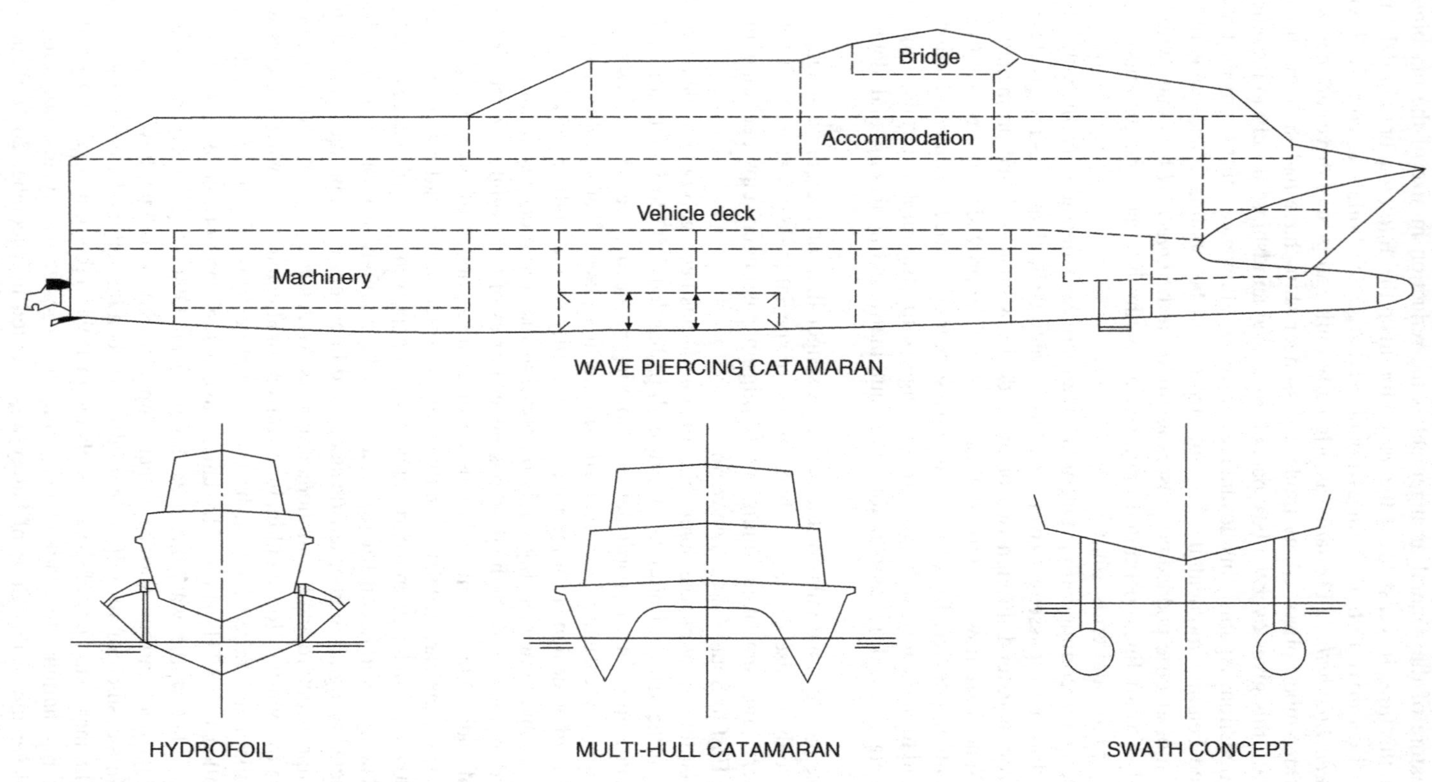

Figure 3.7 Various types of high-speed craft.

illustrates the various types of high-speed craft. Also see Figure 17.11, which shows the midship section of a high-speed wave-piercing catamaran.

Further reading

Barge carriers—A revolution in marine transport, *The Naval Architect*, April 1973.

Bhave, Roy G: Special trade passenger ships, *The Naval Architect*, January 1975.

Burrows: The North Sea platform supply vessel, *ImarEST Trans*. Part 1, 1997.

Code of Safety for Special Purpose Ships. IMO publication (IMO-820E).

Design and operation of bulk carriers. 2005 Conference Proceedings. Royal Institution of Naval Architects Publications.

Design and operation of container ships. 2003 Conference Proceedings. Royal Institution of Naval Architects Publications.

Design and operation of double hull tankers. 2004 Conference Proceedings. Royal Institution of Naval Architects Publications.

Design and operation of gas carriers. 2004 Conference Proceedings. Royal Institution of Naval Architects Publications.

Farell: Chemical tankers—The quiet evolution, *The Naval Architect*, July 1975.

Guidelines for the Design and Construction of Offshore Supply Vessels. IMO publication (IMO-807E).

Guidelines on Early Assessment of Hull Damage and Possible Need for Abandonment of Bulk Carriers. IMO—MSC/Circ. 1143 dated 13 December 2004.

High speed craft. 2004 Conference Proceedings. Royal Institution of Naval Architects Publications.

IMO: *International Code of Safety for High Speed Craft (HSC Code)*, 1994.

Meek: The first OCL container ship, *Trans. RINA*, 1970.

Modern car ferry design and development, *The Naval Architect*, January 1980.

Murray: Merchant ships 1860–1960, *Trans. RINA*, 1960.

Payne: The evolution of the modern cruise liner, *The Naval Architect*, 1990.

Payne: From *Tropicale* to *Fantasy*: A decade of cruiseship development, *Trans. RINA*, 1993.

Payne: The return of the true liner—A design critique of the modern fast cruise ship, *The Naval Architect*, September 1994.

Safety of passenger ro-ro vessels. 1996 Conference Proceedings. Royal Institution of Naval Architects Publications.

Part Two

Materials and Strength of Ships

4 Classification societies

Chapter Outline

Rules and regulations 38
Lloyd's register 38
Lloyd's register classification symbols 39
Classification of ships operating in ice 40
Structural design programs 40
Periodical surveys 41
Annual surveys 41
Intermediate surveys 41
Docking surveys 41
In-water surveys 42
Special surveys 42
Hull planned maintenance scheme 43
Damage repairs 43
Further reading 43
Some useful websites 43

A cargo shipper and the underwriter requested to insure a maritime risk require some assurance that any particular vessel is structurally fit to undertake a proposed voyage. To enable the shipper and underwriter to distinguish the good risk from the bad, a system of classification has been formulated over a period of more than 200 years. During this period reliable organizations have been created for the initial and continuing inspection of ships so that classification may be assessed and maintained.

Recent amendment to the requirements of the International Convention for the Safety of Life at Sea (SOLAS—see Chapter 29) have required ships to which that convention applies to be designed, constructed, and maintained in compliance with the structural, mechanical, and electrical requirements of a classification society that is recognized by the flag administration or with applicable national standards of that administration that provide an equivalent level of safety. In general, flag administrations recognize specific classification societies for this purpose rather than maintaining such national standards.

Whilst there are reported to be more than 50 ship classification organizations worldwide, the 13 major classification societies that claim to class over 90% of all commercial tonnage involved in international trade worldwide are members of the

Ship Construction. DOI: 10.1016/B978-0-08-097239-8.00004-0

International Association of Classification Societies (IACS). These members of the IACS are:

American Bureau of Shipping (ABS)	USA
Bureau Veritas (BV)	France
China Classification Society (CCS)	China
Croatian Register of Shipping	Croatia
Det Norske Veritas (DNV)	Norway
Germanischer Lloyd (GL)	Germany
Indian Register of Shipping	India
Korean Register (KR)	Korea
Lloyd's Register (LR)	Great Britain
Nippon Kaiji Kyokai (Class NK)	Japan
Polish Register of Shipping	Poland
Registro Italiano Navale (RINA)	Italy
Russian Maritime Register of Shipping (RS)	Russia

Rules and regulations

The classification societies each publish rules and regulations that are principally concerned with the strength and structural integrity of the ship, the provision of adequate equipment, and the reliability of the machinery. Ships may be built in any country to a particular classification society's rules and they are not restricted to classification by the relevant society of the country where they are built or owned.

In recent years, under the auspices of the IACS, member societies have been engaged in the development of common structural rules for ships. The first two of these common structural rules, for bulk carriers of 90 meters or more in length and for oil tankers of 150 meters or more in length came into force on 1 April 2006. These common rules will be incorporated into each member society's rule book. In November 2008 the IACS launched the IACS CSR Tracking Database (www.iacs-csrtrack.org.uk) to provide users easy and quick access to full revision history of CSR rules on a paragraph by paragraph basis.

These and other common rules to be developed by IAC members anticipate the nature of future standards to be made under the International Maritime Organization's proposed Goal-Based New Ship Construction Standards (see Chapter 29).

Lloyd's register

Only the requirements of Lloyd's Register, which is the oldest of the classification societies, are dealt with in detail in this chapter. The requirements of other classification societies that are members of the IACS are not greatly different.

Founded in 1760 and reconstituted in 1834, Lloyd's Register was amalgamated with the British Corporation, the only other British classification society in existence at that time, in 1949. Ships built in accordance with Lloyd's Register rules or equivalent standards are assigned a class in the Register Book, and continue to be classed so long as they are maintained in accordance with the rules.

Lloyd's register classification symbols

All ships classed by Lloyd's Register are assigned one or more character symbols. The majority of ships are assigned the characters 100A1 or ✠ 100A1.

The character figure 100 is assigned to all ships considered suitable for sea-going service. The character letter A is assigned to all ships that are built in accordance with or accepted into class as complying with the society's rules and regulations. The character figure 1 is assigned to ships carrying on board anchor and/or mooring equipment complying with the society's rules and regulations. Ships that the society agree need not be fitted with anchor and mooring equipment may be assigned the character letter N in lieu of the character figure 1. The Maltese cross mark is assigned to new ships constructed under the society's special survey, i.e. a surveyor has been in attendance during the construction period to inspect the materials and workmanship.

There may be appended to the character symbols, when considered necessary by the society or requested by the owner, a number of class notations. These class notations may consist of one or a combination of the following: type notation, cargo notation, special duties notation, special features notation, service restriction notation. Type notation indicates that the ship has been constructed in compliance with particular rules applying to that type of ship, e.g. 100A1 'Bulk carrier'. Cargo notation indicates the ship has been designed to carry one or more specific cargoes, e.g. 'Sulfuric acid'. This does not preclude it from carrying other cargoes for which it might be suitable. Special duties notation indicates that the ship has been designed for special duties other than those implied by type or cargo notation, e.g. 'research'. Special features notation indicates the ship incorporates special features that significantly affect the design, e.g. 'movable decks'. Service restriction notation indicates the ship has been classed on the understanding it is operated only in a specified area and/or under specified conditions, e.g. 'Great Lakes and St Lawrence'.

The class notation ✠ LMC indicates that the machinery has been constructed, installed, and tested under the society's special survey and in accordance with the society's rules and regulations. Various other notations relating to the main and auxiliary machinery may also be assigned.

Vessels with a refrigerated cargo installation constructed, installed, and tested under the society's special survey and in accordance with its rules and regulations may be assigned the notation ✠ Lloyd's RMC. A classed liquefied gas carrier or tanker in which the cargo reliquefaction or cargo refrigeration equipment is approved, installed, and tested in accordance with the society's rules and regulations may be assigned the notation ✠ Lloyd's RMC (LG).

Where additional strengthening is fitted for navigation in ice conditions an appropriate notation may be assigned.

Classification of ships operating in ice

Classification societies such as Lloyd's Register and a number of administrations whose waters experience icing have for many years had regulations defining and categorizing ice conditions and specifying design and standard requirements for ships operating in ice. Lloyd's Register have assigned special features notations to many existing ships for operation in first-year ice and for operation in multi-year ice. First-year ice notations are for additional strengthening where waters ice up in winter only and multi-year ice for service in Arctic and Antarctic waters.

The increasing maritime trading within Arctic waters in the past decade and the desire to ship oil, gas, and other commodities from there all year round appears to have resulted in the class societies adopting to some extent the ice strengthening requirements of the 'Finnish–Swedish Ice Class Rules 1985' developed for vessels trading in winter and for which the keel was laid after 1 November 1986. These requirements were intended primarily for vessels operating in the Northern Baltic in winter are given for four different ice classes:

- Ice Class 1AA
- Ice Class 1A
- Ice Class 1B
- Ice Class 1C.

The hull scantling requirements determined under these rules are based on certain assumptions concerning the nature of the ice load the ship's structure may be subjected to. These assumptions have been determined from full-scale observations made in the Northern Baltic.

This increased trading in Arctic waters has also created particular interest in the establishment of universal requirements for ships operating in ice.

Both the IMO and IACS have been involved in this work, with the IMO producing guidelines in December 2002 for ships operating in Arctic ice-covered waters for which they prescribe seven 'Polar Class' descriptions. These range from PC 1 for year-round operation in all Arctic ice covered waters to PC 7 for summer/autumn operation in thin first-year ice that may include old ice inclusions. Subsequently, the IACS set up a working group to develop Unified Requirements for Polar Ships that would cover:

a. Polar class descriptions and applications
b. Structural requirements for Polar class ships
c. Machinery requirements for Polar class ships.

It was intended that with the completion of these Uniform Requirements for Polar Ships and their adoption by the IACS Council, the IACS member societies will have one year in which to implement these common standards for ships operating in ice.

Structural design programs

In recent years the principal classification societies have developed software packages for use by shipyards that incorporate dynamic-based criteria for the scantlings, structural arrangements, and details of ship structures. This was a response to a perception that the traditional semi-empirical published classification rules based on experience could be inadequate for new and larger vessel trends. The computer programs made available to shipyards incorporate a realistic representation of the dynamic loads likely to be experienced by the ship and are used to determine the scantlings and investigate the structural responses of critical areas of the ship's structure.

Lloyd's Register's 'Ship Right Procedures for the Design, Construction and Lifetime Care of Ships' incorporates programs for structural design assessment (SDA) and fatigue design assessment (FDA). Also incorporated are construction monitoring (CM) procedures that ensure the identified critical locations on the ship are built to acceptable standards and approved construction procedures. (These provisions are mandatory for classification of tankers of more than 190 meters in length and for other ships where the type, size, and structural configuration demand.)

Periodical surveys

To maintain the assigned class the vessel has to be examined by the society surveyors at regular periods.

The major hull items to be examined at these surveys only are indicated below.

Annual surveys

All steel ships are required to be surveyed at intervals of approximately one year. These annual surveys are, where practicable, held concurrently with statutory annual or other load-line surveys. At the survey the surveyor is to examine the condition of all closing appliances covered by the conditions of assignment of minimum free-board, the freeboard marks, and auxiliary steering gear. Watertight doors and other penetrations of watertight bulkheads are also examined and the structural fire protection verified. The general condition of the vessel is assessed, and anchors and cables are inspected where possible at these annual surveys. Dry bulk cargo ships are subject to an inspection of a forward and after cargo hold.

Intermediate surveys

Instead of the second or third annual survey after building or special survey, an intermediate survey is undertaken. In addition to the requirements for annual survey, particular attention is paid to cargo holds in vessels over 15 years of age and the operating systems of tankers, chemical carriers, and liquefied gas carriers.

Docking surveys

Ships are to be examined in dry dock at intervals not exceeding 2½ years. At the dry-docking survey particular attention is paid to the shell plating, stern frame and rudder, external and through hull fittings, and all parts of the hull particularly liable to corrosion and chafing, and any unfairness of bottom.

In-water surveys

The society may accept in-water surveys in lieu of any one of the two dockings required in a five-year period. The in-water survey is to provide the information normally obtained for the docking survey. Generally, consideration is only given to an in-water survey where a suitable high-resistance paint has been applied to the underwater hull.

Special surveys

All steel ships classed with Lloyd's Register are subject to special surveys. These surveys become due at five-yearly intervals, the first five years from the date of build or date of special survey for classification and thereafter five years from the date of the previous special survey. Special surveys may be carried out over an extended period commencing not before the fourth anniversary after building or previous special survey, but must be completed by the fifth anniversary.

The hull requirements at a special survey, the details of the compartments to be opened up, and the material to be inspected at any special survey are listed in detail in the rules and regulations (Part 1, Chapter 3). Special survey hull requirements are divided into four ship age groups as follows:

1. Special survey of ships—five years old
2. Special survey of ships—10 years old
3. Special survey of ships—15 years old
4. Special survey of ships—20 years old and at every special survey thereafter.

In each case the amount of inspection required increases and more material is removed so that the condition of the bare steel may be assessed. It should be noted that where the surveyor is allowed to ascertain by drilling or other approved means the thickness of material, nondestructive methods such as ultrasonics are available in contemporary practice for this purpose. Additional special survey requirements are prescribed for oil tankers, dry bulk carriers, chemical carriers, and liquefied gas carriers.

When classification is required for a ship not built under the supervision of the society's surveyors, details of the main scantlings and arrangements of the actual ship are submitted to the society for approval. Also supplied are particulars of manufacture and testing of the materials of construction, together with full details of the equipment. Where details are not available, the society's surveyors are allowed to lift the relevant information from the ship. At the special survey for classification, all the hull requirements for special surveys (1), (2), and (3) are to be carried out. Ships over 20 years old are also to comply with the hull requirements of special survey (4), and oil tankers must comply with the additional requirements stipulated in the rules and regulations. During this survey, the surveyor assesses the standard of the workmanship, and verifies the scantlings and arrangements submitted for approval. It should be noted that the special survey for classification will receive special consideration from Lloyd's Register in the case of a vessel transferred from another recognized classification society. Periodical surveys where the vessel is classed are subsequently held as in the case of ships built under survey, being dated from the date of special survey for classification.

Hull planned maintenance scheme

Along with other classification societies, Lloyd's Register offers a hull planned maintenance scheme (HPMS) that may significantly reduce the scope of the periodical surveys of the hull. The classification society works closely with the shipowner to set up an inspection program that integrates classification requirements with the shipowner's own planned maintenance program. Ship staff trained and accredited by Lloyd's Register are authorized to inspect selected structural items according to an approved schedule. Compliance is verified by Lloyd's Register surveyors at an annual audit.

Damage repairs

When a vessel requires repairs to damaged equipment or to the hull, it is necessary for the work to be carried out to the satisfaction of Lloyd's Register surveyors. In order that the ship maintains its class, approval of the repairs undertaken must be obtained from the surveyors either at the time of the repair or at the earliest opportunity.

Further reading

Lloyd's Register, Rules and Regulations for the Classification of Ships, Part 1, Regulations, Chapters 2 and 3.

Some useful websites

www.iacs.org.uk IACS website—see in particular *Classification Societies—What, Why and How?*

www.lr.org Lloyd's Register website.

www.eagle.org American Bureau of Shipping website.

5 Steels

Chapter Outline

Manufacture of steels 46
- Open hearth process 46
- Electric furnaces 46
- Oxygen process 47
- Chemical additions to steels 47

Heat treatment of steels 48
- Annealing 48
- Normalizing 48
- Quenching (or hardening) 48
- Tempering 48
- Stress relieving 48

Steel sections 48
Shipbuilding steels 49
High tensile steels 50
Corrosion-resistant steels 50
Steel sandwich panels 50
Steel castings 52
Steel forgings 52
Further reading 52
Some useful websites 52

The production of all steels used for shipbuilding purposes starts with the smelting of iron ore and the making of pig-iron. Normally the iron ore is smelted in a blast furnace, which is a large, slightly conical structure lined with a refractory material. To provide the heat for smelting, coke is used and limestone is also added. This makes the slag formed by the incombustible impurities in the iron ore fluid, so that it can be drawn off. Air necessary for combustion is blown in through a ring of holes near the bottom, and the coke, ore, and limestone are charged into the top of the furnace in rotation. Molten metal may be drawn off at intervals from a hole or spout at the bottom of the furnace and run into molds formed in a bed of sand or into metal molds.

The resultant pig-iron contains 92–97% iron, the remainder being carbon, silicon, manganese, sulfur, and phosphorus. In the subsequent manufacture of steels the pig-iron is refined; in other words the impurities are reduced.

Ship Construction. DOI: 10.1016/B978-0-08-097239-8.00005-2

Manufacture of steels

Steels may be broadly considered as alloys of iron and carbon, the carbon percentage varying from about 0.1% in mild steels to about 1.8% in some hardened steels. These may be produced by one of four different processes: the open hearth process, the Bessemer converter process, the electric furnace process, or an oxygen process. Processes may be either an acid or basic process according to the chemical nature of the slag produced. Acid processes are used to refine pig-iron low in phosphorus and sulfur that are rich in silicon and therefore produce an acid slag. The furnace lining is constructed of an acid material so that it will prevent a reaction with the slag. A basic process is used to refine pig-iron that is rich in phosphorus and low in silicon. Phosphorus can be removed only by introducing a large amount of lime, which produces a basic slag. The furnace lining must then be of a basic refractory to prevent a reaction with the slag. About 85% of all steel produced in Britain is of the *basic* type, and with modern techniques is almost as good as the *acid* steels produced with superior ores.

Only the open hearth, electric furnace, and oxygen processes are described here as the Bessemer converter process is not used for shipbuilding steels.

Open hearth process

The open hearth furnace is capable of producing large quantities of steel, handling 150–300 tonnes in a single melt. It consists of a shallow bath, roofed in, and set above two brick-lined heating chambers. At the ends are openings for heated air and fuel (gas or oil) to be introduced into the furnace. Also, these permit the escape of the burned gas, which is used for heating the air and fuel. Every 20 minutes or so the flow of air and fuel is reversed.

In this process a mixture of pig-iron and steel scrap is melted in the furnace, carbon and the impurities being oxidized. Oxidization is produced by the oxygen present in the iron oxide of the pig-iron. Subsequently carbon, manganese, and other elements are added to eliminate iron oxides and give the required chemical composition.

Electric furnaces

Electric furnaces are generally of two types: the arc furnace and the high-frequency induction furnace. The former is used for refining a charge to give the required composition, whereas the latter may only be used for melting down a charge whose composition is similar to that finally required. For this reason only the arc furnace is considered in any detail. In an arc furnace melting is produced by striking an arc between electrodes suspended from the roof of the furnace and the charge itself in the hearth of the furnace. A charge consists of pig-iron and steel scrap, and the process enables consistent results to be obtained and the final composition of the steel can be accurately controlled.

Electric furnace processes are often used for the production of high-grade alloy steels.

Oxygen process

This is a modern steelmaking process by which a molten charge of pig-iron and steel scrap with alloying elements is contained in a basic lined converter. A jet of high-purity gaseous oxygen is then directed onto the surface of the liquid metal in order to refine it.

Steel from the open hearth or electric furnace is tapped into large ladles and poured into ingot molds. It is allowed to cool in these molds until it becomes reasonably solidified, permitting it to be transferred to 'soaking pits' where the ingot is reheated to the required temperature for rolling.

Chemical additions to steels

Additions of chemical elements to steels during the above processes serve several purposes. They may be used to deoxidize the metal, to remove impurities and bring them out into the slag, and finally to bring about the desired composition.

The amount of deoxidizing elements added determines whether the steels are 'rimmed steels' or 'killed steels'. Rimmed steels are produced when only small additions of deoxidizing material are added to the molten metal. Only those steels having less than 0.2% carbon and less than 0.6% manganese can be rimmed. Owing to the absence of deoxidizing material, the oxygen in the steel combines with the carbon and other gases present and a large volume of gas is liberated. So long as the metal is molten, the gas passes upwards through the molten metal. When solidification takes place in ingot form, initially from the sides and bottom and then across the top, the gases can no longer leave the metal. In the central portion of the ingot a large quantity of gas is trapped, with the result that the core of the rimmed ingot is a mass of blow holes. Normally the hot rolling of the ingot into thin sheet is sufficient to weld the surfaces of the blow holes together, but this material is unsuitable for thicker plate.

The term 'killed' steel indicates that the metal has solidified in the ingot mold with little or no evolution of gas. This has been prevented by the addition of sufficient quantities of deoxidizing material, normally silicon or aluminum. Steel of this type has a high degree of chemical homogeneity, and killed steels are superior to rimmed steels. Where the process of deoxidation is only partially carried out by restricting the amount of deoxidizing material, a 'semi-killed' steel is produced.

In the ingot mold the steel gradually solidifies from the sides and base, as mentioned previously. The melting points of impurities like sulfides and phosphides in the steel are lower than that of the pure metal and these will tend to separate out and collect towards the center and top of the ingot, which is the last to solidify. This forms what is known as the 'segregate' in the way of the noticeable contraction at the top of the ingot. Owing to the high concentration of impurities at this point, this portion of the ingot is often discarded prior to rolling plate and sections.

Heat treatment of steels

The properties of steels may be altered greatly by the heat treatment to which the steel is subsequently subjected. These heat treatments bring about a change in the mechanical properties principally by modifying the steel's structure. Those heat treatments that concern shipbuilding materials are described.

Annealing

This consists of heating the steel at a slow rate to a temperature of say 850–950 °C, and then cooling it in the furnace at a very slow rate. The objects of annealing are to relieve any internal stresses, to soften the steel, or to bring the steel to a condition suitable for a subsequent heat treatment.

Normalizing

This is carried out by heating the steel slowly to a temperature similar to that for annealing and allowing it to cool in air. The resulting faster cooling rate produces a harder, stronger steel than annealing, and also refines the grain size.

Quenching (or hardening)

Steel is heated to temperatures similar to that for annealing and normalizing, and then quenched in water or oil. The fast cooling rate produces a very hard structure with a higher tensile strength.

Tempering

Quenched steels may be further heated to a temperature somewhat between atmospheric and 680 °C, and some alloy steels are then cooled fairly rapidly by quenching in oil or water. The object of this treatment is to relieve the severe internal stresses produced by the original hardening process and to make the material less brittle but retain the higher tensile stress.

Stress relieving

To relieve internal stresses the temperature of the steel may be raised so that no structural change of the material occurs and then it may be slowly cooled.

Steel sections

A range of steel sections are rolled hot from ingots. The more common types associated with shipbuilding are shown in Figure 5.1. It is preferable to limit the sections required for shipbuilding to those readily available, i.e. the standard types; otherwise

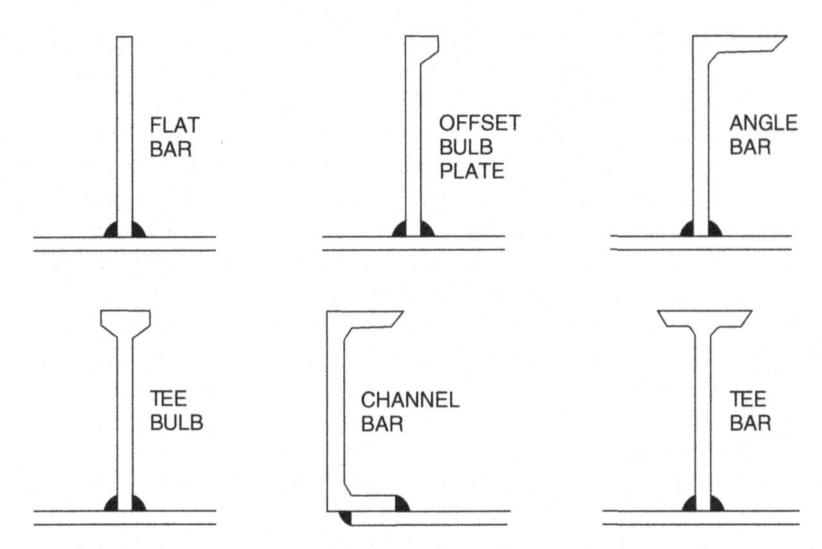

Figure 5.1 Steel sections of shipbuilding.

a steel mill is required to set up rolls for a small amount of material, which is not very economic.

Shipbuilding steels

Steel for hull construction purposes is usually mild steel containing 0.15–0.23% carbon and a reasonably high manganese content. Both sulfur and phosphorus in the mild steel are kept to a minimum (less than 0.05%). Higher concentrations of both are detrimental to the welding properties of the steel, and cracks can develop during the rolling process if the sulfur content is high.

Steel for a ship classed with Lloyd's Register is produced by an approved manufacturer, and inspection and prescribed tests are carried out at the steel mill before dispatch. All certified materials are marked with the society's brand and other particulars as required by the rules.

Ship classification societies originally had varying specifications for steel. However, in 1959, the major societies agreed to standardize their requirements in order to reduce the required grades of steel to a minimum. There are now five different qualities of steel employed in merchant ship construction and now often referred to as IACS steels. These are graded A, B, C, D, and E, Grade A being an ordinary mild steel to Lloyd's Register requirements and generally used in shipbuilding. Grade B is a better quality mild steel than Grade A and specified where thicker plates are required in the more critical regions. Grades C, D, and E possess increasing notch-tough characteristics, Grade C being to American Bureau of Shipping requirements. Lloyd's Register requirements for Grades A, B, D, and E steels may be found in Chapter 3 of Lloyd's Rules for the Manufacture, Testing and Certification of Materials.

High tensile steels

Steels having a higher strength than that of mild steel are employed in the more highly stressed regions of large tankers, container ships, and bulk carriers. Use of higher strength steels allows reductions in thickness of deck, bottom shell, and framing where fitted in the midships portion of larger vessels; it does, however, lead to larger deflections. The weldability of higher tensile steels is an important consideration in their application in ship structures and the question of reduced fatigue life with these steels has been suggested. Also, the effects of corrosion with lesser thicknesses of plate and section may require more vigilant inspection.

Higher tensile steels used for hull construction purposes are manufactured and tested in accordance with Lloyd's Register requirements. Full specifications of the methods of manufacture, chemical composition, heat treatment, and mechanical properties required for the higher tensile steels are given in Chapter 3 of Lloyd's Rules for the Manufacture, Testing and Certification of Materials. The higher strength steels are available at three strength levels, 32, 36, and 40 (kg/mm^2), when supplied in the as-rolled or normalized condition. Provision is also made for material with six higher strength levels, 42, 46, 50, 55, 62, and 69 (kg/mm^2), when supplied in the quenched and tempered condition. Each strength level is subdivided into four grades, AH, DH, EH, and FH, depending on the required level of notch toughness.

Corrosion-resistant steels

Steels with alloying elements that give them good corrosion resistance and colloquially referred to as stainless steels are not commonly used in ship structures, primarily because of their higher initial and fabrication costs. Only in the fabrication of cargo tanks containing highly corrosive cargoes might such steels be found.

For oil tankers the inner surfaces, particularly the deckhead and bottom, are generally protected by high-cost corrosion-resistant coatings that require vigilant inspection and maintenance (see Chapter 27). A recent development in the manufacture of an alloyed shipbuilding steel with claimed improved corrosion resistance properties and its approval by Lloyd's Register for use in certain cargo tanks of a 105,000 dwt tanker indicate that in the future the need to coat oil cargo tanks might be dispensed with.

Steel sandwich panels

As an alternative to conventional shipyard-fabricated stiffened steel plate structures, proprietary manufactured steel sandwich panels have become available and used on ships where their lighter weight was important. Such panels consist of a steel core in the form of a honeycomb with flanges to which the external steel sheets are resistance (spot) or laser (stake) welded. Early use of these bought-in steel sandwich panels was primarily for nonhull structures in naval construction, where their light weight was important. Such panels have also been used for the superstructures of passenger ships, where lightness can allow additional decks and hence increased passenger accommodation. Also, when fabricated using stainless steel their corrosion resistance and low maintenance properties have been utilized.

A proprietary steel sandwich plate system (SPS) has been developed that consists of an elastomer core between steel face plates. Elastomers are a specific class of polyurethane that has a high tolerance to mechanical stress, i.e. it rapidly recovers from deformation. The SPS elastomer also has a high resistance to most common chemical species. Initial application of SPS in shipbuilding has been in passenger ship superstructures, where the absence of stiffening has increased the space available and provided factory-finished surfaces with built-in vibration damping, acoustic insulation, and fire protection. SPS structures have been approved with an A 60 fire-resistance rating (see Chapter 27). The main use of this system has been for repair of ships, especially decks. A single steel panel is used to secure the elastomer core to the existing deck. This creates a sandwich panel that is structurally acceptable. The major benefit is in the reduced time required to complete a major steelwork repair, compared to removing and replacing existing, corroded structure. SPS structures can be fabricated using joining technologies presently used in the shipbuilding industry, but the design of all joints must take into account the structural and material characteristics of the metal–elastomer composite. The manufacturer envisages the use of SPS panels throughout the hull and superstructure of ships, providing a simpler construction with greater carrying capacity and less corrosion, maintenance, and inspection. In association with the manufacturer, Lloyd's Register in early 2006 published provisional rules for the use of this sandwich plate system for new construction and ship repair. The rules cover construction procedures, scantling determination for primary supporting structures, framing arrangements, and methods of scantling determination for steel sandwich panels.

The Norwegian classification society, Det Norske Veritas (DNV), have developed proposals for ship hulls using a lightweight concrete/steel sandwich. They envisaged a steel/concrete/steel composite structure for the cargo hold area of some 600 mm width for the side shell but somewhat greater width for the double bottom area. This sandwich would be much narrower than for a comparable steel-only double-skin bulk carrier, thus increasing the potential carrying capacity, although water ballast may have to be carried in some designated holds as the double skin would not be available for this purpose. DNV consider the other advantages of the concrete/steel sandwich to be reduced stress concentrations with less cracking in critical areas, considerable elimination of corrosion, and elimination of local buckling. The conclusions from the study, a report on which is available for download, are that the sandwich has potential applications for small vessels, for short sea or rivers, and for some offshore structures. It does not appear viable for larger, ocean-going ships at present. Similar panels have been adopted in some offshore applications.

Steel castings

Molten steel produced by the open hearth, electric furnace, or oxygen process is poured into a carefully constructed mold and allowed to solidify to the shape required. After removal from the mold, a heat treatment is required, for example annealing, or normalizing and tempering to reduce brittleness. Stern frames, rudder frames, spectacle frames for bossings, and other structural components may be produced as castings.

Steel forgings

Forging is simply a method of shaping a metal by heating it to a temperature where it becomes more or less plastic and then hammering or squeezing it to the required form. Forgings are manufactured from killed steel made by the open hearth, electric furnace, or oxygen process, the steel being in the form of ingots cast in chill molds. Adequate top and bottom discards are made to ensure no harmful segregations in the finished forgings and the sound ingot is gradually and uniformly hot worked. Where possible the working of the metal is such that metal flow is in the most favorable direction with regard to the mode of stressing in service. Subsequent heat treatment is required, preferably annealing or normalizing and tempering to remove effects of working and non-uniform cooling.

Further reading

Lloyd's Register, Rules and Regulations for the Classification of Ships 2011, Rules for the Manufacture, Testing and Certification of Materials.

Some useful websites

http://lloydsregister.axinteractive.com/category/2-marine.aspx Lloyd's Register rules for classification of ships, including downloadable information on materials.

http://eagle.org Website for ABS with downloadable rules, including materials, welding, and testing.

www.ie-sps.com For more information about SPS systems.

http://www.dnv.com/industry/maritime/ News about sandwich panels.

6 Other shipbuilding materials

Chapter Outline

Aluminum alloy 53
Production of aluminum 54
Aluminum alloys 55
Riveting 55
Aluminum alloy sandwich panels 57
Fire protection 57
Fiber-reinforced composites (FRCs) 58
Some useful websites 59

Aluminum alloy

There are three advantages that aluminum alloys have over mild steel in the construction of ships. Firstly, aluminum is lighter than mild steel (approximate weights being 2.723 tonnes/m^3, mild steel 7.84 tonnes/m^3), and with an aluminum structure it has been suggested that up to 60% of the weight of a steel structure may be saved. This is in fact the principal advantage as far as merchant ships are concerned, the other two advantages of aluminum being a high resistance to corrosion and its nonmagnetic properties. The nonmagnetic properties can have advantages in warships and locally in the way of the magnetic compass, but they are generally of little importance in merchant vessels. Good corrosion properties can be utilized, but correct maintenance procedures and careful insulation from the adjoining steel structure are necessary. A major disadvantage of the use of aluminum alloys is their higher initial and fabrication costs. The higher costs must be offset by an increased earning capacity of the vessel, resulting from a reduced lightship weight or increased passenger accommodation on the same ship dimensions. Experience with large passenger liners on the North Atlantic service has indicated that maintenance costs of aluminum alloy structures can be higher for this type of ship and service.

Although aluminum was first used for small craft in 1891 and for experimental naval vessels in 1894, it has not been a significant material for ships until comparatively recently.

A significant number of larger ships have been fitted with superstructures of aluminum alloy and, apart from the resulting reduction in displacement, benefits have been obtained in improving the transverse stability. Since the reduced weight of

Ship Construction. DOI: 10.1016/B978-0-08-097239-8.00006-4

superstructure is at a position above the ship's center of gravity, this ensures a lower center of gravity than that obtained with a comparable steel structure. For example, on the *Queen Elizabeth 2* with a limited beam to transit the Panama Canal, the top five decks constructed of aluminum alloy enabled the ship to support one more deck than would have been possible with an all-steel construction.

Only in those vessels having a fairly high speed and hence power, also ships where the deadweight/lightweight ratio is low, are appreciable savings to be expected. Such ships are moderate- and high-speed passenger liners having a low deadweight. It is interesting to note, however, that for the *Queen Mary 2*, not having a beam limitation, the owners decided to avoid aluminum alloy as far as possible to ensure ease of maintenance over a life cycle of 40 years. A very small number of cargo liners have been fitted with an aluminum alloy superstructure, principally to clear a fixed draft over a river bar with maximum cargo.

Smaller naval vessels have often had aluminum superstructures on a steel hull as a weight saving measure. A difficulty is the joining of the two metals, in such a way as to avoid a corrosion cell being set up. Either bolted connections with washers to separate the two metals, or an explosively bonded steel/aluminum transition piece (trade name 'Kelomet' and produced by Nobles explosives) can be used. In the explosively bonded material the interface between the metals remains corrosion free.

The total construction in aluminum alloy of a large ship is not considered an economic proposition and it is only in the construction of smaller multi-hull and other high-speed craft where aluminum alloys of higher strength-to-weight ratio are fully used to good advantage. Aluminum has been used for specialized craft, including hydrofoils, and is currently particularly used for high-speed ferries where weight is critical. The material is also used for small, high-speed military vessels.

One advantage of aluminum is that the material can be extruded to produce a very wide range of profiles. These can have benefits in design of efficient structures. An example is a plate that is extruded with stiffeners as part of the profile. The plates can be joined to assemble deck panels, with minimum welding and hence a lower risk of distortion. The production process is also faster. Specialized extrusions can be expensive so the economic benefits have to be considered carefully.

Production of aluminum

For aluminum production at the present time the ore, bauxite, is mined containing roughly 56% aluminum. The actual extraction of the aluminum from the ore is a complicated and costly process involving two distinct stages. Firstly, the bauxite is purified to obtain pure aluminum oxide, known as alumina; the alumina is then reduced to metallic aluminum. The metal is cast in pig or ingot forms and alloys are added where required before the metal is cast into billets or slabs for subsequent rolling, extrusion, or other forming operations.

Sectional material is mostly produced by the extrusion process. This involves forcing a billet of the hot material through a die of the desired shape. More intricate shapes are produced by this method than are possible with steel, where the sections are rolled. However, the range of thickness of section may be limited since each thickness requires a different die. Typical sections are shown in Figure 6.1.

Aluminum alloys

Pure aluminum has a low tensile strength and is of little use for structural purposes; therefore, the pure metal is alloyed with small percentages of other materials to give greater tensile strengths (see Table 6.1). There are a number of aluminum alloys in use, but these may be separated into two distinct groups, nonheat-treated alloys and heat-treated alloys. The latter, as implied, are subjected to a carefully controlled heating and cooling cycle in order to improve the tensile strength.

Cold working of the nonheat-treated plate has the effect of strengthening the material and this can be employed to advantage. However, at the same time the plate becomes less ductile, and if cold working is considerable the material may crack; this places a limit on the amount of cold forming possible in ship building. Cold-worked alloys may be subsequently subjected to a slow heating and cooling annealing or stabilizing process to improve their ductility.

With aluminum alloys a suitable heat treatment is necessary to obtain a high tensile strength. A heat-treated aluminum alloy that is suitable for shipbuilding purposes is one having as its main alloying constituents magnesium and silicon. These form a compound Mg_2Si and the resulting alloy has very good resistance to corrosion and a higher ultimate tensile strength than that of the nonheat-treated alloys. Since the material is heat treated to achieve this increased strength, subsequent heating, for example welding or hot forming, may destroy the improved properties locally.

Aluminum alloys are generally identified by their Aluminum Association numeric designation, the 5000 alloys being nonheat treated and the 6000 alloys being heat treated. The nature of any treatment is indicated by additional lettering and numbering. Lloyd's Register prescribes the following commonly used alloys in shipbuilding:

5083-0	annealed
5083-F	as fabricated
5083-H321	strain hardened and stabilized
5086-0	annealed
5086-F	as fabricated
5086-H321	strain hardened and stabilized
6061-T6	solution heat treated and artificially aged
6082-T6	solution heat treated and artificially aged

Riveting

Riveting may be used to attach stiffening members to light aluminum alloy plated structures where appearance is important and distortion from the heat input of welding is to be avoided.

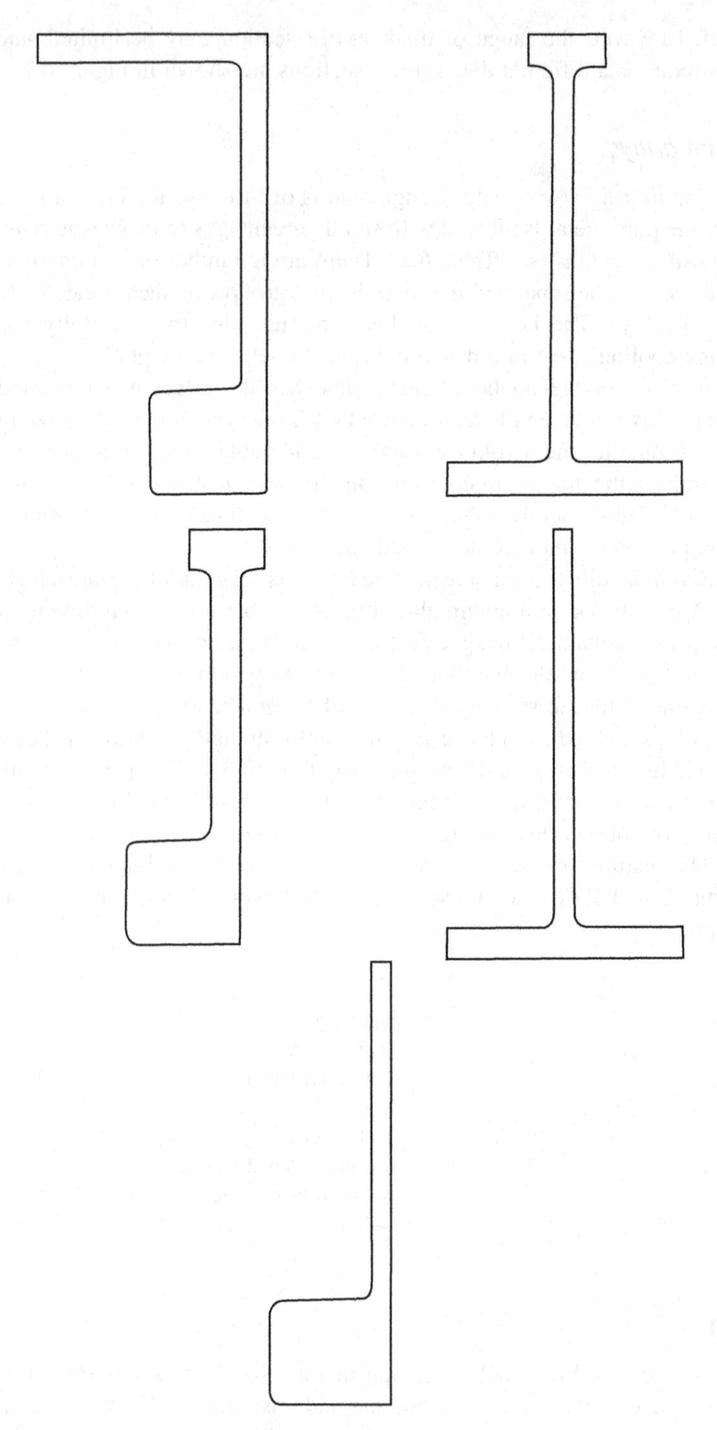

Figure 6.1 Typical aluminum alloy sections.

Table 6.1 Alloying elements

Element	5083	5086	6061	6082
Copper	0.10 max	0.10 max	0.15–0.40	0.10 max
Magnesium	4.0–4.9	3.5–4.5	0.8–1.2	0.6–1.2
Silicon	0.40 max	0.40 max	0.4–0.8	0.7–1.3
Iron	0.40 max	0.50 max	0.70 max	0.50 max
Manganese	0.4–1.0	0.2–0.7	0.15 max	0.4–1.0
Zinc	0.25 max	0.25 max	0.25 max	0.20 max
Chromium	0.05–0.25	0.05–0.25	0.04–0.35	0.25 max
Titanium	0.15 max	0.15 max	0.15 max	0.10 max
Other elements				
Each	0.05 max	0.05 max	0.05 max	0.05 max
Total	0.15 max	0.15 max	0.15 max	0.15 max

The commonest stock for forging rivets for shipbuilding purposes is a nonheat-treatable alloy NR5 (R for rivet material) that contains 3–4% magnesium. Nonheat-treated alloy rivets may be driven cold or hot. In driving the rivets cold relatively few heavy blows are applied and the rivet is quickly closed to avoid too much cold work, i.e. becoming work hardened so that it cannot be driven home. Where rivets are driven hot the temperature must be carefully controlled to avoid metallurgical damage. The shear strength of hot driven rivets is slightly less than that of cold driven rivets.

Aluminum alloy sandwich panels

As with steel construction, proprietary aluminum alloy honeycomb sandwich panels are now available to replace fabricated plate and stiffener structures and can offer extremely low weight options for the superstructures of high-speed craft.

Fire protection

It was considered necessary to mention when discussing aluminum alloys that fire protection is more critical in ships in which this material is used because of the low melting point of aluminum alloys. During a fire the temperatures reached may be sufficient to cause a collapse of the structure unless protection is provided. The insulation on the main bulkheads in passenger ships will have to be sufficient to make the aluminum bulkhead equivalent to a steel bulkhead for fire purposes.

For the same reason it is general practice to fit steel machinery casings through an aluminum superstructure on cargo ships.

Fiber-reinforced composites (FRCs)

Composite materials combine two elements with very different characteristics to provide a material with good structural capability. The fiber provides the strength and the matrix in which it is contained, usually a plastic, holds the fiber in place. The fiber can be arranged to provide directional strength so the composite can be tailored to very specific structural requirements.

Composites have a lengthy history, for example mud bricks reinforced with straw are a composite, as is reinforced concrete. For marine applications glass fiber-reinforced plastic (GRP), first introduced in the 1950s, is the main material for small boats. It can be considered to be the most forgiving of all construction media. There can be problems with poor construction technique, and this can result in expensive repairs, for example due to hull and deck cores becoming water saturated and laminates blistered. However, well-built GRP boats can still be operational after 50 years.

FRC is used extensively in leisure craft, where it provides advantages of lightness and durability. FRC has also been used for naval vessels, in particular mine countermeasures vessels, where the attraction of the material is that it is nonmagnetic. The vessels are relatively large, typically around 50 meters loa. The 'Visby' class corvettes, built for the Swedish navy by Kockums AB, have an overall length of 72.7 meters and a beam of 10.4 meters.

The largest composite vessel is the superyacht 'Mirabella V', which has an overall length of 75.2 meters and a beam of 14.82 meters. The full load displacement is 700 tonnes. For both a mine countermeasures ship and a luxury yacht the proportion of cost represented by the hull is small. Such large vessels would probably not be economic for a commercial application, due to the labor-intensive nature of FRP construction.

Major advantages of FRP for small vessels include low weight, combined with high strength and stiffness. The material is resistant to harsh marine environments. The hull is laid up as a continuous structure, with no seams, which avoids any possible leakage problems. FRP is durable and has low maintenance requirements.

For specialist applications the ability of FRP to absorb high energy can be valuable, and the material can be tailored for specific loadings. This gives excellent design flexibility and leads to efficient structures.

The material consists of fibers, usually glass but including Kevlar and carbon fibers, for very high strength and low weight applications, which are bound into a resin. The higher strength fibers are used in such applications as masts, where strength and light weight are important considerations. For most FRP applications the fibers may be randomly arranged in a chopped strand mat or woven into a specific material to provide additional strength.

The hull is laid up in a mold. The mold is built using a former, known as the 'plug', which is an initial hull form. Once the mold is made it can be used many times, which is one reason for the popularity of the material for small boats. The mold has to be carefully supported to ensure it does not deform under the loads imposed by the FRP and the production activities. The surface finish of the mold is important as this will determine the smoothness of the finished hull.

A coating of PV (a clear odorless liquid), known as the releasing agent, is applied to the mold to prevent the mold from sticking to the plug or finished hull. Then a gel coat is applied, which once cured will form the smooth, water-resistant outer surface of the hull. At this stage the gel coat can be colored. The glass mat or woven roving is then laid onto the mold and the resin applied. The mat can be cut to match the hull shape. It is important that the fibers are thoroughly wetted by the resin and that all air is excluded. The hull is built up to the required thickness with several layers of glass. Additional glass is used to build up thicker material where additional strength is required.

Stiffeners can be incorporated in the hull. These are usually a solid or hollow core former. These are positioned on the hull in the mold and then covered with the glass and resin. When covered with several layers of fiberglass mat they create a closed box or semicircular section. These are known as 'top-hat' and 'half-round' stiffeners. Internal fittings, machinery mountings, and locations for bulkheads are also built into the hull as it progresses. Once the hull, and fittings incorporated into it, are complete the hull is left to cure until the resin is fully set.

The mold is often in two halves for small vessels and this allows it to be split to release the hull.

Some useful websites

www.austal.com Construction of aluminum fast vessels.

http://www.sciencedirect.com/science/article/pii/S0950061889800029 Makes available an article on Kelomet explosively bonded steel/aluminum joining pieces.

http://www.nationalcompositescentre.co.uk/ General information on composite use and developments.

http://www.reinforcedplastics.com/ Also general information on composites and specific reports on boats.

7 Testing of materials

Chapter Outline

Classification society tests for hull materials 63

Tensile test 63

Impact tests 63

Aluminum alloy tests 65

Metals are tested to ensure that their strength, ductility, and toughness are suitable for the function they are required to perform.

Material properties are important to the capability of the structure. The properties outlined below are appropriate in determining the suitability of a material for ship construction.

The strength of the material is its ability to resist deformation. Yield stress and ultimate tensile strength measure the ability to resist forces on the structure.

Hardness of a material describes its ability to resist abrasion. Hardness is important, for example, in bulk carriers where the cargo handling produces abrasive action on the cargo hold structures. Hardness is usually measured on a scale (Rockwell or Brinell), based on test results.

Ductility is the ability of a material to be deformed before it fails.

Brittleness is the opposite of ductility and describes a material that fails under stress because it cannot deform. Softer metals, such as aluminum, are ductile. Hard materials such as cast iron are strong but brittle.

Toughness is the ability of a material to absorb energy.

In comparing the strengths of various metals, stresses and strains are often referred to and require to be defined. Stress is a measure of the ability of a material to transmit a load, and the intensity of stress in the material, which is the load per unit area, is often stated. The load per unit area is simply obtained by dividing the applied load by the cross-sectional area of the material, e.g. if a tensile load of P kg is applied to a rod having a cross-sectional area of A mm^2, then the tensile stress in the material of the rod is P/A kg/mm^2 (see Figure 7.1).

Total strain is defined as the total deformation that a body undergoes when subjected to an applied load. The strain is the deformation per unit length or unit volume, e.g. if the tensile load P applied to the rod of original length l produces an elongation, or extension, of the rod of amount dl, then the tensile

Ship Construction. DOI: 10.1016/B978-0-08-097239-8.00007-6

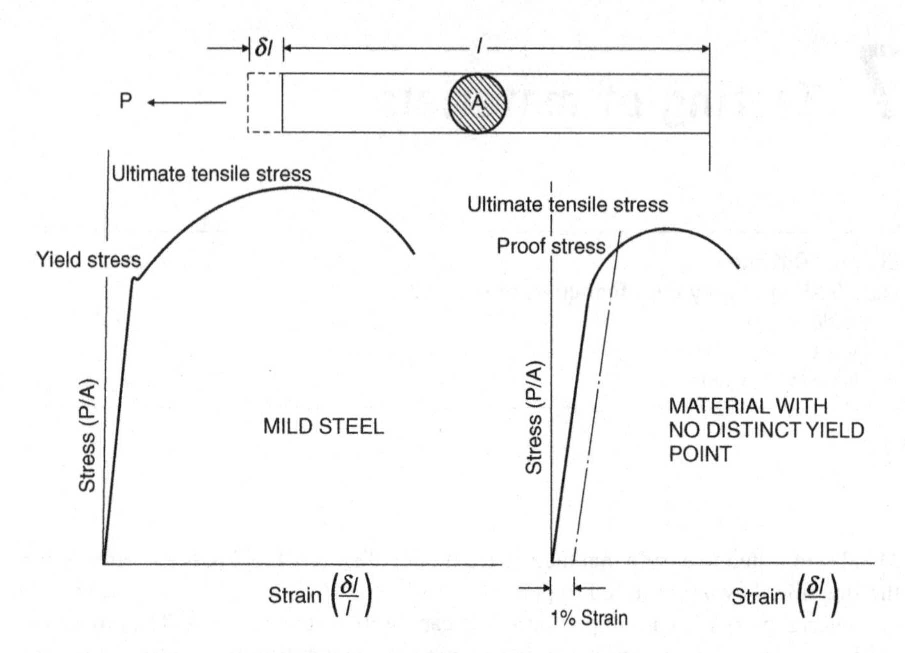

Figure 7.1 Stress–strain relationship of shipbuilding materials.

strain to which the material of the rod is subjected is given by the extension per unit length, i.e.

$$\frac{\text{Extension}}{\text{Original length}} \quad \text{or} \quad \frac{dl}{l}$$

It can be shown that the load on the rod may be increased uniformly and the resulting extension will also increase uniformly until a certain load is reached. This indicates that the load is proportional to extension, and hence stress and strain are proportional since the cross-sectional area and original length of the rod remain constant. For most metals this direct proportionality holds until what is known as the 'elastic limit' is reached. The metal behaves elastically to this point, the rod for example returning to its original length if the load is removed before the 'elastic limit' is reached.

If a mild steel bar is placed in a testing machine and the extensions are recorded for uniformly increasing loads, a graph of load against extension, or stress against strain, may be plotted as in Figure 7.1. This shows the straight-line relationship (i.e. direct proportionality) between stress and strain up to the elastic limit.

Since stress is directly proportional to strain, the stress is equal to a constant, which is in fact the slope of the straight-line part of the graph, and is given by:

$$\text{A constant} = \text{stress} \div \text{strain}$$

This constant is referred to as the modulus of elasticity for the metal and is denoted E (for mild steel its value is approximately 21,100 kg/mm^2 or 21.1 tonnes/mm^2).

The yield stress for a metal corresponds to the stress at the 'yield point'; that is, the point at which the metal no longer behaves elastically. Ultimate tensile stress is the maximum load to which the metal is subjected, divided by the original cross-sectional area. Beyond the yield point the metal behaves plastically, which means that the metal deforms at a greater, unproportional, rate when the yield stress is exceeded, and will not return to its original dimensions on removal of the load. It becomes deformed or is often said to be permanently 'set'.

Many metals do not have a clearly defined yield point, for example aluminum, having a stress–strain curve over its lower range that is a straight line becoming gradually curved without any sharp transformation on yielding, as shown by mild steel (see Figure 7.1). A 'proof stress' is quoted for the material and this may be obtained by setting off on the base some percentage of the strain, say 0.2%, and drawing a line parallel to the straight portion of the curve. The intersection of this line with the actual stress–strain curve marks the proof stress.

It is worth noting at this stage that the ship's structure is designed for working stresses that are within the elastic range and much lower than the ultimate tensile strength of the material to allow a reasonable factor of safety.

Classification society tests for hull materials

Both mild steel and higher tensile steel plates and sections built into a ship are to be produced at works approved by the appropriate classification society. During production an analysis of the material is required and so are prescribed tests of the rolled metal. Similar analyses and tests are required by the classification societies for steel forgings and steel castings, in order to maintain an approved quality.

Destructive tests are made on specimens obtained from the same product as the finished material in accordance with the societies' requirements, which may be found in the appropriate rules. These tests usually take the form of a tensile test and impact test.

Tensile test

The basic principle of this test has already been described, a specimen of given dimensions being subject to an axial pull and a minimum specified yield stress, ultimate tensile stress, and elongation must be obtained. In order to make comparisons between the elongation of tensile test pieces of the same material the test pieces must have the same proportions of sectional area and gage length. Therefore, a standard gage length equal to 5.65 times the square root of the cross-sectional area, which is equivalent to a gage length of five times the diameter, is adopted by the major classification societies.

Impact tests

There are several forms of impact test, but the Charpy V-notch test or Charpy U-notch test is commonly specified and therefore described in this text. The object of the

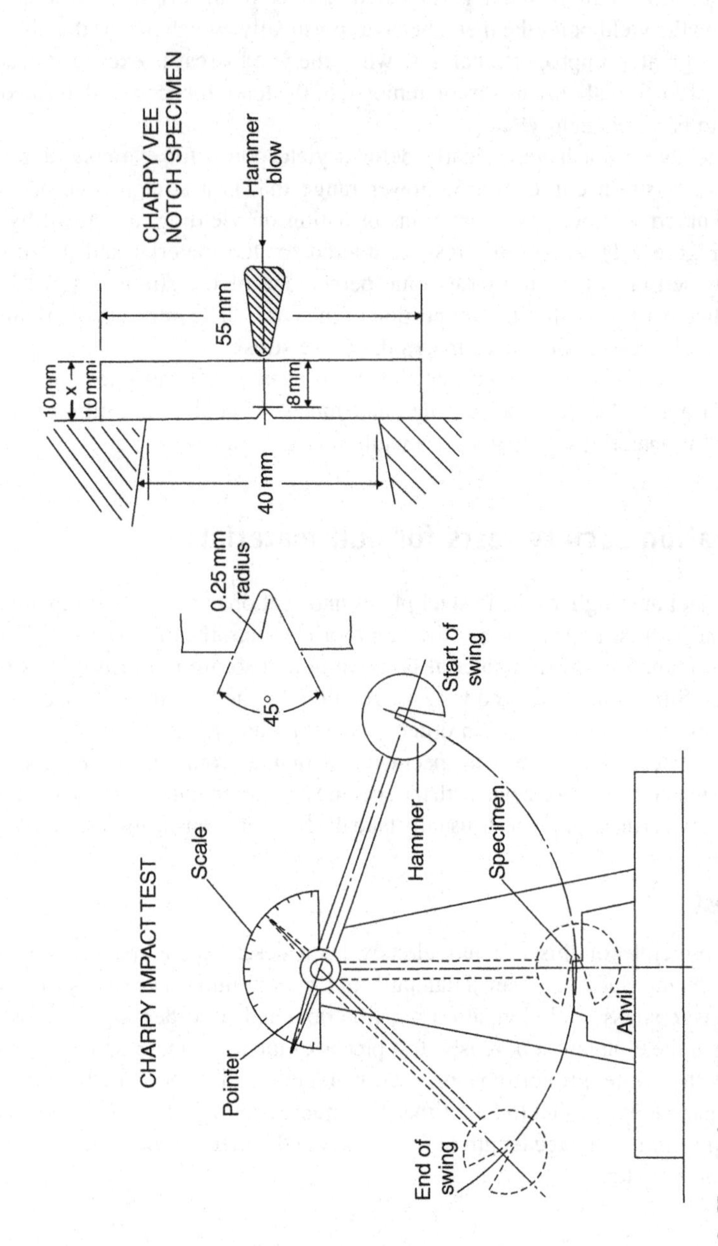

Figure 7.2 Charpy impact test.

impact test is to determine the toughness of the material; that is, its ability to withstand fracture under shock loading. In Figure 7.2 the principle of the Charpy test machine is illustrated, along with the standard test specimen for a Charpy V-notch test. This specimen is placed on an anvil and the pendulum is allowed to swing so that the striker hits the specimen opposite the notch and fractures it. Energy absorbed in fracturing the specimen is automatically recorded by the machine. Basically, making allowances for friction, the energy absorbed in fracturing the specimen is the difference between the potential energy the pendulum possesses before being released, and that which it attains in swinging past the vertical after fracturing the specimen. A specified average impact energy for the specimens tested must be obtained at the specified test temperature, fracture energy being dependent on temperature, as will be illustrated in Chapter 8.

Aluminum alloy tests

Aluminum alloy plate and section material is subject to specified tensile tests. Bar material for aluminum alloy rivets is subject to a tensile test and also a dump test. The latter test requires compression of the bar until its diameter is increased to 1.6 times the original diameter without cracking occurring. Selected manufactured rivets are also subjected to the same dump test.

8 Stresses to which a ship is subject

Chapter Outline

Vertical shear and longitudinal bending in still water 67
Bending moments in a seaway 68
Longitudinal shear forces 68
Bending stresses 68
The ship as a beam 71
Strength deck 71
Transverse stresses 73
Racking 73
Torsion 73
Local stresses 73
Panting 73
Pounding 74
Other local stresses 74
Brittle fracture 75
Fatigue failures 76
Buckling 76
Monitoring ship stresses at sea 78
Further reading 78
Some useful websites 78

The stresses experienced by the ship floating in still water and when at sea may conveniently be considered separately.

Vertical shear and longitudinal bending in still water

If a homogeneous body of uniform cross-section and weight is floating in still water, at any section the weight and buoyancy forces are equal and opposite. Therefore, there is no resultant force at a section and the body will not be stressed or deformed. A ship floating in still water has an unevenly distributed weight owing to both cargo distribution and structural distribution. The buoyancy distribution is also non-uniform since the underwater sectional area is not constant along the length. Total weight and total buoyancy are of course balanced, but at each section there will be a resultant force or load, either an excess of buoyancy or excess of load. Since the vessel remains intact there are vertical upward and downward forces tending to distort the vessel

Ship Construction. DOI: 10.1016/B978-0-08-097239-8.00008-8

(see Figure 8.1), which are referred to as vertical shearing forces, since they tend to shear the vertical material in the hull.

The ship shown in Figure 8.1 will be loaded in a similar manner to the beam shown below it, and will tend to bend in a similar manner owing to the variation in vertical loading. It can be seen that the upper fibers of the beam will be in tension, as will the material forming the deck of the ship with this loading. Conversely, the lower fibers of the beam, and likewise the material forming the bottom of the ship, will be in compression. A vessel bending in this manner is said to be 'hogging' and if it takes up the reverse form with excess weight amidships is said to be 'sagging'. When sagging the deck will be in compression and the bottom shell in tension. Lying in still water the vessel is subjected to bending moments either hogging and sagging depending on the relative weight and buoyancy forces, and it will also be subjected to vertical shear forces.

Bending moments in a seaway

When a ship is in a seaway the waves with their troughs and crests produce a greater variation in the buoyant forces and therefore can increase the bending moment, vertical shear force, and stresses. Classically the extreme effects can be illustrated with the vessel balanced on a wave of length equal to that of the ship. If the crest of the wave is amidships the buoyancy forces will tend to 'hog' the vessel; if the trough is amidships the buoyancy forces will tend to 'sag' the ship (see Figure 8.2). In a seaway the overall effect is an increase of bending moment from that in still water when the greater buoyancy variation is taken into account.

Longitudinal shear forces

When the vessel hogs and sags in still water and at sea, shear forces similar to the vertical shear forces will be present in the longitudinal plane (see Figure 8.2). Vertical and longitudinal shear stresses are complementary and exist in conjunction with a change of bending moment between adjacent sections of the hull girder. The magnitude of the longitudinal shear force is greatest at the neutral axis and decreases towards the top and bottom of the girder.

Bending stresses

From classic bending theory the bending stress (σ) at any point in a beam is given by:

$$\sigma = \frac{M}{I} \times y$$

where M = applied bending moment, y = distance of point considered from neutral axis, and I = second moment of area of cross-section of beam about the neutral axis.

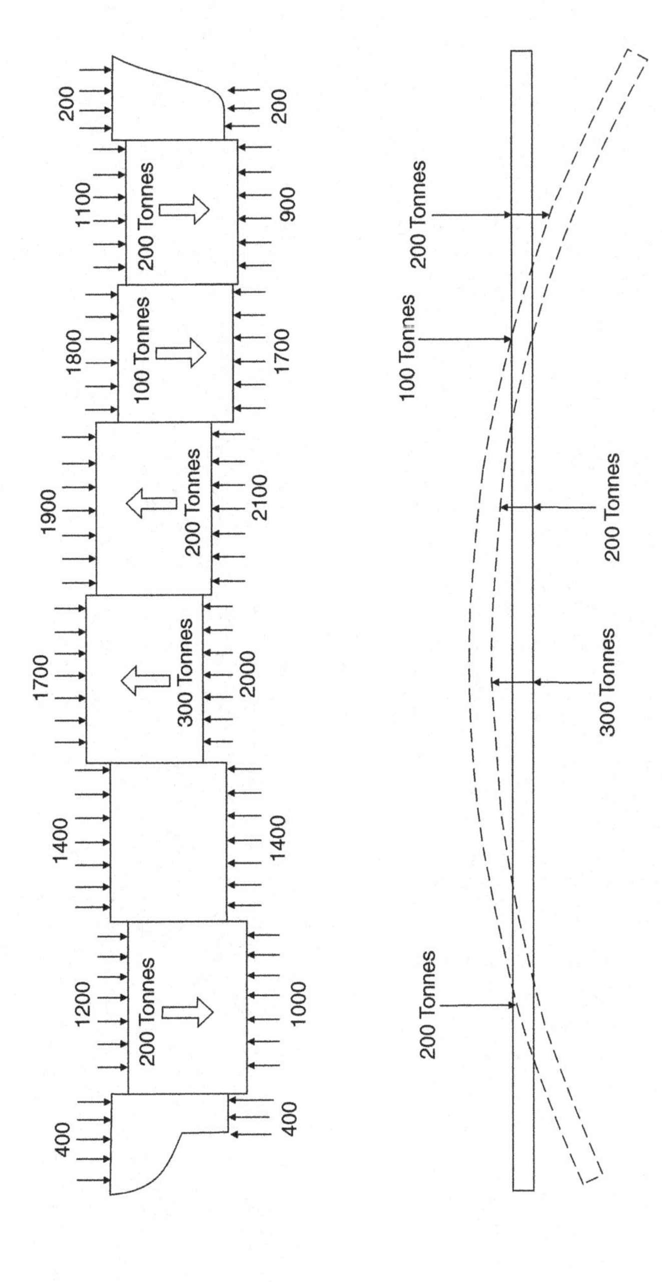

Figure 8.1 Vertical shear and longitudinal bending in still water.

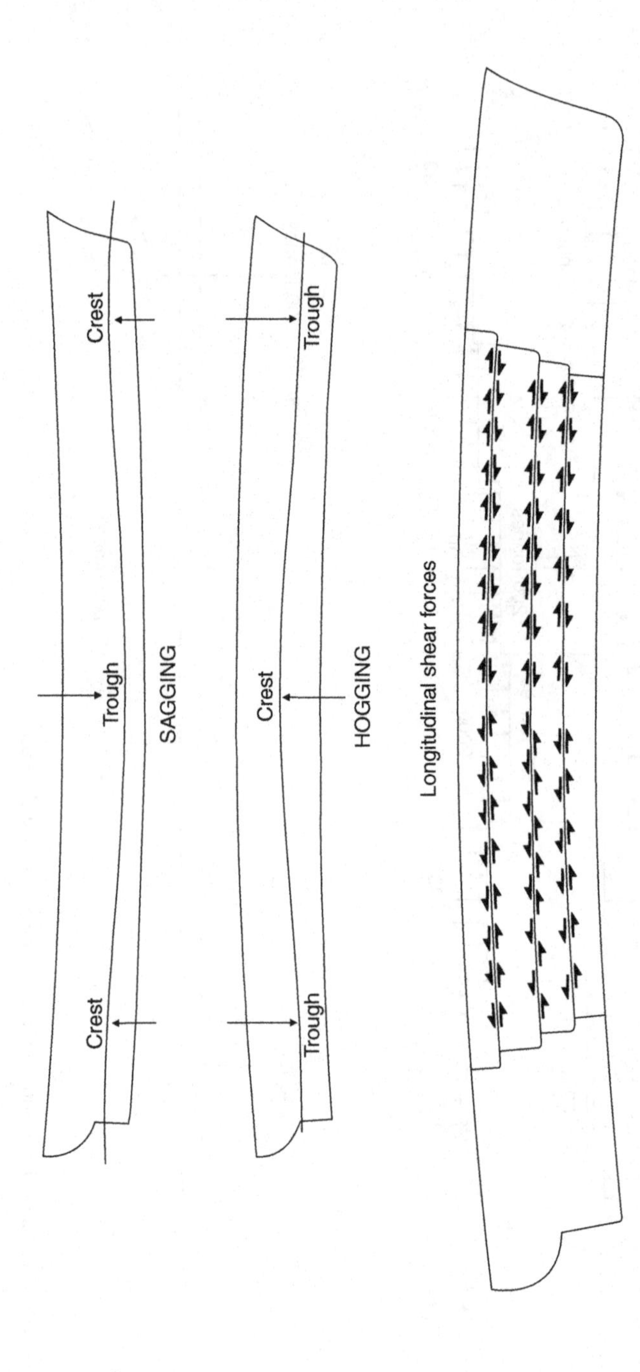

Figure 8.2 Wave bending moments.

When the beam bends it is seen that the extreme fibers are, say in the case of hogging, in tension at the top and in compression at the bottom. Somewhere between the two there is a position where the fibers are neither in tension nor compression. This position is called the *neutral axis*, and at the furthest fibers from the neutral axis the greatest stress occurs for plane bending. It should be noted that the neutral axis always contains the center of gravity of the cross-section. In the equation the second moment of area (I) of the section is a divisor; therefore, the greater the value of the second moment of area, the less the bending stress will be. This second moment of area of section varies as the depth squared and therefore a small increase in depth of section can be very beneficial in reducing the bending stress. Occasionally reference is made to the sectional modulus (Z) of a beam; this is simply the ratio between the second moment of area and the distance of the point considered from the neutral axis, i.e. $I/y = Z$.

The bending stress (σ) is then given by $\sigma = M/Z$.

The ship as a beam

It was seen earlier that the ship bends like a beam, and in fact the hull can be considered as a box-shaped girder for which the position of the neutral axis and second moment of area may be calculated. The deck and bottom shell form the flanges of the hull girder, and are far more important to longitudinal strength than the sides that form the web of the girder and carry the shear forces. The box-shaped hull girder and a conventional I girder are compared in Figure 8.3.

In a ship the neutral axis is generally nearer the bottom, since the bottom shell will be heavier than the deck, having to resist water pressure as well as the bending stresses. In calculating the second moment of an area of the cross-section all longitudinal material is of greatest importance and the further the material is from the neutral axis, the greater will be its second moment of area about the neutral axis. However, at greater distances from the neutral axis, the sectional modulus will be reduced and correspondingly higher stress may occur in extreme hull girder plates such as the deck stringer, sheer strake, and bilge. These strakes of plating are generally heavier than other plating.

Bending stresses are greater over the middle portion of the length and it is owing to this variation that Lloyd's give maximum scantlings over 40% of the length amidships. Other scantlings may taper towards the ends of the ship, apart from locally highly stressed regions where other forms of loading are encountered.

Strength deck

The deck forming the uppermost flange of the main hull girder is often referred to as the strength deck. This is to some extent a misleading term since all continuous decks are in fact strength decks if properly constructed. Along the length of the ship the top flange of the hull girder, i.e. the strength deck, may step from deck to deck where large superstructures are fitted or there is a natural break, for instance in way of a raised quarter deck. Larger superstructures tend to deform with the main hull and stresses of

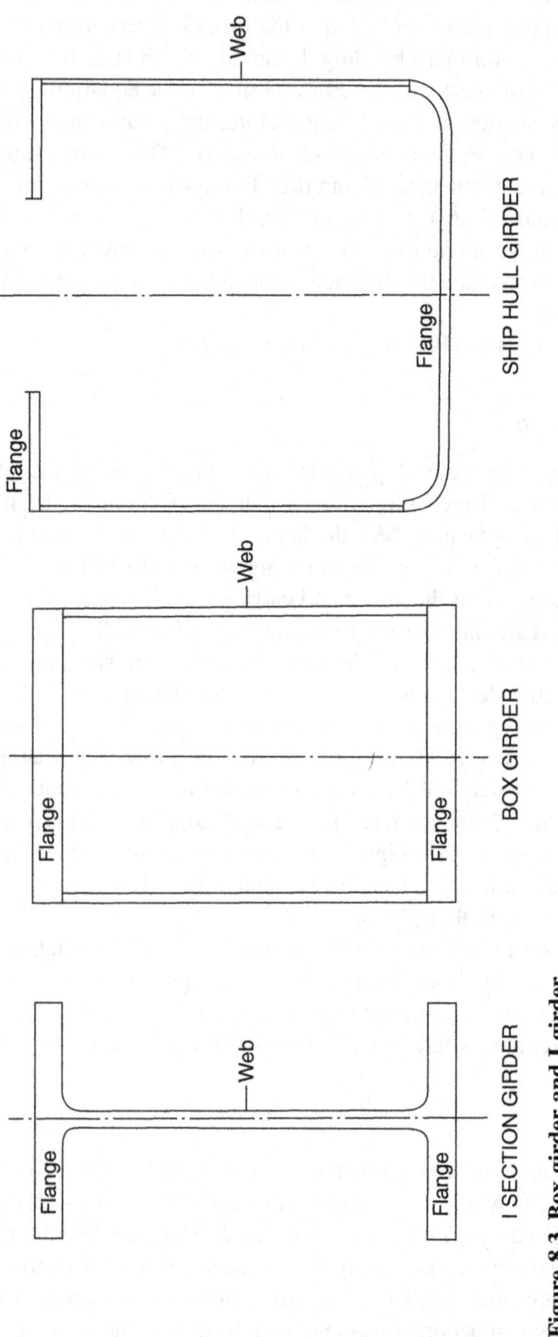

Figure 8.3 Box girder and I girder.

appreciable magnitude will occur in the structure. Early vessels fitted with large superstructures of light construction demonstrated this to their cost. Attempts to avoid fracture have been made by fitting expansion joints, which made the light structure discontinuous. These were not entirely successful and the expansion joint may itself form a stress concentration at the strength deck, which one would wish to avoid. In modern construction the superstructure is usually made continuous and of such strength that its sectional modulus is equivalent to that which the strength deck would have if no superstructure were fitted (see Chapter 19).

Transverse stresses

When a ship experiences transverse forces these tend to change the shape of the vessel's cross-sections and thereby introduce transverse stresses. These forces may be produced by hydrostatic loads and impact of seas or cargo and structural weights, both directly and as the result of reactions due to change of ship motion.

Racking

When a ship is rolling, the deck tends to move laterally relative to the bottom structure, and the shell on one side to move vertically relative to the other side. This type of deformation is referred to as 'racking'. Transverse bulkheads primarily resist such transverse deformation, the side frames' contribution being insignificant provided the transverse bulkheads are at their usual regular spacings. Where transverse bulkheads are widely spaced, deep web frames and beams may be introduced to compensate.

Torsion

When any body is subject to a twisting moment, which is commonly referred to as torque, that body is said to be in 'torsion'. A ship heading obliquely (45°) to a wave will be subjected to righting moments of opposite direction at its ends, twisting the hull and putting it in 'torsion'. In most ships these torsional moments and stresses are negligible but in ships with extremely wide and long deck openings they are significant. A particular example is the larger container ship, where at the topsides a heavy torsion box girder structure including the upper deck is provided to accommodate the torsional stresses (see Figures 8.4 and 17.9).

Local stresses

Panting

Panting refers to a tendency for the shell plating to work 'in' and 'out' in a bellows-like fashion, and is caused by the fluctuating pressures on the hull at the ends when the ship is amongst waves. These forces are most severe when the vessel is running into

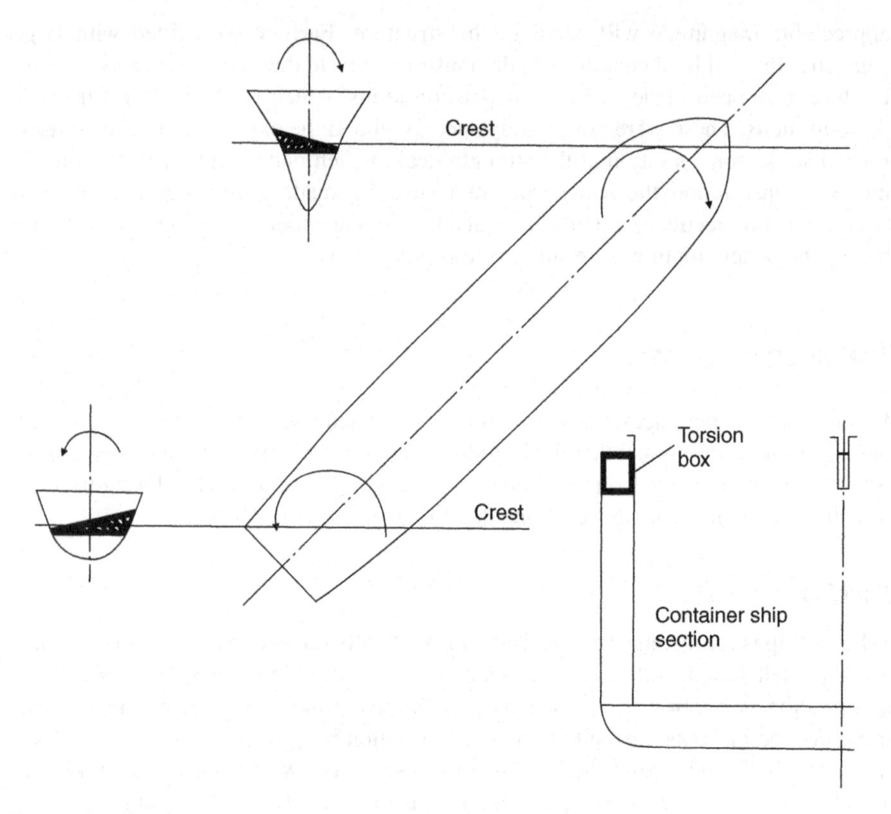

Figure 8.4 Torsion.

waves and is pitching heavily, the large pressures occurring over a short time cycle. Strengthening to resist panting both forward and aft is covered in Chapter 17.

Pounding

Severe local stresses occur in way of the bottom shell and framing forward when a vessel is driven into head seas. These pounding stresses, as they are known, are likely to be most severe in a lightly ballasted condition, and occur over an area of the bottom shell aft of the collision bulkhead. Additional stiffening is required in this region; this is dealt with in Chapter 16.

Other local stresses

Ship structural members are often subjected to high stresses in localized areas, and great care is required to ensure that these areas are correctly designed. This is particularly the case where various load-carrying members of the ship intersect, examples being where longitudinals meet at transverse bulkheads and at intersections of longitudinal and transverse bulkheads. Another highly stressed area occurs where there is a discontinuity of the hull girder at ends of deck house structures, also at hatch and other opening corners, and where there are sudden breaks in the bulwarks.

Brittle fracture

With the large-scale introduction of welding in ship construction, much consideration has been given to the correct selection of materials and structural design to prevent the possibility of brittle fracture occurring. During the Second World War the incidence of this phenomenon was high amongst tonnage hastily constructed, whilst little was known about the mechanics of brittle fracture. Although instances of brittle fracture were recorded in riveted ships, the consequences were more disastrous in the welded vessels because of the continuity of metal provided by the welded joint as opposed to the riveted lap, which tended to limit the propagating crack.

Brittle fracture occurs when an otherwise elastic material fractures without any apparent sign or little evidence of material deformation prior to failure. Fracture occurs instantaneously with little warning and the vessel's overall structure need not be subject to a high stress at the time. Mild steel used extensively in ship construction is particularly prone to brittle fracture given the conditions necessary to trigger it off. The subject is too complex to be dealt with in detail in this text, but it is known that the following factors influence the possibility of brittle fracture and are taken into consideration in the design and material selection of modern ships:

1. A sharp notch is present in the structure from which the fracture initiates.
2. A tensile stress is present.
3. There is a temperature above which brittle fracture will not occur.
4. The metallurgical properties of the steel plate.
5. Thick plate is more prone.

A brittle fracture is distinguishable from a ductile failure by the lack of deformation at the edge of the tear, and its bright granular appearance. A ductile failure has a dull gray appearance. The brittle fracture is also distinguished by the apparent chevron marking, which aids location of the fracture initiation point since these tend to point in that direction.

Factors that are known to exist where a brittle fracture may occur must be considered if this is to be avoided. Firstly, the design of individual items of ship structure must be such that sharp notches where cracks may be initiated are avoided. With welded structures as large as a ship the complete elimination of crack initiation is not entirely possible owing to the existence of small faults in the welds, a complete weld examination not being practicable. Steel specified for the hull construction should therefore have good 'notch ductility' at the service temperatures, particularly where thick plate is used. Provision of steel having good 'notch ductility' properties has the effect of making it difficult for a crack to propagate. Notch ductility is a measure of the relative toughness of the steel, which has already been seen to be determined by an impact test. Steels specified for ship construction have elements added (particularly manganese with a carbon limit), and may also be subjected to a controlled heat treatment, which will enhance the notch toughness properties. Figure 8.5 illustrates the improved notch ductility of a manganese/carbon steel against a plain carbon steel. Grade D and Grade E steels, which have higher notch ductility, are employed where thick plate is used and in way of higher stressed regions, as will be seen when the ship structural details are considered later.

In association with the problem of brittle fracture it was not uncommon at one time to hear reference to the term 'crack arrester'. The term related to the now outdated practice of introducing riveted seams in cargo ships to subdivide the vessel into welded substructures so that any possible crack propagation was limited to the substructure. In particular, such a 'crack arrester' was usually specified in the sheerstrake/stringer plate area of larger ships. Today strakes of higher notch toughness steel are required to be fitted in such areas. Lloyd's Register, for example, require the mild steel sheerstrake and stringer plate at the strength deck over the midship portion of vessels of more than 250 meters length to be Grade D if less than 15 mm thick and Grade E if of greater thickness (see Chapter 17).

Fatigue failures

Unlike brittle fracture, fatigue fracture occurs very slowly and can in fact take years to propagate. The greatest danger with fatigue fractures is that they occur at low stresses that are applied to a structure repeatedly over a period of time (Figure 8.5). A fatigue crack, once initiated, may grow unnoticed until the load-bearing member is reduced to a cross-sectional area that is insufficient to carry the applied load. Fatigue failures are associated with sharp notches or discontinuities in structures, and are especially prevalent at 'hard spots', i.e. regions of high rigidity in ship structures.

With the growth in size of oil tankers, bulk carriers, and container ships there has been increasing use of higher yield strength steels in their hull structures. The classification societies have subsequently placed special emphasis on analysis of the fatigue performance of these larger structures, usually over a 25-year life cycle, as part of their approval process.

Buckling

With the substantial increase in size of oil tankers, bulk carriers, and container ships in recent years, greater attention has had to be given to the buckling strength of the stiffened plate panels constituting the shell.

Buckling of a structural member loaded in compression may occur at a stress level that is substantially lower than the material's yield stress (see Figure 7.1). The load at which buckling occurs is a function of the structural member's geometry and the material's modulus of elasticity rather than the material's strength. The most common example of buckling failure is the collapse of a pillar under a compressive load. A stiffened plate panel in compression will also have a critical buckling load

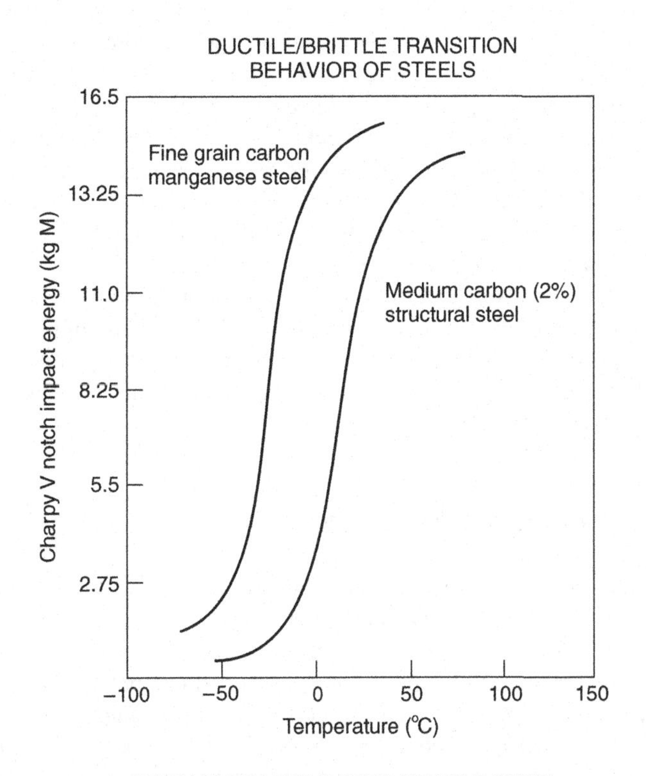

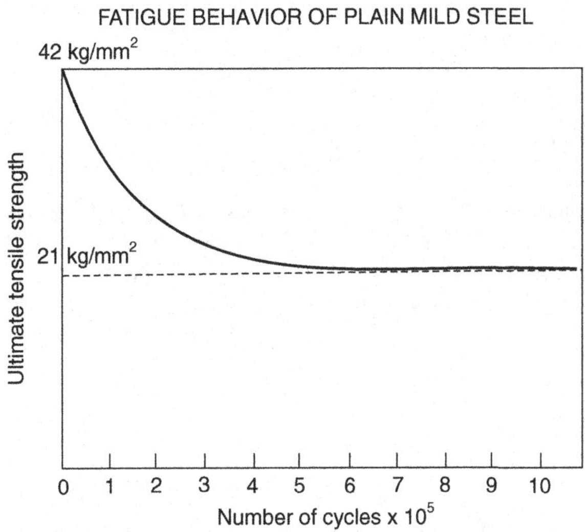

Figure 8.5 Benefit of manganese/carbon steel.

whose value depends on the plate thickness, unsupported dimensions, edge support conditions, and the material's modulus of elasticity. Unlike the pillar, however, slightly exceeding this load will not necessarily result in collapse of the plate but only in elastic deflection of the center portion of the plate from its initial plane. After removal of the load, the plate will return to its original undeformed state. The ultimate load that may be carried by a buckled plate is determined by the onset of yielding (i.e. when the yield point of the material is reached) at some point in the plate or in the stiffeners. Once begun, this yielding may propagate rapidly throughout the stiffened panel with further increase in load until failure of the plate or stiffeners occurs.

In recent years Lloyd's Register has introduced for oil tankers over 150 meters in length rules that contain formulae to check the buckling capacity of plate panels and their main supporting stiffeners. Where further buckling assessment is required a computer-based general and local stiffened plate panel ultimate buckling strength evaluation assessment procedure is used.

Monitoring ship stresses at sea

In order to enhance safety during shipboard operations, real-time motion and stress monitoring information equipment can be supplied by classification societies such as Lloyd's Register to a ship at the owner's request. This entails the fitting of strain gages to the deck structure, an accelerometer, and a personal computer with software that displays ship stress and motion readings on the bridge. An alarm is activated if the safety limits are exceeded, enabling remedial action to be taken. Where this equipment is installed the notation SEA is assigned and if coupled to a data recorder the assigned notation is SEA(R).

Further reading

Fatigue in ship structures—New light on an eternal issue, *The Naval Architect*, January 2003.

Rawson, Tupper: *Basic Ship Theory*, ed 5, vol 1, Chapter 6: The ship girder; Chapter 7: Structural design and analysis, 2001, Butterworth Heinemann.

Refined ABS rules for very large container ships, *The Naval Architect*, October 2005.

Stress monitoring system for NASSCO's Alaska-class tankers, *The Naval Architect*, September 2003.

Some useful websites

www.iacs.org.uk See Guidelines and Recommendations—Recommendation 46 'Bulk Carriers—Guidance and Information on Bulk Cargo Loading and Discharging to Reduce the Likelihood of Over-stressing the Hull Structure' and Recommendation 56 'Fatigue Assessment of Ship Structures'.

Part Three

Welding and Cutting

9 Welding and cutting processes used in shipbuilding

Chapter Outline

Gas welding 82
Electric arc welding 84
Slag-shielded processes 84
Manual welding electrodes 86
Automatic welding with cored wires 86
Submerged arc welding 86
Stud welding 88
Gas-shielded arc welding processes 89
Tungsten inert gas (TIG) welding 89
Metal inert gas (MIG) welding 89
Plasma welding 93
Other welding processes 94
Electro-slag welding 94
Electro-gas welding 94
Laser welding 95
Thermit welding 96
Friction stir welding[1] 96
Cutting processes 98
Gas cutting 98
Plasma-arc cutting 98
Gouging 100
Laser cutting 100
Water jet cutting 101
Further reading 101
Some useful websites 101

Initially welding was used in ships as a means of repairing various metal parts. During the First World War various authorities connected with shipbuilding, including Lloyd's Register, undertook research into welding and in some cases prototype welded structures were built. However, riveting remained the predominant method employed for joining ship plates and sections until the time of the Second World War. During and after this war the use and development of welding for shipbuilding purposes was widespread, and welding totally replaced riveting in the latter part of the twentieth century.

Ship Construction. DOI: 10.1016/B978-0-08-097239-8.00009-X

There are many advantages to be gained from employing welding in ships as opposed to having a riveted construction. These may be considered as advantages in both building and in operating the ship.

For the shipbuilder the advantages are:

1. Welding lends itself to the adoption of prefabrication techniques.
2. It is easier to obtain watertightness and oiltightness with welded joints.
3. Joints are produced more quickly.
4. Less skilled labor is required.

For the shipowner the advantages are:

1. Reduced hull steel weight, therefore more deadweight.
2. Less maintenance from slack rivets, etc.
3. The smoother hull with the elimination of overlapping plate joints leads to reduced skin friction resistance, which can reduce fuel costs.

Other than some blacksmith work involving solid-phase welding, the welding processes employed in shipbuilding are of the fusion welding type. Fusion welding is achieved by means of a heat source that is intense enough to melt the edges of the material to be joined as it is traversed along the joint. Gas welding, arc welding, laser welding, and resistance welding all provide heat sources of sufficient intensity to achieve fusion welds.

Gas welding

A gas flame was probably the first form of heat source to be used for fusion welding, and a variety of fuel gases with oxygen have been used to produce a high-temperature flame. The most commonly used gas in use is acetylene, which gives an intense concentrated flame (average temperature 3000 °C) when burnt in oxygen.

An oxyacetylene flame has two distinct regions: an inner cone, in which the oxygen for combustion is supplied via the torch; and a surrounding envelope, in which some or all the oxygen for combustion is drawn from the surrounding air. By varying the ratio of oxygen to acetylene in the gas mixture supplied by the torch, it is possible to vary the efficiency of the combustion and alter the nature of the flame (Figure 9.1). If the oxygen supply is slightly greater than the supply of acetylene by volume, what is known as an 'oxidizing' flame is obtained. This type of flame may be used for welding materials of high thermal conductivity, e.g. copper, but not steels, as the steel may be decarburized and the weld pool depleted of silicon. With equal amounts of acetylene and oxygen a 'neutral' flame is obtained, and this would normally be used for welding steels and most other metals. Where the acetylene supply exceeds the oxygen by volume a 'carburizing' flame is obtained, the excess acetylene decomposing and producing submicroscopic particles of carbon. These readily go into solution in the molten steel, and can produce metallurgical problems in service.

The outer envelope of the oxyacetylene flame by consuming the surrounding oxygen to some extent protects the molten weld metal pool from the surrounding air. If unprotected the oxygen may diffuse into the molten metal and produce porosity

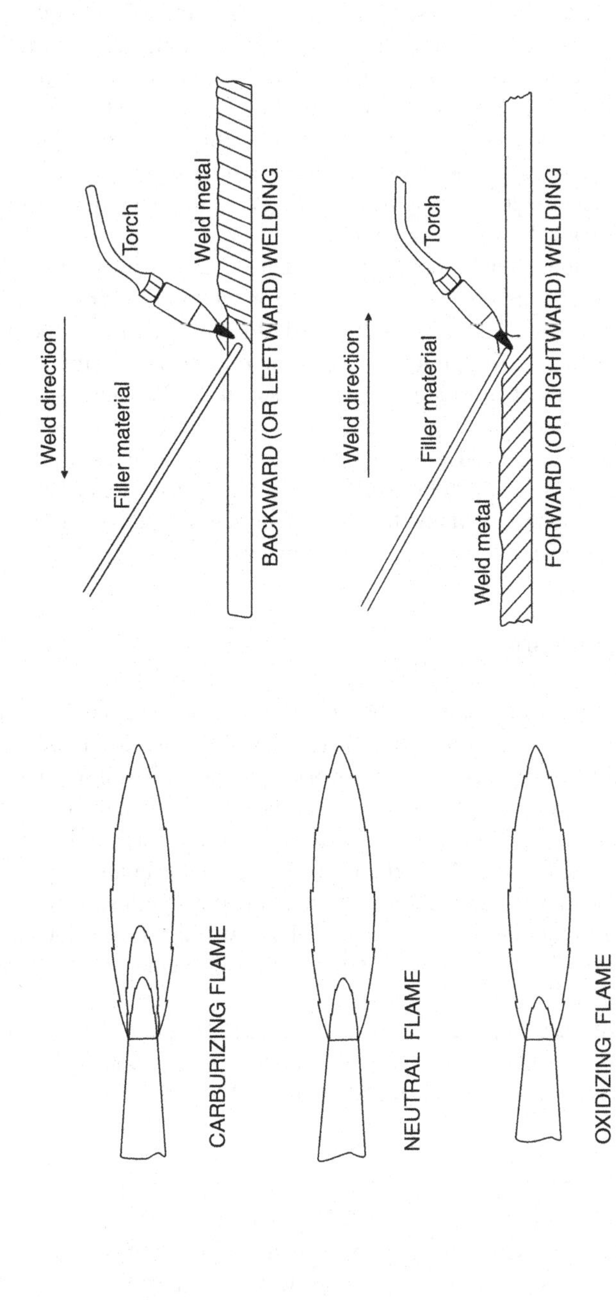

Figure 9.1 Gas welding.

when the weld metal cools. With metals containing refractory oxides, such as stainless steels and aluminum, it is necessary to use an active flux to remove the oxides during the welding process.

Both oxygen and acetylene are supplied in cylinders, the oxygen under pressure and the acetylene dissolved in acetone since it cannot be compressed. Each cylinder, which is distinctly colored (red—acetylene, black—oxygen), has a regulator for controlling the working gas pressures. The welding torch consists of a long thick copper nozzle, a gas mixer body, and valves for adjusting the oxygen and acetylene flow rates. Usually a welding rod is used to provide filler metal for the joint, but in some cases the parts to be joined may be fused together without any filler metal. Gas welding techniques are shown in Figure 9.1.

Oxyacetylene welding tends to be slower than other fusion welding processes because the process temperature is low in comparison with the melting temperature of the metal, and because the heat must be transferred from the flame to the plate. The process is therefore only really applicable to thinner mild steel plate, thicknesses up to 7 mm being welded using this process with a speed of 3–4 meters per hour. In shipbuilding oxyacetylene welding has almost disappeared but can be employed in the fabrication of ventilation and air-conditioning trunking, cable trays, and light steel furniture; some plumbing and similar work may also make use of gas welding. These trades may also employ the gas flame for brazing purposes, where joints are obtained without reaching the fusion temperature of the material being joined.

Electric arc welding

The basic principle of electric arc welding is that a wire or electrode is connected to a source of electrical supply with a return lead to the plates to be welded. If the electrode is brought into contact with the plates an electric current flows in the circuit. By removing the electrode a short distance from the plate, so that the electric current is able to jump the gap, a high-temperature electrical arc is created. This will melt the plate edges and the end of the electrode if this is of the consumable type.

Electrical power sources vary, DC generators or rectifiers with variable or constant voltage characteristics being available, as well as AC transformers with variable voltage characteristics for single or multiple operation. The latter are most commonly used in shipbuilding.

Illustrated in Figure 9.2 are the range of manual, semi-automatic, and automatic electric arc welding processes that might be employed in shipbuilding. Each of these electric arc welding processes is discussed below with its application.

Slag-shielded processes

Metal arc welding started as bare wire welding, the wire being attached to normal power lines. This gave unsatisfactory welds, and subsequently it was discovered that by dipping the wire in lime a more stable arc was obtained. As a result of further developments many forms of flux are now available for coating the wire or for

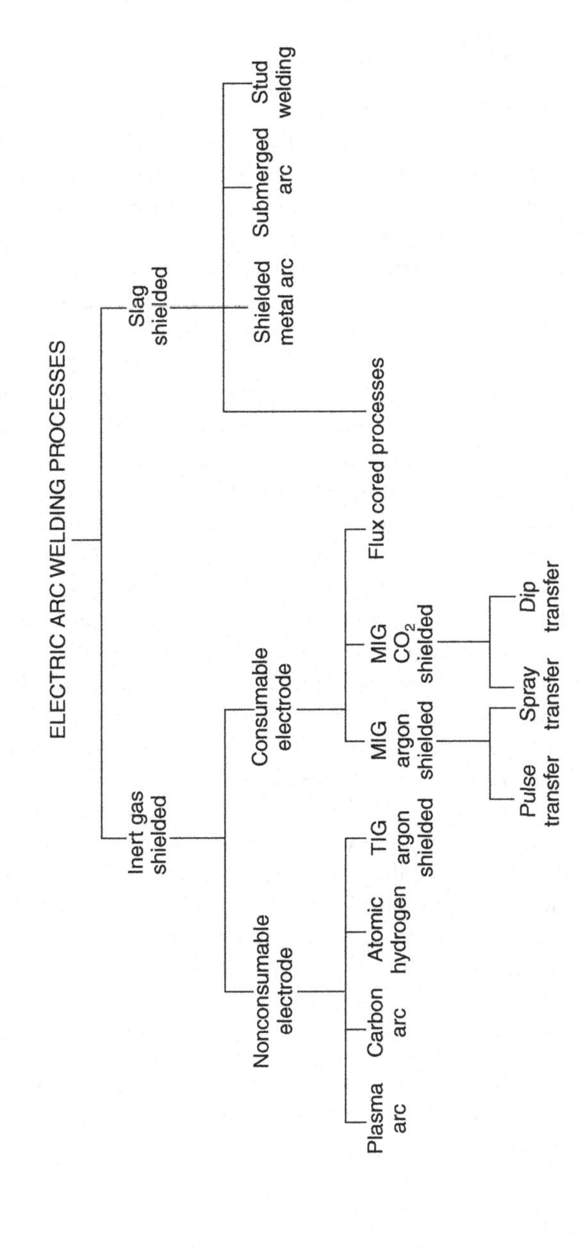

Figure 9.2 Electric arc welding processes.

deposition on the joint prior to welding. Other developments include a hollow wire for continuous welding with the flux within the hollow core. The flux melts, then solidifies during the welding process, forming a solid slag that protects the weld from atmospheric oxygen and nitrogen.

Manual welding electrodes

The core wire normally used for mild steel electrodes is rimming steel. This is ideal for wire-drawing purposes, and elements used to 'kill' steel such as silicon or aluminum tend to destabilize the arc, making 'killed' steels unsuitable. Coatings for the electrodes normally consist of a mixture of mineral silicates, oxides, fluorides, carbonates, hydrocarbons, and powdered metal alloys plus a liquid binder. After mixing, the coating is then extruded onto the core wire and the finished electrodes are dried in batches in ovens.

Electrode coatings should provide gas shielding for the arc, easy striking and arc stability, a protective slag, good weld shape, and most important of all a gas shield consuming the surrounding oxygen and protecting the molten weld metal. Various electrode types are available, the type often being defined by the nature of the coating. The more important types are the rutile and basic (or low-hydrogen) electrodes. Rutile electrodes have coatings containing a high percentage of titania, and are general-purpose electrodes that are easily controlled and give a good weld finish with sound properties. Basic or low-hydrogen electrodes, the coating of which has a high lime content, are manufactured with the moisture content of the coating reduced to a minimum to ensure low-hydrogen properties. The mechanical properties of weld metal deposited with this type of electrode are superior to those of other types, and basic electrodes are generally specified for welding the higher tensile strength steels. Where high restraint occurs, for example at the final erection seam weld between two athwartships rings of unit structure, low-hydrogen electrodes may also be employed. An experienced welder is required where this type of electrode is used since it is less easily controlled. Welding with manual electrodes may be accomplished in the downhand position, for example welding at the deck from above, also in the horizontal vertical, or vertical positions, for example across or up a bulkhead, and in the overhead position, for example welding at the deck from below (Figure 9.3). Welding in any of these positions requires selection of the correct electrode (positional suitability stipulated by the manufacturer), correct current, correct technique, and inevitably experience, particularly for the vertical and overhead positions.

Automatic welding with cored wires

Flux-cored wires (FCAW) are often used in mechanized welding, allowing higher deposition rates and improved quality of weld. Basic or rutile flux-cored wires are commonly used for one-sided welding with a ceramic backing.

Submerged arc welding

This is an arc welding process in which the arc is maintained within a blanket of granulated flux (see Figure 9.4). A consumable filler wire is employed and the arc is

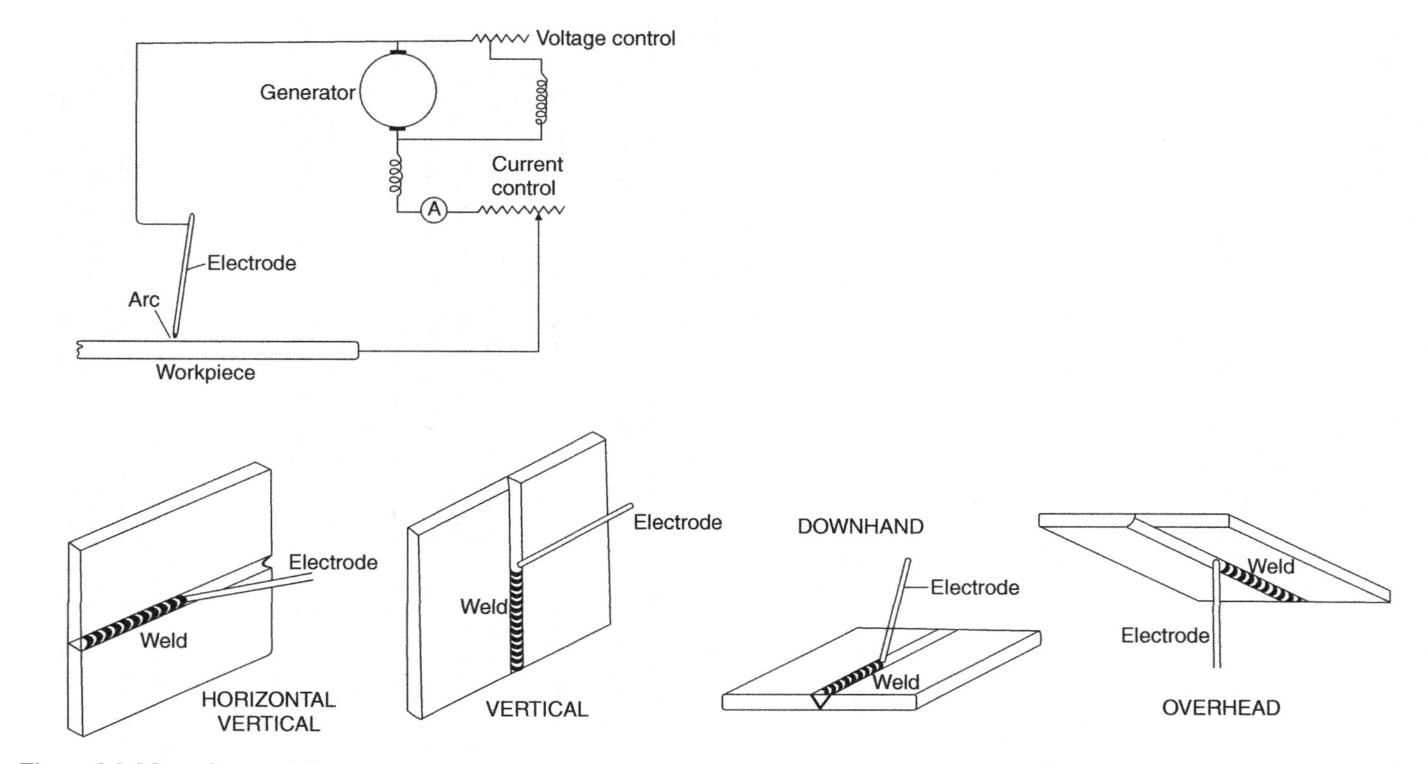

Figure 9.3 Manual arc welding.

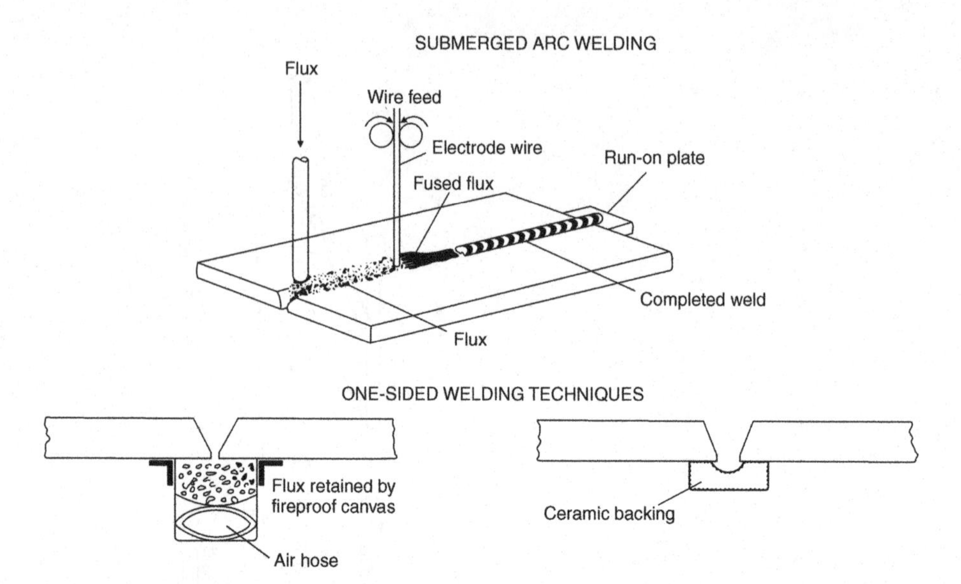

Figure 9.4 Automatic arc welding.

maintained between this wire and the parent plate. Around the arc the granulated flux breaks down and provides some gases, and a highly protective thermally insulating molten container for the arc. This allows a high concentration of heat, making the process very efficient and suitable for heavy deposits at fast speeds. After welding the molten metal is protected by a layer of fused flux, which together with the unfused flux may be recovered before cooling.

This is the most commonly used process for downhand mechanical welding in the shipbuilding industry, in particular for joining plates for ship shell, decks, and bulkheads. Metal powder additions that result in a 30–50% increase in metal deposition rate without incurring an increase in arc energy input may be used for the welding of joint thicknesses of 25 mm or more. Submerged arc multi-wire and twin-arc systems are also used to give high productivity.

With shipyards worldwide adopting one-side welding in their ship panel lines for improved productivity, the submerged arc process is commonly used with a fusible backing, using either flux or glass fiber materials to contain and control the weld penetration bead.

Stud welding

Stud welding may be classed as a shielded arc process, the arc being drawn between the stud (electrode) and the plate to which the stud is to be attached. Each stud is inserted into a stud welding gun chuck, and a ceramic ferrule is slipped over it before the stud is placed against the plate surface. On depressing the gun trigger the stud is automatically retracted from the plate and the arc established, melting the end of the stud and the local plate surface. When the arcing period is complete, the current is automatically shut off and the stud driven into a molten pool of weld metal, so attaching stud to plate.

Apart from the stud welding gun the equipment includes a control unit for timing the period of current flow. Granular flux is contained within the end of each stud to create a protective atmosphere during arcing. The ceramic ferrule that surrounds the weld area restricts the access of air to the weld zone; it also concentrates the heat of the arc and confines the molten metal to the weld area (see Figure 9.5).

Stud welding is often used in shipbuilding, generally for the fastening of stud bolts to secure supports for pipe hangars, electric cable trays and other fittings, also insulation to bulkheads and wood sheathing to decks, etc. Apart from various forms of stud bolts, items like stud hooks and rings are also available.

Gas-shielded arc welding processes

The application of bare wire welding with gas shielding was developed in the 1960s, and was quickly adopted for the welding of lighter steel structures in shipyards, as well as for welding aluminum alloys. Gas-shielded processes are principally of an automatic or semi-automatic nature.

Tungsten inert gas (TIG) welding

In the TIG welding process the arc is drawn between a water-cooled nonconsumable tungsten electrode and the plate (Figure 9.6). An inert gas shield is provided to protect the weld metal from the atmosphere, and filler metal may be added to the weld pool as required. Ignition of the arc is obtained by means of a high-frequency discharge across the gap, since it is not advisable to strike an arc on the plate with the tungsten electrode. Normally in Britain the inert gas shield used for welding aluminum and steel is argon. Only plate thicknesses of less than 6 mm would normally be welded by this process, and in particular aluminum sheet, a skilled operator being required for manual work. This may also be referred to as TAGS welding, i.e. tungsten arc gas-shielded welding.

Metal inert gas (MIG) welding

This is in effect an extension of TIG welding, the electrode in this process becoming a consumable metal wire.

Basically the process is as illustrated in Figure 9.6, a wire feed motor supplying wire via guide rollers through a contact tube in the torch to the arc. An inert gas is supplied to the torch to shield the arc, and electrical connections are made to the contact tube and workpiece. Welding is almost always done with a DC source and electrode positive for regular metal transfer, and when welding aluminum to remove the oxide film by the action of the arc cathode. Although the process may be fully automatic, semi-automatic processes as illustrated with hand gun are now in greater use, and are particularly suitable in many cases for application to shipyard work.

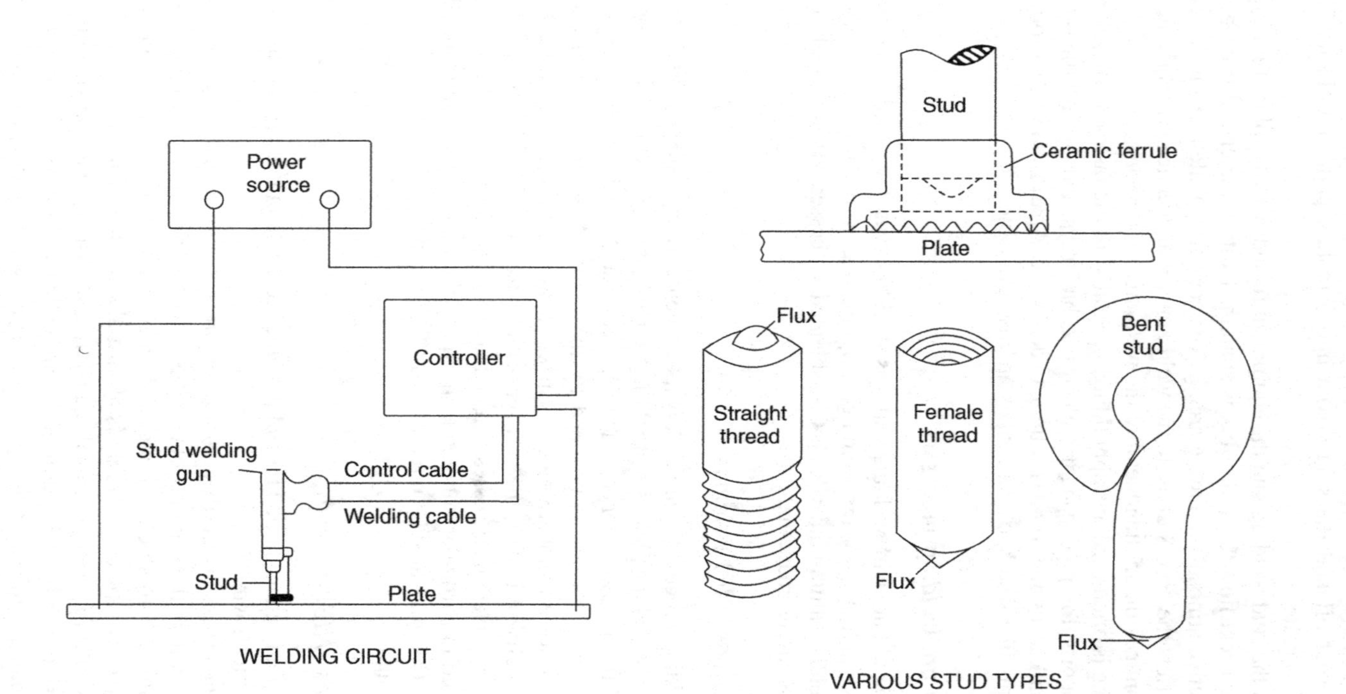

Figure 9.5 Stud welding.

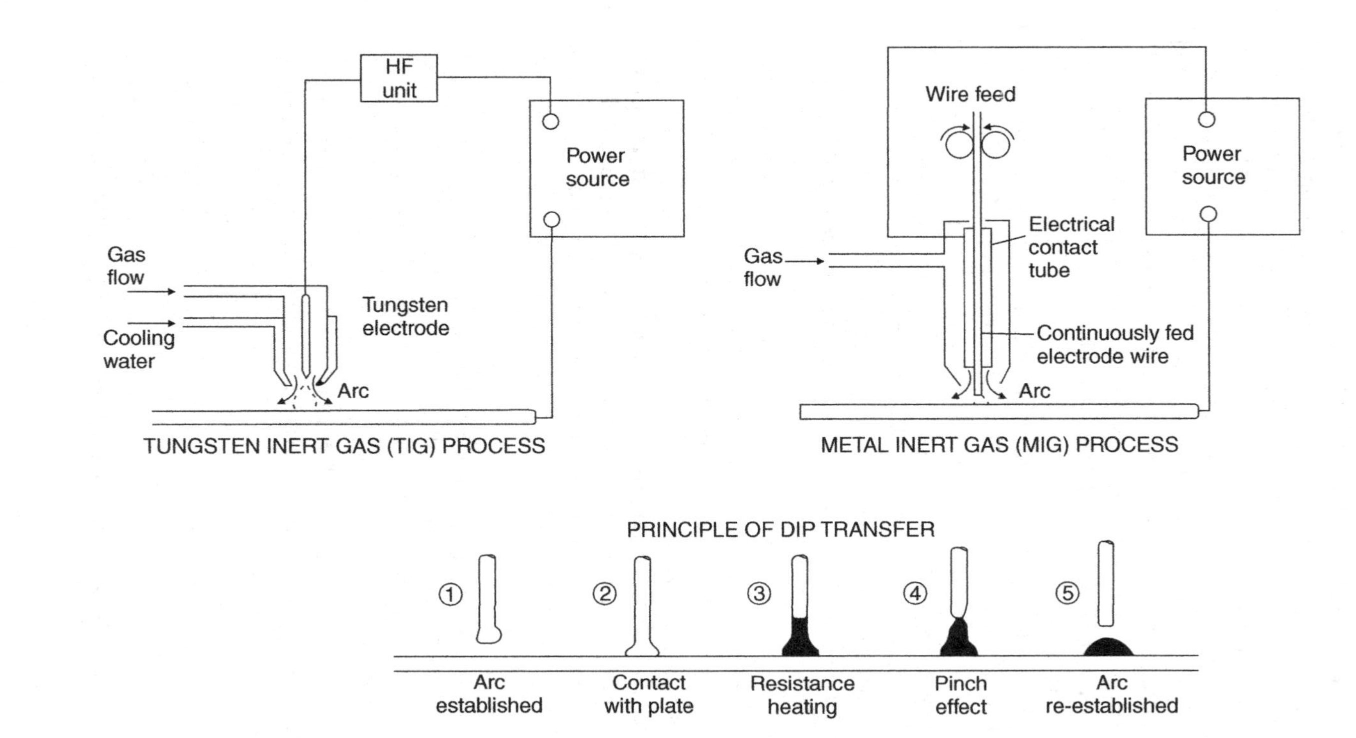

Figure 9.6 Metal inert gas welding.

Initially aluminum accounted for most of the MIG welding, with argon being used as the inert shielding gas. Much of the welding undertaken on aluminum deckhouses, and liquid methane gas tanks of specialized carriers, has made use of this process. Generally larger wire sizes and heavier currents have been employed in this work, metal transfer in the arc being by means of a spray transfer, i.e. metal droplets being projected at high speed across the arc. At low currents metal transfer in the arc is rather difficult and very little fusion of the plate results, which has made the welding of light aluminum plate rather difficult with the MIG/argon process. The introduction of the 'pulsed arc' process has to some extent overcome this problem and made positional welding easier. Here a low-level current is used with high-level pulses of current that detach the metal from the electrode and accelerate it across the arc to give good penetration.

Early work on the welding of mild steel with the metal inert gas process made use of argon as a shielding gas, but as this gas is rather expensive, and satisfactory welding could only be accomplished in the downhand position, an alternative shielding gas was sought. Research in this direction was concentrated on the use of CO_2 as the shielding gas, and the MIG/CO_2 process is now widely used for welding mild steel. Using higher current values with thicker steel plate a fine spray transfer of the metal from the electrode across the arc is achieved, with a deep penetration. Wire diameters in excess of 1.6 mm are used, and currents above about 350 amps are required to obtain this form of transfer. Much of the higher current work is undertaken with automatic machines, but some semi-automatic torches are available to operate in this range in the hands of skilled welders. Welding is downhand only.

On thinner plating where lower currents would be employed, a different mode of transfer of metal in the arc is achieved with the MIG/CO_2 process. This form of welding is referred to as the dip transfer (or short-circuiting) process. The sequence of metal transfer is (see Figure 9.6):

1. Establish the arc.
2. Wire fed into arc until it makes contact with plate.
3. Resistance heating of wire in contact with plate.
4. Pinch effect, detaching heated portion of wire as droplet of molten metal.
5. Re-establish the arc.

To prevent a rapid rise of current and 'blast off' of the end of the wire when it short-circuits on the plate, variable inductance is introduced in the electrical circuit. Smaller wire diameters, 0.8 and 1.2 mm, are used where the dip transfer method is employed on lighter plate at low currents. The process is suitable for welding light mild steel plate in all positions. It may be used in shipbuilding as a semi-automatic process, particularly for welding deckhouses and other light steel assemblies.

The pulsed MIG/argon process, developed for positional welding of light aluminum plate, may be used for positional welding of light steel plate but is likely to prove more expensive.

Use of the MIG semi-automatic processes can considerably increase weld output and lower costs.

This form of welding may also be collectively referred to as MAGS welding, i.e. metal arc gas-shielded welding.

Plasma welding

This is very similar to TIG welding as the arc is formed between a pointed tungsten electrode and the plate. But, with the tungsten electrode positioned within the body of the torch, the plasma arc is separated from the shielding gas envelope (see Figure 9.7). Plasma is forced through a fine-bore copper nozzle that constricts the arc. By varying the bore diameter and plasma gas flow rate, three different operating modes can be achieved:

1. Microplasma—the arc is operated at very low welding currents (0.1–15 amps) and used for welding thin sheets (down to 0.1 mm thickness).
2. Medium current—the arc is operated at currents from 15 to 200 amps. Plasma welding is an alternative to conventional TIG welding, but with the advantage of achieving deeper penetration and having greater tolerance to surface contamination. Because of the bulkiness of the torch, it is more suited to mechanized welding than hand welding.
3. Keyhole plasma—the arc is operated at currents above 100 amps and by increasing the plasma flow a very powerful plasma beam is created. This can penetrate thicknesses up to 10 mm, but when using a single-pass technique is normally limited to a thickness of 6 mm. This operating mode is normally used for welding sheet metal (over 3 mm) in the downhand position.

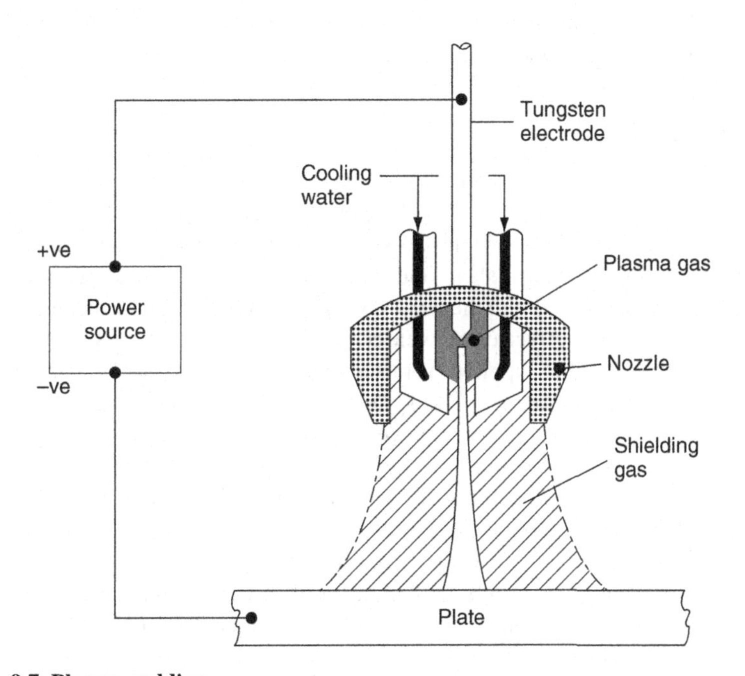

Figure 9.7 Plasma welding.

Other welding processes

There are some welding processes that cannot strictly be classified as gas or arc welding processes and these are considered separately.

Electro-slag welding

The electro-slag welding process is used for welding heavy casting structure components such as stern frames and was also used at an earlier stage to make vertical welds in heavier side shell when fabricated hull units where joined at the berth. With the development of the electro-gas welding process, electro-slag welding is no longer used for the latter purpose.

To start the weld an arc is struck, but welding is achieved by resistance path heating through the flux, the initial arcing having been discouraged once welding is started. In Figure 9.8 the basic electro-slag process is illustrated; the current passes into the weld pool through the wire, and the copper water-cooled shoes retain the molten pool of weld metal. These may be mechanized so that they move up the plate as the weld is completed, flux being fed into the weld manually by the operator. A square edge preparation is used on the plates, and it is found that the final weld metal has a high plate dilution. 'Run-on' and 'run-off' plates are required for stopping and starting the weld, and it is desirable that the weld should be continuous. If a stoppage occurs it will be impossible to avoid a major slag inclusion in the weld, and it may then be necessary to cut out the original metal and start again. If very good weld properties are required with a fine grain structure (electro-slag welds tend to have a coarse grain structure) it is necessary to carry out a local normalizing treatment.

Electro-gas welding

Of greater interest to the shipbuilder is a further development, electro-gas welding (see Figure 9.8). This is in fact an arc welding process that combines features of gas-shielded welding with those of electro-slag welding. Water-cooled copper shoes similar to those for the electro-slag welding process are used, but a flux-cored wire rather than a bare wire is fed into the weld pool. Fusion is obtained by means of an arc established between the surface of the weld pool and the wire, and the CO_2 or CO_2 with argon mixture gas shield is supplied from separate nozzles or holes located centrally near the top of the copper shoes. The system is mechanized utilizing an automatic vertical-up welding machine fed by a power source and having a closed-loop cooling circuit and a level sensor that automatically adjusts the vertical travel speed.

The process is more suitable for welding plates in the thickness range of 13–50 mm with square or vee edge preparations and is therefore used for shipbuilding purposes in the welding of vertical butts when erecting side shell panels or for the vertical shell butt joints when joining building blocks on the berth or dock. For this purpose the use of a single- or double-vee butt with the electro-gas process is preferable since this permits completion of the weld manually if any breakdown occurs. A square butt with appreciable gap would be almost impossible to bridge manually.

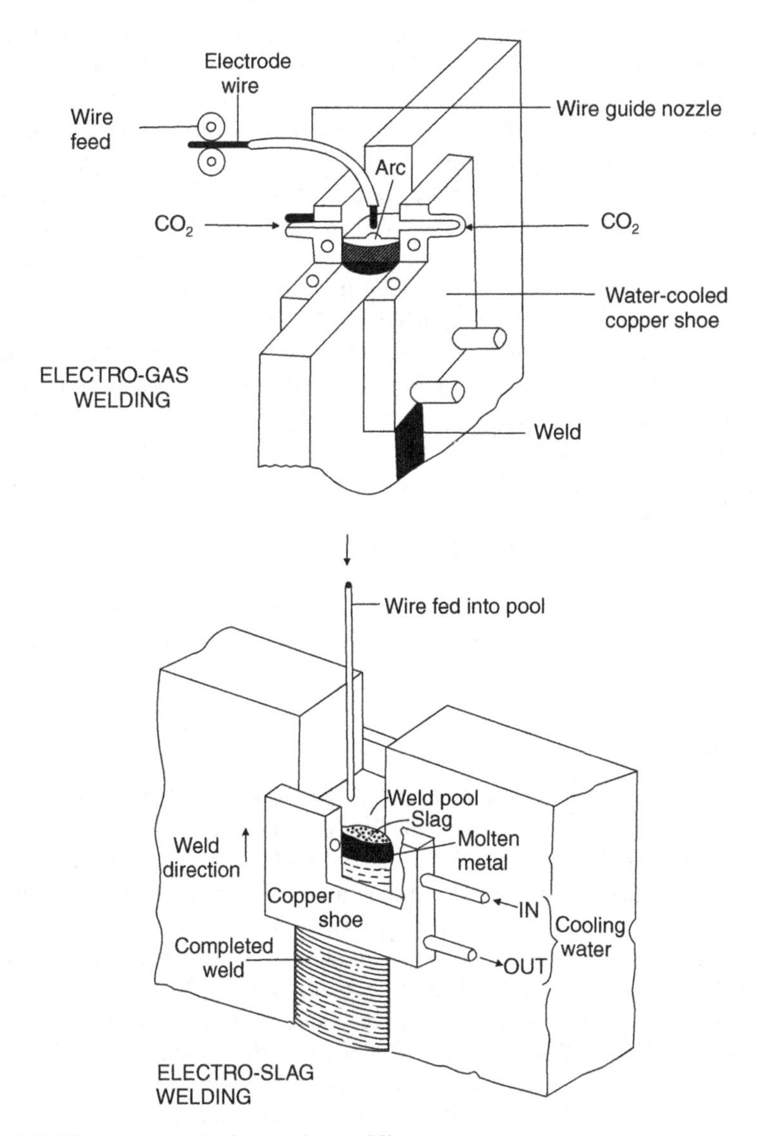

Figure 9.8 Electro-gas and electro-slag welding.

Laser welding

Laser welding is being used in the shipbuilding industry and shows much promise as a welding process that offers low heat input and therefore minimum distortion of welded plates and stiffeners.

For shipbuilding welding applications the laser source is either CO_2 (see 'Laser cutting' section below) or Nd:YAG (neodimium–yttrium–aluminum–garnet) crystals. Because of the wide range of applied powers and power densities available from

Nd:YAG lasers, different welding methods are possible. If the laser is in pulsed mode and the surface temperature is below boiling point, heat transfer is predominantly by conduction and a conduction limited weld is produced. If the applied power is higher (for a given speed), boiling begins in the weld pool and a deep penetration weld can be formed. After the pulse, the material flows back into the cavity and solidifies. Both these methods can be used to produce spot welds or 'stake' welds. Laser 'stake welding' has been used in shipbuilding, stiffening members being welded to plate panels from the plate side only. A seam weld is produced by a sequence of overlapping deep penetration 'spot' welds or by the formation of a continuous molten weld pool. Pulsed laser welding is normally used at thicknesses below about 3 mm. Higher power (4 to 10 kW) continuous-wave Nd:YAG lasers are capable of keyhole-type welding materials from 0.8 to 15 mm thickness. Nd:YAG lasers can be used on a wide range of steels and aluminum alloys. Also, because of the possibility of using fiber-optic beam delivery, Nd:YAG lasers are often used in conjunction with articulated arm robots for welding fabricated units of complex shape. Since the beam can burn the skin or severely damage the eyes, Nd:YAG lasers require enclosures within the fabrication shop that are fully opaque to the Nd:YAG laser wavelength.

Hybrid laser–arc welding is also used in the shipbuilding industry, this being a combination of laser and arc welding that produces deep penetration welds with good tolerance to poor joint fit-up. The Nd:YAG laser is combined with a metal arc welding gas (laser-MAG).

Thermit welding

This is a very useful method of welding that may be used to weld together large steel sections, for example parts of a stern frame. It is in fact often used to repair castings or forgings of this nature. Thermit welding is basically a fusion process, the required heat being evolved from a mixture of powdered aluminum and iron oxide. The ends of the part to be welded are initially built into a sand or graphite mold, whilst the mixture is poured into a refractory lined crucible. Ignition of this mixture is obtained with the aid of a highly inflammable powder consisting mostly of barium peroxide. During the subsequent reaction within the crucible the oxygen leaves the iron oxide and combines with the aluminum producing aluminum oxide, or slag, and superheated thermit steel. This steel is run into the mold, where it preheats and eventually fuses and mixes with the ends of the parts to be joined. On cooling a continuous joint is formed and the mold is removed.

Friction stir welding[1]

Friction stir welding is a relatively new materials joining process that has been used in the shipbuilding industry and is likely to be more widely used.

Friction stir welding is a solid-state process that offers advantages over fusion welding for certain applications. In producing butt joints it uses a nonconsumable rotating tool, the profiled pin of which is plunged into the butted joint of two plates

[1] Friction stir welding was invented and patented by TWI Ltd., Cambridge, UK.

and then moves along the joint. The plate material is softened in both plates and forced around the rotating profiled pin, resulting in a solid-state bond between the two plates (see Figure 9.9). To contain the softened material in the line of the joint, a backing bar is used and the tool shoulder under pressure retains material at the upper surface. Both plates and the backing bar require substantial clamping because of the forces involved. Plates of different thickness may be butt welded by inclining the rotating tool (Figure 9.9).

The process is currently used for welding aluminum alloy plates and such plates to aluminum alloy extrusions or castings. Dissimilar aluminum alloys may be joined by the process. A suitable material for a rotating tool to permit friction stir welding of steel has yet to be developed.

Typical applications of friction stir welding are the construction of aluminum alloy deck panels for high-speed craft from extruded sections and aluminum alloy honeycomb panels for passenger ship cabin bulkheads.

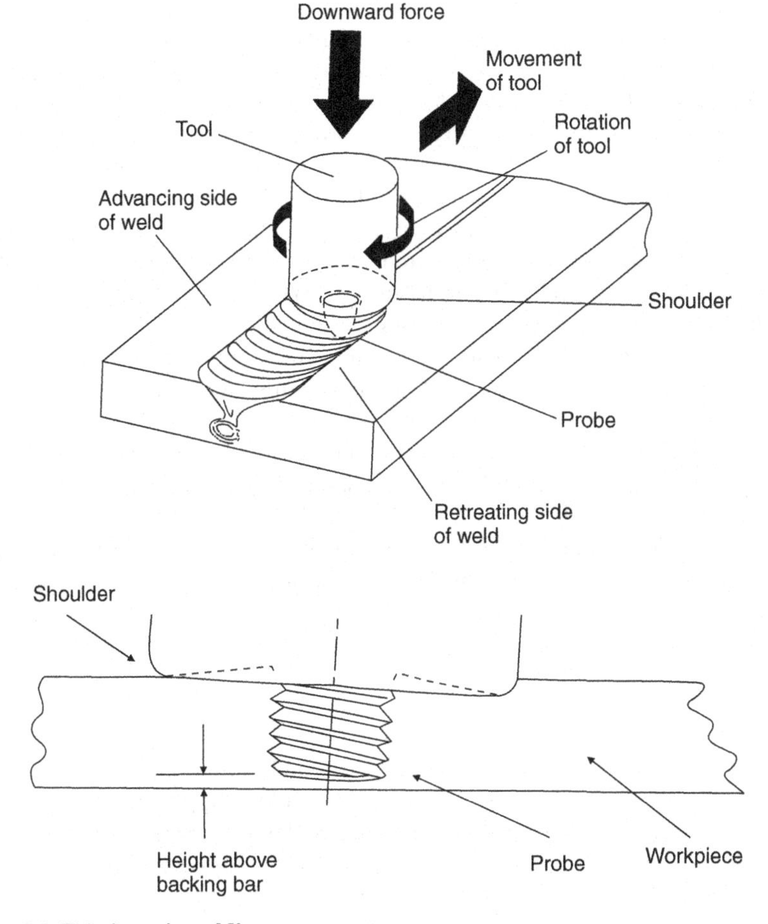

Figure 9.9 Friction stir welding.

Cutting processes

Steel plates and sections were mostly cut to shape in shipyards using a gas cutting technique, but the introduction of competitive plasma-arc cutting machines has led to their widespread use in shipyards today.

Gas cutting

Gas cutting is achieved by what is basically a chemical/thermal reaction occurring with iron and iron alloys only. Iron or its alloys may be heated to a temperature at which the iron will rapidly oxidize in an atmosphere of high purity oxygen.

The principle of the process as applied to the cutting of steel plates and sections in shipbuilding is as follows. Over a small area the metal is preheated to a given temperature, and a confined stream of oxygen is then blown onto this area. The iron is then oxidized in a narrow band, and the molten oxide and metal are removed by the kinetic energy of the oxygen stream. A narrow parallel sided gap is then left between the cut edges. Throughout the cutting operation the preheat flame is left on to heat the top of the cut since most of the heat produced by the reaction at the cutting front is not instantaneous, and tends to be liberated at the lower level of the cut only. Alloying elements in small amounts are dissolved in the slag and removed when cutting steel. However, if they are present in large quantities, alloying elements, especially chromium, will retard and even prevent cutting. The reason for this is that they either decrease the fluidity of the slag or produce a tenacious oxide film over the surface which prevents further oxidation of the iron. This may be overcome by introducing an iron rich powder into the cutting area, a process often referred to as 'powder cutting'. When cutting stainless steels which have a high chromium content 'powder cutting' would be employed.

Generally acetylene is used with oxygen to provide the preheat flame but other gases can be used: propane for example or hydrogen which is used for underwater work because of its compressibility. Apart from the torch, the equipment is similar to that for gas welding. The torch has valves for controlling the volume of acetylene and oxygen provided for the preheat flame, and it has a separate valve for controlling the oxygen jet (see Figure 9.10).

The oxyacetylene cutting process has been highly automated for use in shipyards; these developments are considered in Chapter 13. Hand burning with an oxyacetylene flame is used extensively for small jobbing work, and during the fabrication and erection of units.

Plasma-arc cutting

Plasma in this sense is a mass of ionized gas which will conduct electricity. An electrode is connected to the negative terminal of a DC supply and a gas shield is supplied for the arc from a nozzle which has a bore less than the natural diameter of the arc. As a result a constricted arc is obtained which has a temperature considerably higher than that of an open arc. The arc is established between the electrode and workpiece when the ionized conducting gas comes into contact with the work.

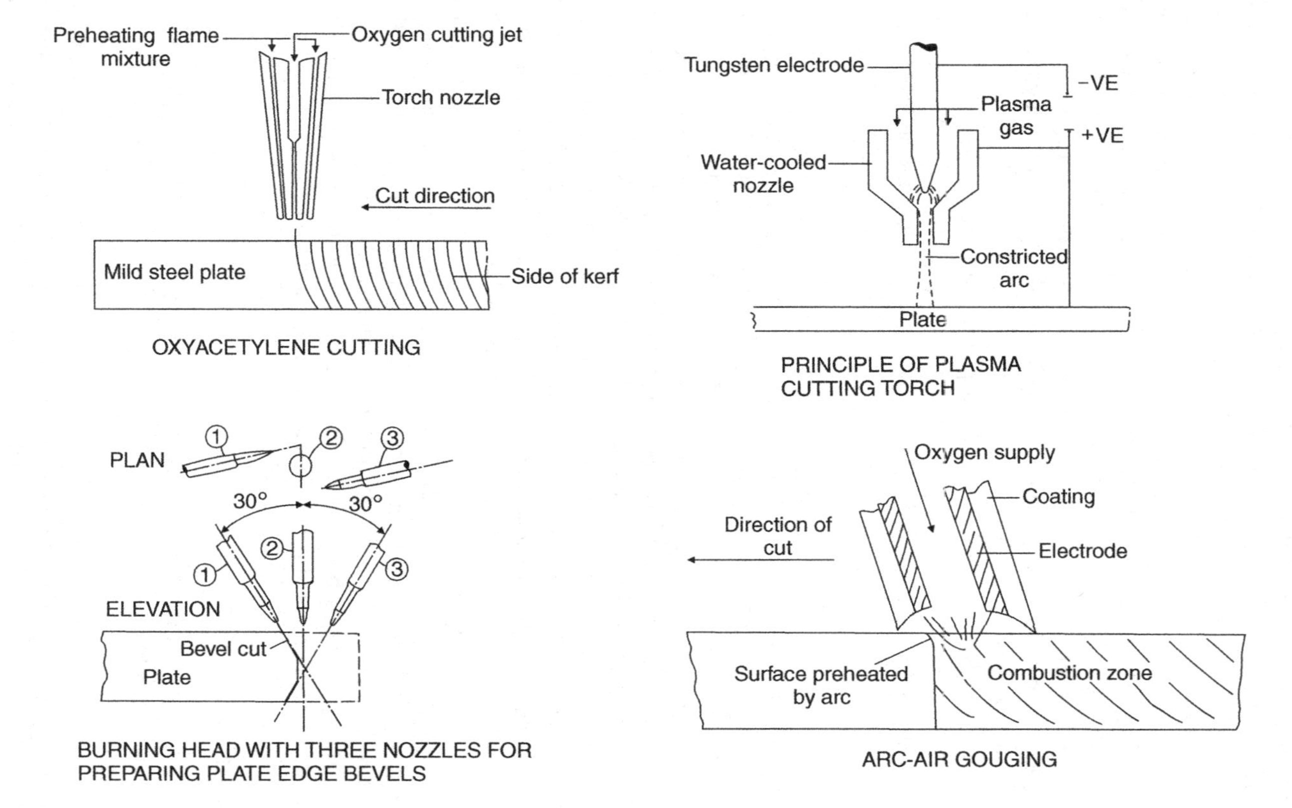

Figure 9.10 Metal cutting processes.

This gas is ionized in the first place by a subsidiary electrical discharge between the electrode and the nozzle. Plates are cut by the high-temperature concentrated arc melting the material locally (Figure 9.10).

The plasma-arc process may be used for cutting all electrically conductive materials. Cutting units are available with cutting currents of 20–1000 amps to cut plates with thicknesses of 0.6–150 mm. The plasma carrier gas may be compressed air, nitrogen, oxygen, or argon/hydrogen to cut mild or high alloy steels, and aluminum alloys, the more expensive argon/hydrogen mixture being required to cut the greater thickness sections. A water-injection plasma-arc cutting system is available for cutting materials up to 75 mm thick using nitrogen as the carrier gas. A higher cutting speed is possible and pollution minimized with the use of water and an exhaust system around the torch.

Water cutting tables were often used with plasma-arc cutting, but more recent systems have dispensed with underwater cutting. Cutting in water absorbed the dust and particulate matter and reduced the plasma noise and ultraviolet radiation of earlier plasma cutters.

Gouging

Both gas and arc welding processes may be modified to produce means of gouging out shallow depressions in plates to form edge preparations for welding purposes where precision is not important. Gouging is particularly useful in shipbuilding for cleaning out the backs of welds to expose clean metal prior to depositing a weld back run. The alternative to gouging for this task is mechanical chipping, which is slow and arduous. Usually, where gouging is applied for this purpose, what is known as 'arc-air' gouging is used. A tubular electrode is employed, the electrode metal conducting the current and maintaining an arc of sufficient intensity to heat the workpiece to incandescence. Whilst the arc is maintained, a stream of oxygen is discharged from the bore of the electrode that ignites the incandescent electrode metal and the combustible elements of the workpiece. At the same time the kinetic energy of the excess oxygen removes the products of combustion, and produces a cut. Held at an angle to the plate, the electrode will gouge out the unwanted material (Figures 9.8, 9.10).

A gas cutting torch may be provided with special nozzles that allow gouging to be accomplished when the torch is held at an acute angle to the plate.

Laser cutting

Profile cutting and planing at high speeds can be achieved with a concentrated laser beam and has increasingly been employed in a mechanized or robotic form in the shipbuilding industry in recent years. In a laser beam the light is of one wavelength, travels in the same direction, and is coherent, i.e. all the waves are in phase. Such a beam can be focused to give high energy densities. For welding and cutting the beam is generated in a CO_2 laser. This consists of a tube filled with a mixture of CO_2, nitrogen, and helium that is made to fluoresce by a high-voltage discharge. The tube emits infrared radiation with a wavelength of about 1.6 μm and is capable of delivering outputs up to 20 kW.

Laser cutting relies on keyholing to penetrate the thickness, and the molten metal is blown out of the hole by a gas jet. A nozzle is fitted concentric with the output from a CO_2 laser so that a gas jet can be directed at the work coaxial with the laser beam. The jet can be an inert gas, nitrogen, or in the case of steel, oxygen. With oxygen there is an exothermic reaction with the steel, giving additional heat as in oxy-fuel cutting. The thermal keyholing gives a narrow straight-sided cut compared with the normal cut obtained by other processes relying on a chemical reaction.

Water jet cutting

The cutting tool employed in this process is a concentrated water jet, with or without abrasive, which is released from a nozzle at 2½ times the speed of sound and at a pressure level of several thousand bar. Water jet cutting can be used on a range of materials such as timber, plastics, rubber, etc., as well as steels and aluminum alloys. Mild steel from 0.25 to 150 mm in thickness and aluminum alloys from 0.5 to 250 mm in thickness can be cut. Being a cold cutting process the heat-affected zone, mechanical stresses, and distortion are left at the cut surface.

Water jet cutting is slower than most thermal cutting processes and is not a portable machine tool.

Further reading

ESAB advances in welding and cutting technology, *The Naval Architect*, July/August 2003.
Exploiting friction-stir welding of aluminium, *The Naval Architect*, July/August 2004.
Modern materials and processes for shipbuilding, *The Naval Architect*, July/August 2005.
Smith BD: *Welding Practice*, 1996, Butterworth Heinemann.

Some useful websites

http://www.rina.org.uk/tna.html Site for *The Naval Architect* magazine, with regular updates on ship and shipbuilding technology, including welding.
www.esab.com Recommended for informative papers on welding processes and practices.
www.controlwaterjet.co.uk Use, benefits, and capabilities of water jet cutting.

10 Welding practice and testing welds

Chapter Outline

Welding practice 103
Welding automation 105
Welding distortion 107
Welding sequences 107
Testing welds 110
Weld faults 110
Nondestructive testing 112
Classification society weld tests 114
Further reading 115
Some useful websites 115

The strongest welded joint that may be produced in two plates subsequently subjected to a tensile pull is the butt joint. A butt joint is one where the two joined plates are in the same plane, and in any welded structure it is desirable that butt joints should be used wherever possible.

In mild steel the weld metal tends to have a higher yield strength than the plate material (see Figure 10.1). Under tension it is found that initial yielding usually occurs adjacent to a butt weld in the plate when the yield strength of the plate material is reached locally. Since a good butt weld in tension has a strength equivalent to that of the mild steel plate, it is not considered as a line of structural weakness.

Lapped joints, where fillet welds are used to connect the plates, should be avoided in strength members of a welded structure. As the fillet welds are in shear when the plates are in tension, the strength of the joint is very much less than that of the plate material or butt joint. Fillet welds are unavoidable where sections or plates are connected at an angle to an adjacent plate, but often there is not the same problem as the loading is different. The fatigue strength of fillet welds is also inferior to that of a butt weld.

Welding practice

In making a butt weld with manual arc welding, where the plate thickness exceeds, say, 5–6 mm it will become necessary to make more than one welding pass to deposit sufficient weld metal to close the joint. With the higher current automatic welding

Ship Construction. DOI: 10.1016/B978-0-08-097239-8.00010-6

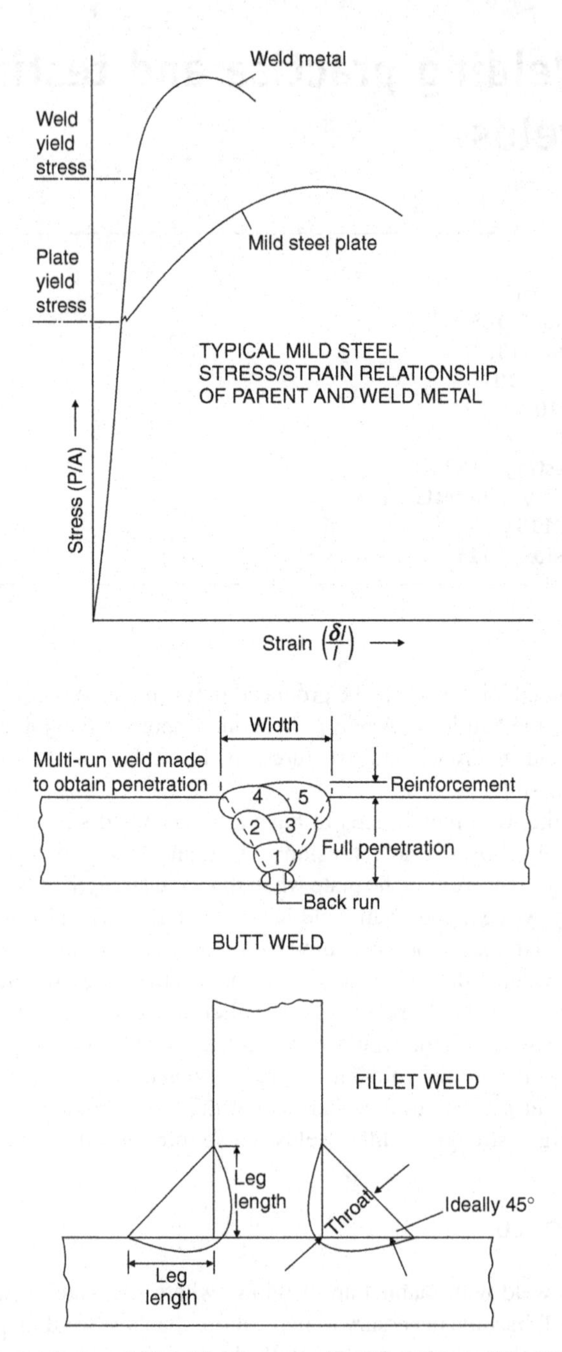

Figure 10.1 Stress–strain relationship for parent/weld metals and butt/fillet weld detail.

processes thicker plates may be welded with a single pass, but at greater thicknesses multi-pass welds become necessary.

In ship work unless a permanent backing bar is used, or a 'one-sided' welding technique or process (see Chapter 9) is used, a back run of weld is required to ensure complete weld penetration. This is made on the reverse side of the joint after cleaning out the slag, etc., by chipping or gouging. Permanent backing bars may conveniently be introduced where it is desired to weld from one side only during erection at the berth. A good example is the use of a cut-down channel bar used as a deck beam, the upper flange providing the backing bar for a deck panel butt weld, made by machine above.

Tack welds are used throughout the construction to hold plates and sections in place after alignment and prior to completion of the full butt or fillet weld. These are short light runs of weld metal, which may be welded over, or cut out in more critical joints during the final welding of the joint.

Fillet welds may be continuous or intermittent depending on the structural effectiveness of the member to be welded. Where fillets are intermittent they may be either staggered or chain welded (see Figure 10.2), the member may also be scalloped to give the same result when continuously welded.

On thicker plates it becomes necessary to bevel the edges of plates that are to be butted together in order to achieve complete penetration of the weld metal (Figure 10.2). This operation may be carried out whilst profiling or trimming the plate edges, which must be aligned correctly. Most edge preparations are made by gas or plasma heads having three nozzles out of phase that can be set at different angles to give the required bevels. Alternatively, the edge preparation may be obtained by mechanical machining methods using either a planing or milling tool. For very high quality welds in thick plate, particularly of the higher tensile types of steel, mechanical machining may well be specified. It is worth noting that there is little to choose between the two as far as metallurgical damage goes, but mechanical methods provide a better finish.

Plates of varying thickness may be butt welded together at different locations, a good example being where heavy insert plates are fitted. Insert plates are preferred to doubling plates in welded construction, and the heavy plate is chamfered to the thickness of the adjacent thinner plate before the butt edge preparation is made.

To ease the assembly of welded units it is common practice to make use of what is known as an 'egg box' construction. Within the double-bottom unit the floors and side girders may be slotted at their intersections so that they fit neatly together prior to construction.

Welding automation

Larger shipyards with a large production line throughput of welded panels use automated welding systems to produce the stiffened panels. To join the plates, high-speed one-sided submerged arc welding (see Figure 9.4) is used. The required welding parameters are set in advance in the operation box and linked to a computer. The operator selects the plate thickness and starts the machine. The machine

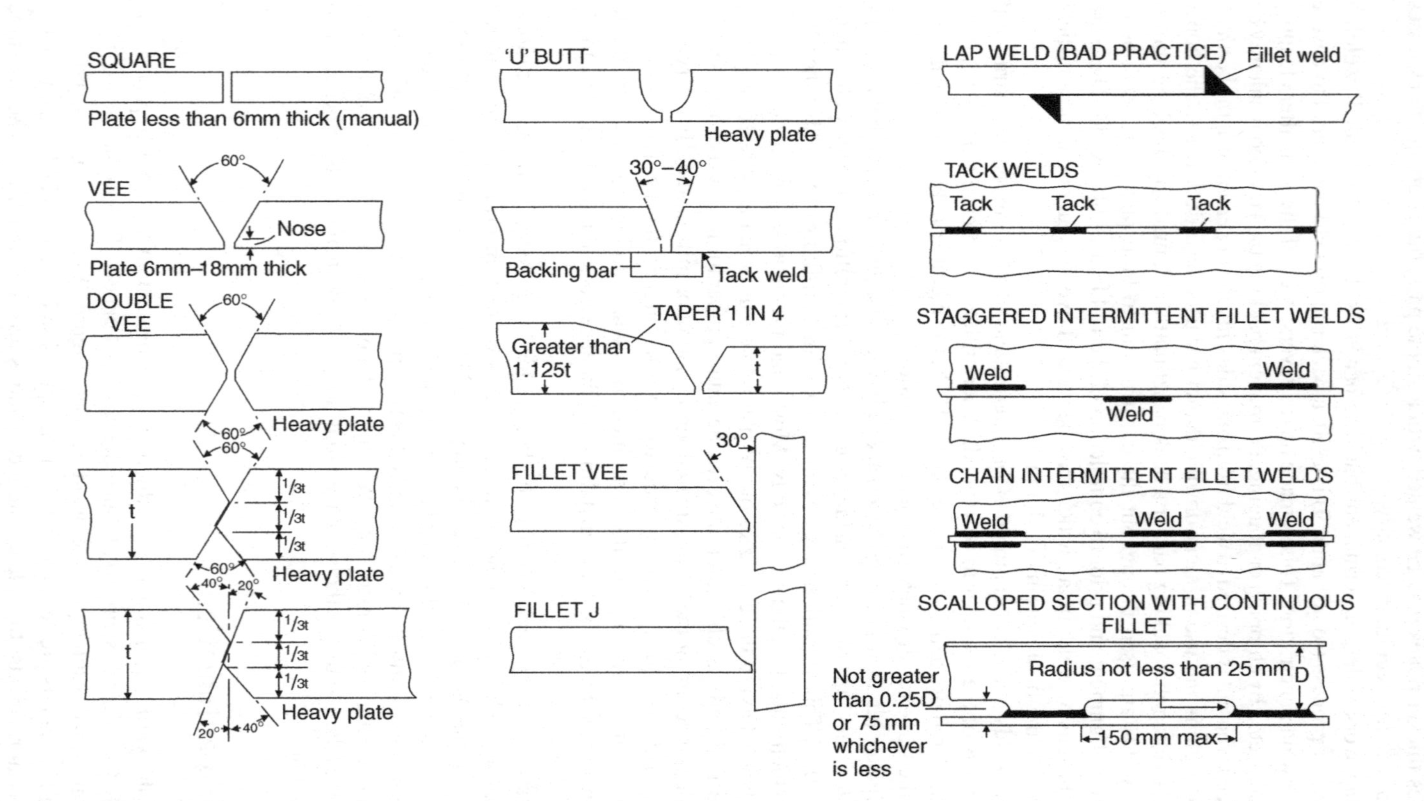

Figure 10.2 Plate edge preparation.

automatically controls the welding parameters for the weld crater and stops when the run-off tab is reached at the end of the plates. The welded plate panel is then moved over floor-mounted rollers to the next stage where the stiffening members are to be attached. Each stiffener is lowered onto the plate panel and tack welded using metal arc gas-shielded welding. The plate panel with tacked stiffeners is then rolled within reach of a fillet welding gantry with double-fillet welding machines. The gantry moves parallel to the stiffeners at the same speed as the welding heads and carries packs containing the welding wires and power sources for the double-fillet welding heads. Welding is carried out using flux-cored wires with CO_2 gas shielding.

The use of robotics for welding is referred to in Chapter 13.

Welding distortion

During the welding process the metal is heated, which causes expansion and the metal then contracts on cooling. The initial rapid heating causes the welded area to expand locally. The slower cooling of the weld causes the plate to move as the weld contracts. The result is a distortion of the part, and this is a major cause of extra work during assembly of units and construction of the ship. The need to adjust distorted parts so that they fit correctly can take considerable time and effort. In-plane distortion is basically shrinkage of the plate. For repeatable processes, which are usual in shipbuilding, the shrinkage can be measured and sufficient data built up to allow the shrinkage to be predicted. Computer-aided design systems can now include an allowance for shrinkage, so that the part as modeled in the system can then be adjusted during the generation of cutting information. The plates are cut oversized and the shrinkage after welding brings it to the correct size. More recently, work has been carried out to model shrinkage of more complex parts, for example structural webs with face flats. Again, these parts can then have their dimensions adjusted prior to cutting so that the effect of shrinkage is to bring them to the correct shape. Out-of-plane distortion is much more difficult to predict and manage. The cause is the same as in-plane shrinkage, but the distortion is often associated with fillet welds used to attach stiffeners to plates. The fillet welds, as they shrink, pull the plate out of plane, resulting in the typical appearance of a welded hull with indentations between the frames. The effect is much more noticeable for thin plate structures, for smaller ships, and for superstructures. Restraining the plate during assembly and welding is one commonly used solution.

The causes of distortion are complex and also include any residual stresses in the steel plate as a result of the steel mill rolling and cooling. Some of the stress may be relieved by rolling the plate prior to production, but distortion remains a significant problem for many shipbuilders.

Welding sequences

In order to minimize distortion in manual welding the 'backstep' and 'wandering' methods of welding are often used, the length of each step being the amount of weld

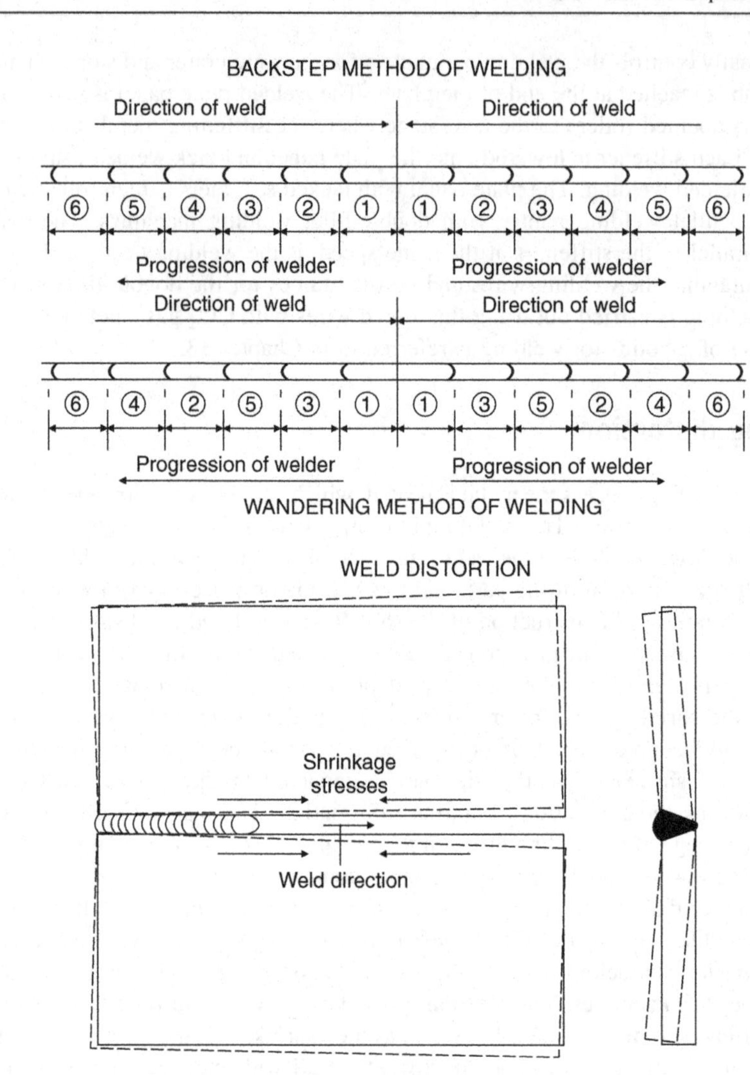

Figure 10.3 Backstep and wandering methods of welding.

metal laid down by an electrode to suit the required cross-section of weld (see Figure 10.3).

To reduce distortion and limit the residual stresses in the structure it is important that a correct welding sequence should be utilized throughout the construction. This applies both during the fabrication of units and at erection and joining on the berth.

Of the more important welds in the construction of the ship the sequences involving welding of butts and seams in plating panels may be considered (see Figure 10.4). At T intersections it is necessary to weld the butt first fully, then gouge

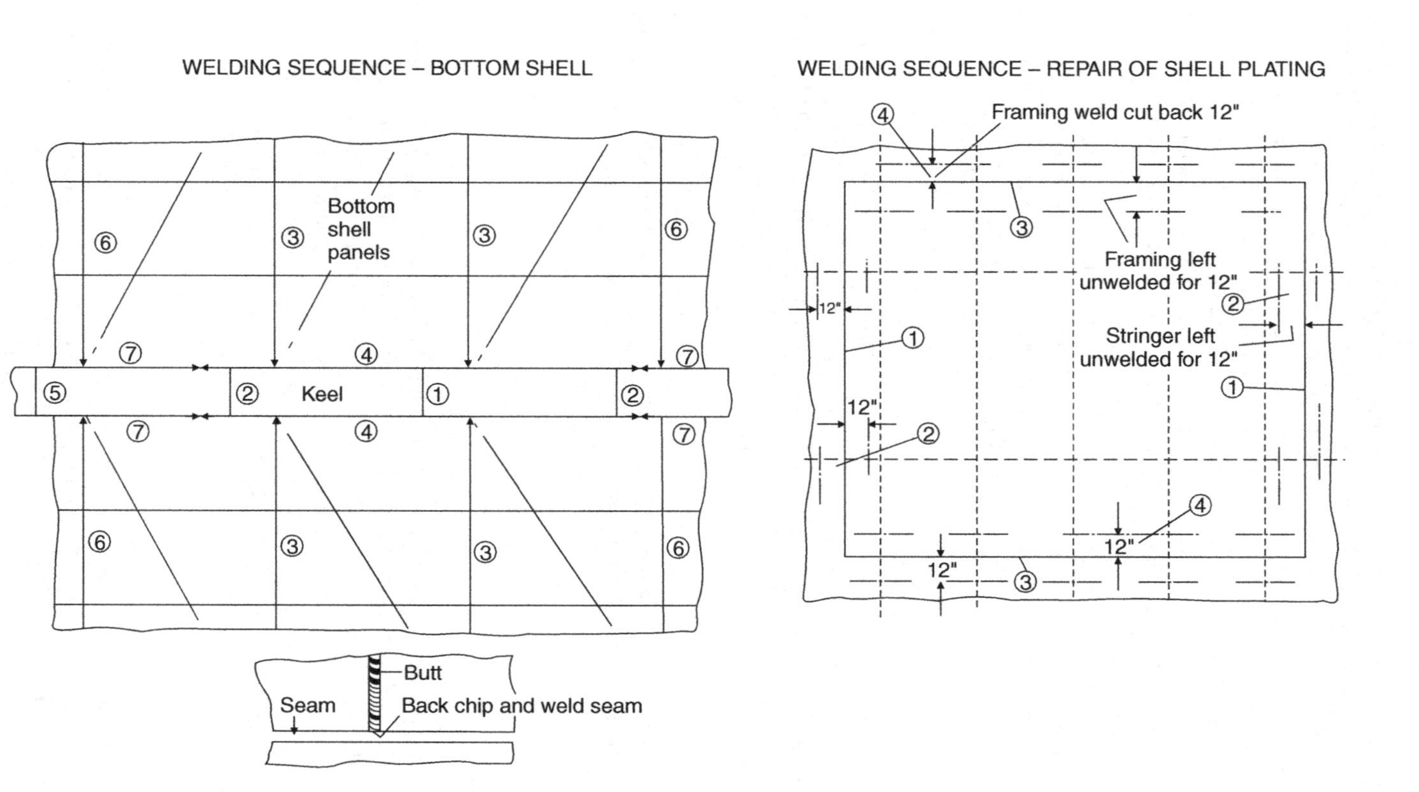

Figure 10.4 Welding sequences.

out the ends to renew the seam edge preparation before welding the seam. Welding the seam first would cause high restraint across the plate strake and when the butt was finished a crack might occur. General practice when welding shell panels is to start by welding the central butts and then adjacent seams, working outwards both transversely and longitudinally. Ships' structural panels have various forms of stiffener attached to the plate panels, these generally being welded to the panel after completing the welding of the panel plates. These stiffening members are left unwelded across the butts and seams of the plates until these are completed, if they are attached at some intermediate stage.

Erection welding sequences generally follow the principles laid down for plating panels. In welded ships the lower side plating seams should not be welded before the upper seams, particularly the deck and gunwale seams. If this sequence of welding the side shell were adopted the upper portion of the hull structure would tend to be shortened, causing the hull to rise from the blocks at the ends. Where in modern construction the side shell and deck plating are erected in blocks and a suitable welding sequence is employed, this problem does not arise.

In repair work correct welding sequences are also important, particularly where new material is fitted into the existing relatively rigid structure. Again, the procedure follows the general pattern for butts and seams in plate panels. If a new shell plate is to be welded in place the seams and butts in the surrounding structure are cut back 300–375 mm from the opening, likewise the connection of the stiffening in way of the opening.

The inserted plate panel is then welded to within 300–375 mm of the free edges, the butts are completed, and then the seams after welding any longitudinal stiffening across the butts. Finally, the vertical framing is welded in way of the seams (Figure 10.4).

Testing welds

For economic reasons much of the weld testing carried out in shipbuilding is done visually by trained inspectors. Spot checks at convenient intervals are made on the more important welds in merchant ship construction, generally using radiographic or ultrasonic equipment. Welding materials are subjected to comprehensive tests before they are approved by Lloyd's Register or the other classification societies for use in ship work. Operatives are required to undergo periodical welder approval tests to ascertain their standard of workmanship.

Weld faults

Various faults may be observed in butt and fillet welds. These may be due to a number of factors: bad design, incorrect welding procedure, use of wrong materials, and bad workmanship. Different faults are illustrated in Figure 10.5. The judgment of the seriousness of the fault rests with the weld inspector and surveyor, and where the weld is considered to be unacceptable it will be cut out and rewelded.

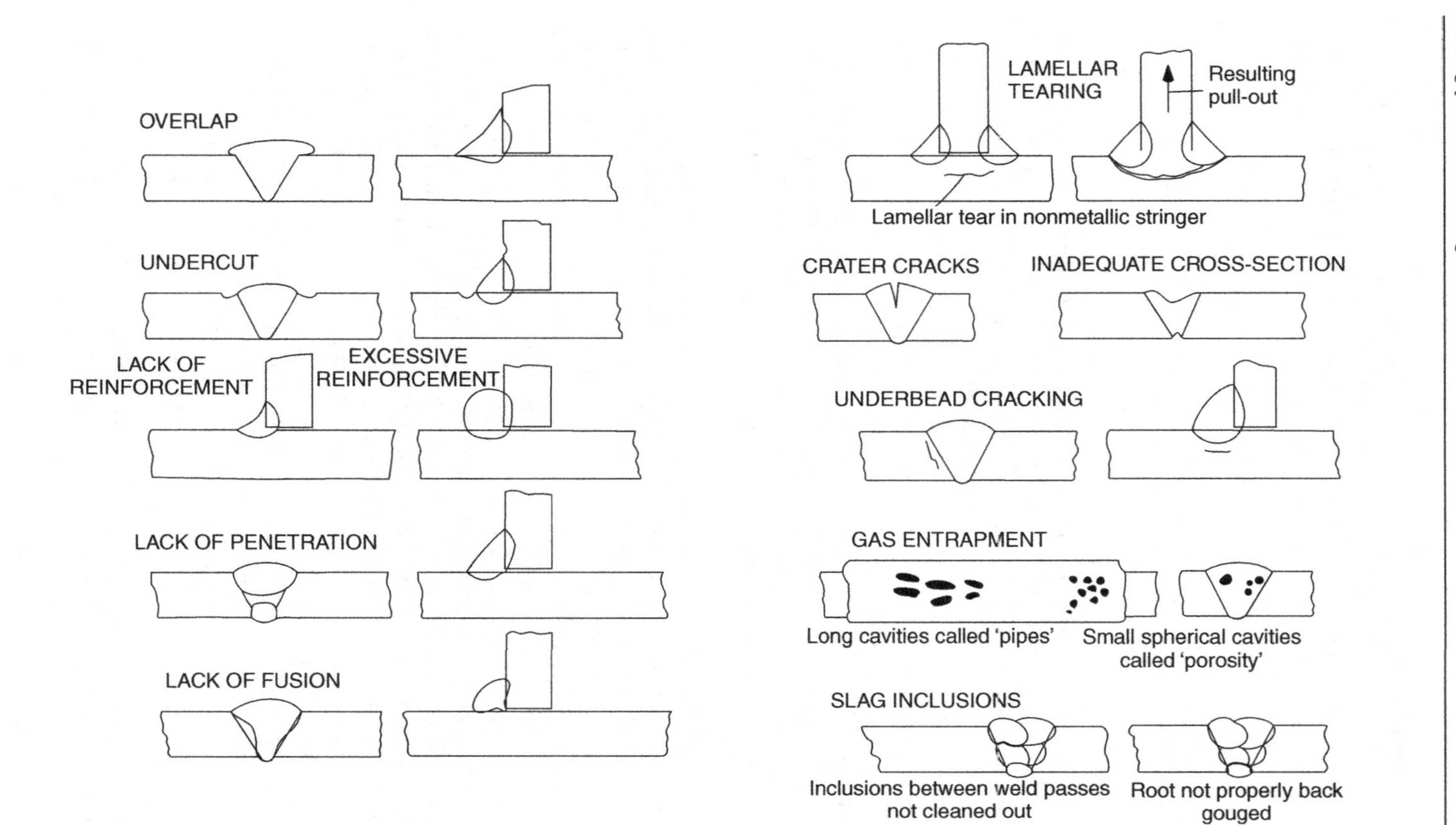

Figure 10.5 Weld faults.

Nondestructive testing

For obvious reasons some form of nondestructive test is required to enable the soundness of ship welds to be assessed. The various available nondestructive testing methods may be summarized as follows:

- Visual examination
- Dye penetrant
- Magnetic particle
- Radiographic
- Ultrasonic.

Of these five methods, the dye penetrant and magnetic particle tests have few applications in ship hull construction, being used for examining surface cracks in stern frames and other castings. Visual, radiographic, and ultrasonic examinations are considered in more detail, as they are in common use.

Magnetic particle testing is carried out by magnetizing the casting and spreading a fluid of magnetic particles (e.g. iron fillings suspended in paraffin) on the surface. Any discontinuity such as a surface crack will show up as the particles will concentrate at this point where there is an alteration in the magnetic field. A dye penetrant will also show up a surface flaw if it remains after the casting has been washed following the application of the dye. To aid the detection of a surface crack the dye penetrant used is often luminous and is revealed under an ultraviolet light.

Visual inspection of welds is routine procedure, and surface defects are soon noticed by the experienced inspector and surveyor. Incorrect bead shape, high spatter, undercutting, bad stop and start points, incorrect alignment, and surface cracks are all faults that may be observed at the surface. Subsurface and internal defects are not observed, but the cost of visual inspection is low, and it can be very effective where examination is made before, during, and after welding.

The principle of radiographic inspection is simply to subject a material to radiation from one side, and record the radiation emitted from the opposite side. Any obstacle in the path of the radiation will affect the radiation density emitted and may be recorded. As radiation will expose photographic plate, for all practical weld test purposes this is used to record the consistency of the weld metal. The photographic plate records changes in radiation density emitted; for example, a void will show up as a darker shadow on the radiograph.

Either X-ray or gamma-ray devices may be used to provide the source of radiation. X-ray equipment consists of a high-voltage power source (50–400 kV), which is used to provide potential between a cathode and target anode in a glass vacuum tube. Only a small percentage of this energy is converted to X-rays, so that large amounts of heat have to be conducted away from the target. From the target the X-rays are projected out of the tube onto the weld surface (see Figure 10.6).

Where gamma-ray devices are used ray emission is produced by decay of a radioactive nucleus, the rate of emission being reduced with time. The radiation given off may be magnetically separated into three parts, α-rays, β-rays and γ-rays,

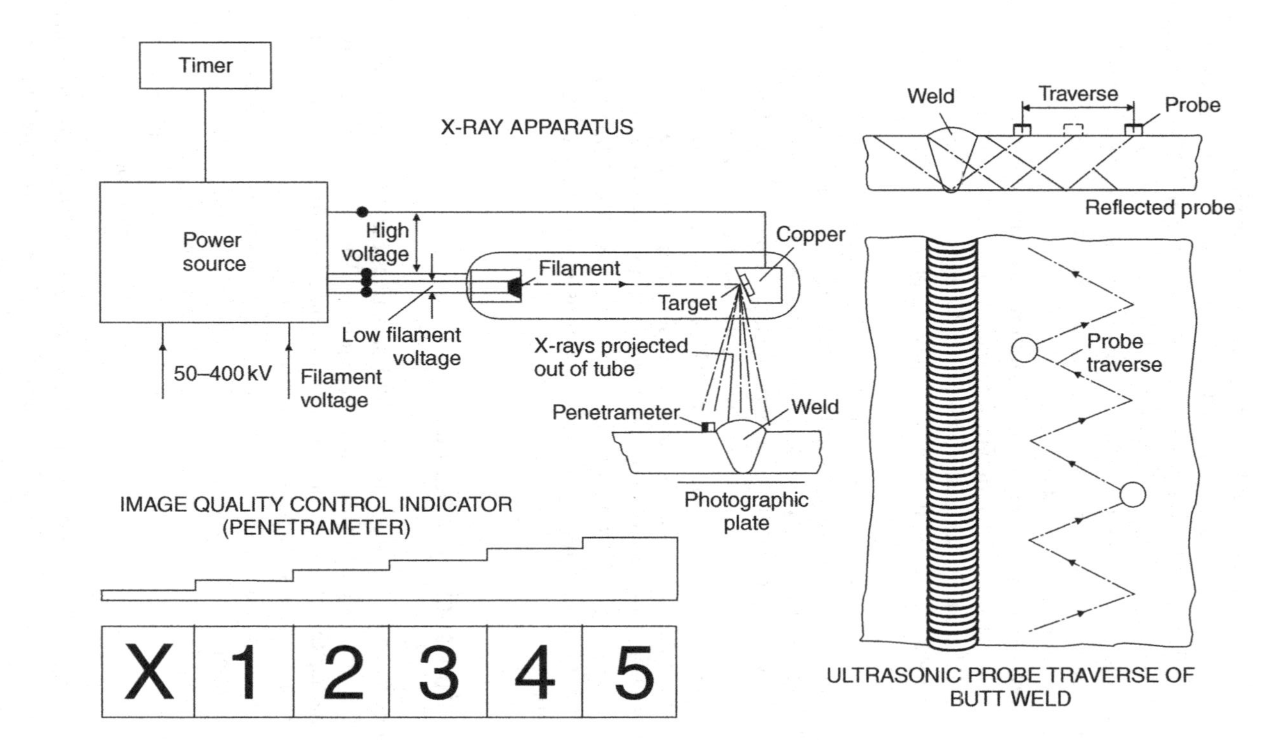

Figure 10.6 **Inspection of welds.**

the γ-rays being similar to X-rays and of most importance since they are very penetrating, but this also means that heavy shielding is required. Since natural radioactive sources are in short supply, great use is made of artificial radioactive sources, namely isotopes.

To interpret the weld radiograph a large amount of experience is required, and a sound knowledge of the welding process. Radiographs usually carry the image of an 'image quality indicator', which shows the minimum change of thickness revealed by the technique. This image quality indicator may have graded steps of metal, each step being identified on the radiograph so that the minimum step thickness discernible is noted and the sensitivity of the radiograph assessed. This indicator is placed adjacent to the weld prior to taking the radiograph.

Ultrasonic energy is commonly used as a tool for locating defects in welds, and has several advantages over radiography, particularly as no health hazard is involved. The technique is particularly useful for locating fine cracks that are often missed by radiography, particularly where they lie perpendicular to the emission source.

The principle of ultrasonic inspection depends on the fact that pulses of ultrasonic energy are reflected from any surface that they encounter. Ultrasonic waves traveling through a plate may be reflected from the surface of the metal and also from the surfaces of any flaws that exist in the metal. Virtually total reflection occurs at an air–metal interface, and therefore to get the ultrasonic wave into the metal a liquid is placed between the source and metal. The pattern of reflection is revealed on a cathode ray tube, which may be calibrated using a standard reference block. An experienced operator is able to recognize flaws from the cathode ray tube display, and to some extent recognition of defect types is possible. Apart from weld inspection, ultrasonic techniques are valuable for assessing the thickness of structural members.

Classification society weld tests

Classification societies specify a number of destructive tests that are intended to be used for initial electrode and weld material approval. These tests are carried out to ascertain whether the electrode or wire-flux combination submitted is suitable for shipbuilding purposes in the category specified by the manufacturer.

Tests are made for conventional electrodes, deep penetration electrodes, wire-gas and wire-flux combinations, consumables for electro-slag and electro-gas welding, and consumables for one-sided welding with temporary backing. Tensile, bend, and impact tests are carried out on the deposited weld metal and welded plate specimens. Other tests are made for the composition of the weld metal deposited and possible cracking.

All works where electrodes, wire-flux and wire-gas combinations, consumables for electro-slag and electro-gas welding, and consumables for one-sided welding with temporary backing are produced, and have been initially approved, are subject to annual inspection.

Further reading

Boekholt R: *Welding Mechanization and Automation in Shipbuilding Worldwide*. Abington Publishing, 1996.

Some useful websites

www.esab.com Includes information on welding equipment.
www.pemamek.fi See 'Shipyards' for details of welding/cutting automation and mechanization, including robotics.

Part Four

Shipyard Practice

11 Shipyard layout

Chapter Outline

Further reading 123
Some useful websites 124
Shipyards 124

Until the advent of steel ships, a shipbuilding operation could be set up almost anywhere close to the sea or a river and to trees for the main construction materials. Iron, rapidly followed by steel, construction resulted in shipbuilding moving to areas where raw materials, primarily coal and iron ore, were available. These were also often areas where the basic metalworking skills were to be found as the basis for the labor force, or in some cases a labor force was moved into the area. Shipyards were usually found along river banks, or in protected harbors, giving sheltered water and their basic arrangement did not vary. Figure 11.1 shows the typical arrangement of a shipyard up to around 1960. The slipways on which the ships are constructed piece by piece are supported by small and simple workshops. There was a relatively small

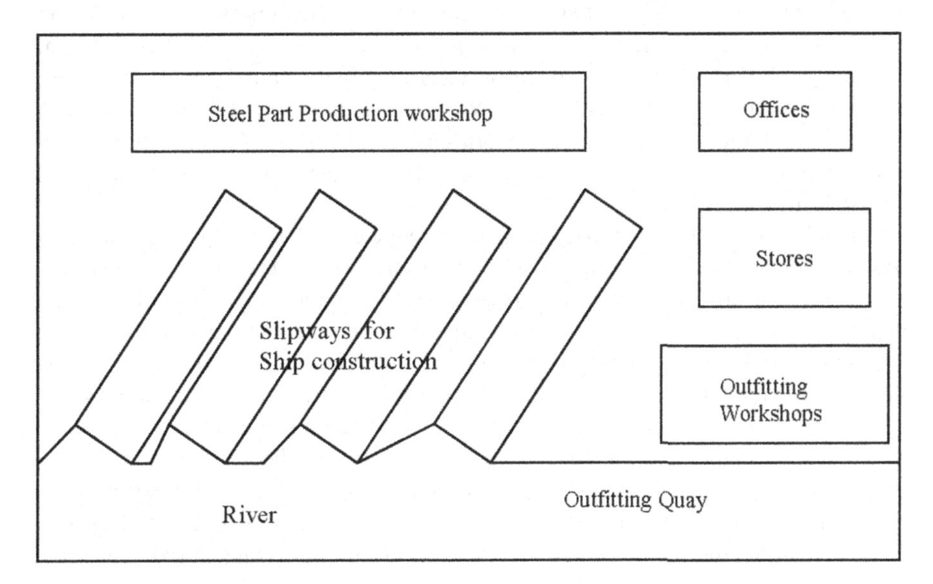

Figure 11.1 'Traditional' riverside layout.

Ship Construction. DOI: 10.1016/B978-0-08-097239-8.00011-8

initial investment and the output could be varied by opening or closing building slipways and taking on or laying off labor.

The layout of a shipyard did not vary significantly until the mid 1950s. A relatively small number of shipyards engaged in capital warship construction or passenger ships, where the product is significantly larger and more complex than average commercial ships. These had extensive outfitting workshops and quays, as well as larger slipways. Large cranes, almost always fixed in position, were available to lift heavy items, perhaps 200 tonnes for large, complex ships. However, for the lifting of hull parts and most outfitting the available lifting capacity was usually below 10 tonnes.

The major change in the shipyards came about initially because of rapid increases in commercial ship size after 1950. At that time a typical cargo ship was of below 10,000 deadweight tonnes, and a tanker of 20,000 deadweight tonnes was considered large. By 1958, the first tanker over 100,000 deadweight tones was in operation and the first over 200,000 tonnes deadweight by 1966. By 1970, 250,000-tonne and larger ships were being built. An important aspect of these newer ships was a tenfold increase in steelweight, from a typical 3000 tonnes to 30,000 tonnes. Also, the largest ships were in excess of 300 meters in length. As such, they were too large for most existing shipyards' slipways and the lifting capacities of the small cranes usually available would have meant an excessive construction time.

The result was first that existing shipyards reduced the number of slipways and increased the size of their cranes as the ship size increased. This allowed them to construct the ships in a shorter time, so keeping the construction time acceptable. A very few European shipyards increased the slipway sizes and cranes to be able to build VLCCs. However, the 1960s saw the emergence of a substantial number of new shipyards, primarily in Asia but also in Europe, which were purpose designed for the new, large ships. The basic arrangement was established by the mid 1960s in Japan, and many shipyards built subsequently, including in Eastern Europe, and more recently in South Korea, China and now Vietnam, have been specifically planned to construct the larger ships. The contemporary shipbuilding practices and production methods have improved but the basic technology and main equipment has been consistent for the last half century.

A number of traditional shipbuilders, which were often based on river banks, also established new yards where they could build larger ships and/or exploit the new technology and production methods. In general, the smaller shipbuilders have been able to reconfigure their site in order to utilize new technology and improve production, whilst continuing to build smaller and medium-sized ships. In many cases the shipyards are constrained as to the size and type of ship that can be built. Many smaller shipyards have adopted total undercover construction, providing a dry, warm (or cool) environment. Undercover ship construction has been extended to some larger ships, including many military ships and also passenger ships, especially in Northern Europe. As in the early days of such large ships, where the shipyards differed from the routine, the complexity and cost of the ship can justify the massive capital expenditure on covered facilities.

An ideal layout for a modern shipyard is based on a production flow basis, as in Figure 11.2, with the shipyard built on a greenfield site and no longer, as with existing

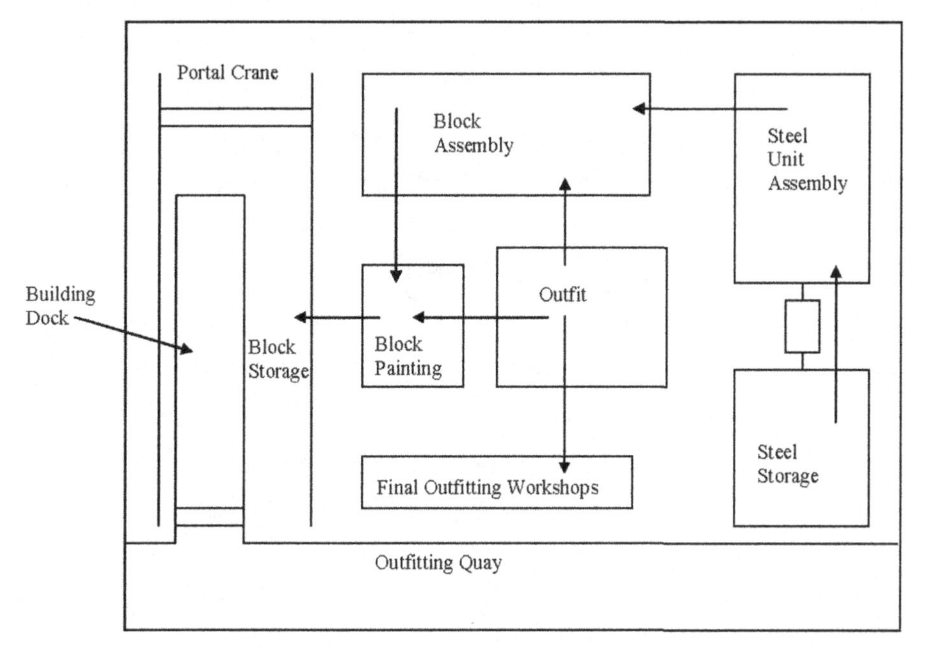

Figure 11.2 Modern, large shipyard layout.

shipyards, having to follow the river bank. This removes typical restrictions on old sites, restricted by their location in a built-up area or the physical river bank slope from extending back from the river, so that modified production flow lines are required.

The sequence of layout development is outlined below. It should be noted that particular locations and circumstances can dictate significantly different arrangements that may not be ideal but that do work.

Planning a new shipyard, or re-planning an existing one, will involve decisions to be made on the following:

- Size and type of ship to be built
- Number of ships per year to be achieved
- Breakdown of the ships into structural blocks and outfit modules (interim products)
- Material handling equipment required for the interim products
- Part production and assembly processes to be installed
- Amount of outfit and engine installation to be undertaken
- Control services to be supplied
- Administration facilities required.

Shipyards usually have a fitting out basin or berth where the virtually completed ship is tied up after launching and the finishing off work and static trails may be carried out.

Some of the facilities identified may be omitted from the shipyard and a subcontractor used instead. This will depend on the location, availability of subcontractors, and the economics of the alternatives.

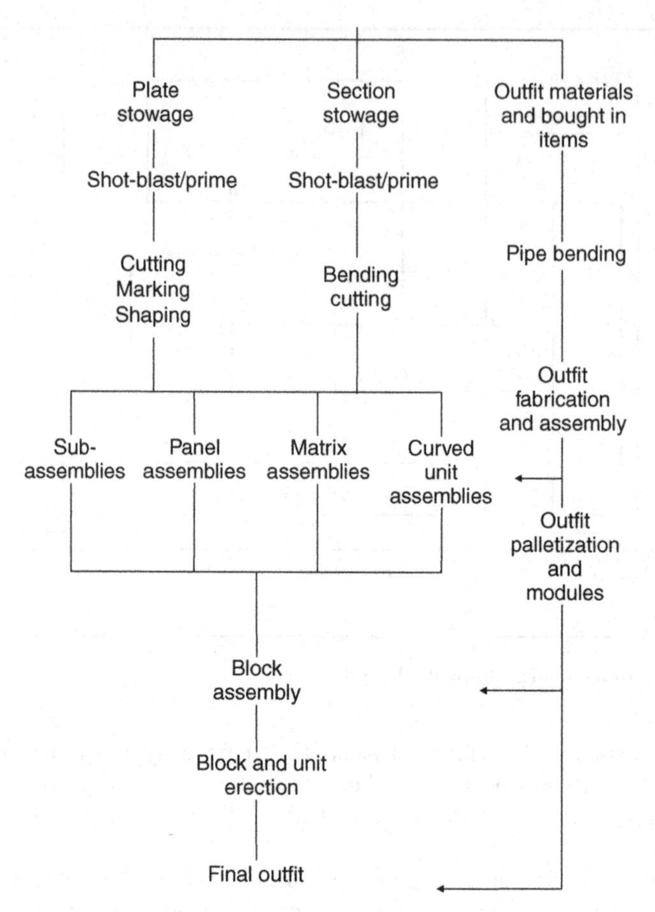

Figure 11.3 Shipbuilding process.

Before considering the actual layout of the shipyard, it is as well to consider the relationship of the work processes involved in building a ship, as illustrated in Figure 11.3.

An idealized layout of a new shipyard is indicated in Figure 11.4, which might be appropriate for a smaller yard specializing in one or two standard type ships with a fairly high throughput so that one covered building dock or berth is sufficient.

At this point it may be convenient to mention the advantages and disadvantages of building docks as opposed to building berths. Building docks can be of advantage in the building of large vessels where launching costs are high, and there is a possibility of structural damage owing to the large stresses imposed by a conventional launch. They also give good crane clearance for positioning units. The greatest disadvantage of the building dock is its high initial cost. However, the dock is the usual choice for new shipyards, especially for larger ships. The level base simplifies the construction process, for example alignment of structural blocks. Also, the dock is more flexible in operation, for example in some cases ships are built in two stages with the outfit and

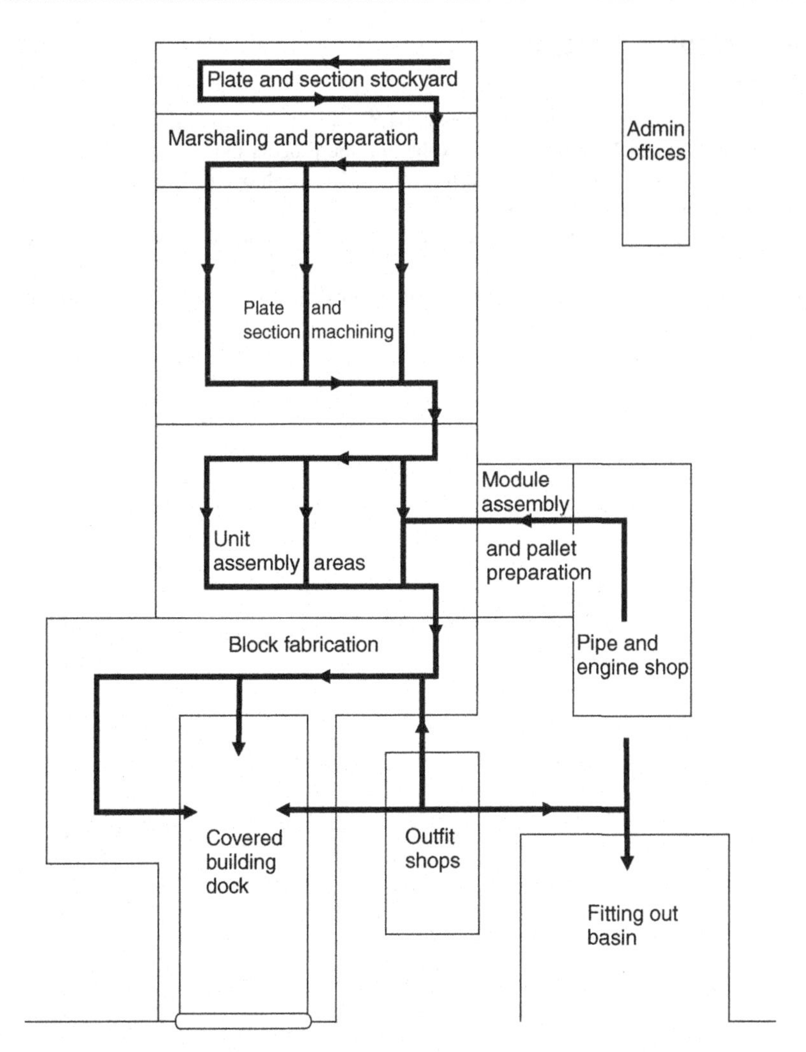

Figure 11.4 Shipyard layout.

labor-intensive stern, containing machinery and accommodation, built as a single structure, then moved along the dock for the cargo-carrying part to be constructed. This doubles the available time for outfitting the ship.

Further reading

Brand-new Portsmouth complex for VT: *The Naval Architect*, July/August 2003.
Chinese shipbuilders on the way to the top: *The Naval Architect*, September 2005.
Covered hall benefit for Bijlsma: *The Naval Architect*, May 2002.
The Naval Architect (published by the RINA) regularly carries articles on shipbuilding developments.

Some useful websites

Shipyards

http://www.stxons.com/service/kor/main.aspx STX is a major group in Korea and Europe. The site includes virtual shipyard tours explaining the process used.

www.shi.samsung.co.kr Take the virtual tour of the Geoje Shipyard. Building large ships on a new site with building docks.

http://www.baesystems.com/Sites/SurfaceShips/ Has information on UK military ship construction.

http://www.fsg-ship.de/198-1-latest-videos.html Shows construction of mainly ro-ro ships on a modernized, covered slipway.

12 Design information for production

Chapter Outline

Ship drawing office 126
Lines plan 126
Three-dimensional representation of shell plating 128
Shell expansion 128
Loftwork following drawing office 128
10:1 scale lofting 130
Computer-aided design (CAD)/computer-aided manufacturing (CAM) 130
Ship product model 131
Further reading 134
Some useful websites 134

In order to produce a ship, it is necessary to develop the initial structural and arrangement design into information usable by production. While the initial design is focused on the operation of the ship once completed, production requires information in a form that relates to the manufacturing, assembly, and construction processes. Historically, this was done by creating a full-size model of the structure, in the mold loft, from which dimensions could be lifted and templates could be made. The relatively simple outfitting of early ships was carried out on the completed ship structure with pipes and other connectors run where space was available. Larger, more complex ships, faster construction times and cost have made this process obsolete, and it has largely been replaced by computer modeling.

This chapter briefly outlines the original functions of the ship drawing office and subsequent full or 10:1 scale lofting of the hull and its structural components. It then concentrates on current use by shipyards of computer-aided design (CAD) for these purposes and the extensive use of computer-aided manufacturing (CAM) in shipbuilding.

In the past, the shipyard design office defined the principal ship dimensions, the scantlings for all major structure, specified major equipment, and satisfied all salutatory requirements. The drawing office developed the design by creating drawings that identified piece parts and critical dimensions. The mold loft then defined the detailed, accurate geometry to manufacture hull piece parts for the ship structure and other trades defied the geometry for their own products, for example pipes and ventilation trunking.

Ship Construction. DOI: 10.1016/B978-0-08-097239-8.00012-X

The planning of production, the flow of material through the manufacturing stages, and the sequence of unit erection in the dock or slipway were determined by the production management.

Ship drawing office

The ship drawing office was traditionally responsible for producing detailed working structural, general arrangement, and outfit drawings for a new ship. It was also common practice for the drawing office to contain a material ordering department that would identify the necessary material requirements from the drawings and progress them.

Structural drawings prepared by the drawing office would be in accordance with Lloyd's or other classification society rules and subject to their approval; also, owner's additional requirements and standard shipyard practices would be incorporated in the drawings. General arrangements of all the accommodation and cargo spaces and stores would also be prepared, incorporating statutory requirements as well as any shipowner's requirements and standards. Outfit plans including piping arrangements, ventilation and air-conditioning (which may be done by an outside contractor), rigging arrangements, furniture plans, etc. were also prepared. Two plans of particular significance were the ship's 'lines plan' and 'shell expansion'.

Lines plan

A preliminary version of this was, in effect, prepared at the time of the conceptual design to give the required capacity, displacement, and propulsive characteristics. It was subsequently refined during the preliminary design stage and following any tank testing or other method of assessing the hull's propulsive and seakeeping characteristics. The lines plan is a drawing, to a suitable scale, typically 1:100, of the molded lines of the vessel in plan, profile, and section. Transverse sections of the vessel at equally spaced stations between the after and forward perpendiculars are drawn to form what is known as the body plan. Usually, 10 equally spaced sections are selected with half ordinates at the ends where a greater change of shape occurs. A half transverse section only is drawn since the vessel is symmetrical about the center line, and forward half sections are drawn to the right of the center line with aft half sections to the left. Preliminary body plans are drawn initially to give the correct displacement, trim, capacity, etc., and must be laid off in plan and elevation to ensure fairness of the hull form. When the final faired body plan is available the full lines plan is completed, showing also the profile or sheer plan of the vessel and the plan of the waterline shapes at different heights above the base.

A lines plan is illustrated in Figure 12.1. The lines of the lateral sections in the sheer plan as indicated are referred to as 'bow lines' forward and 'buttock lines' aft. Bilge diagonals would be drawn with 'offsets' taken along the bilge diagonal to check fairness.

When the lines plan was completed manually the draughtsmen would compile a 'table of offsets', i.e. a list of half breadths, heights of decks and stringer, etc., at each of the drawn stations. These 'offsets' and the lines plan were then passed to

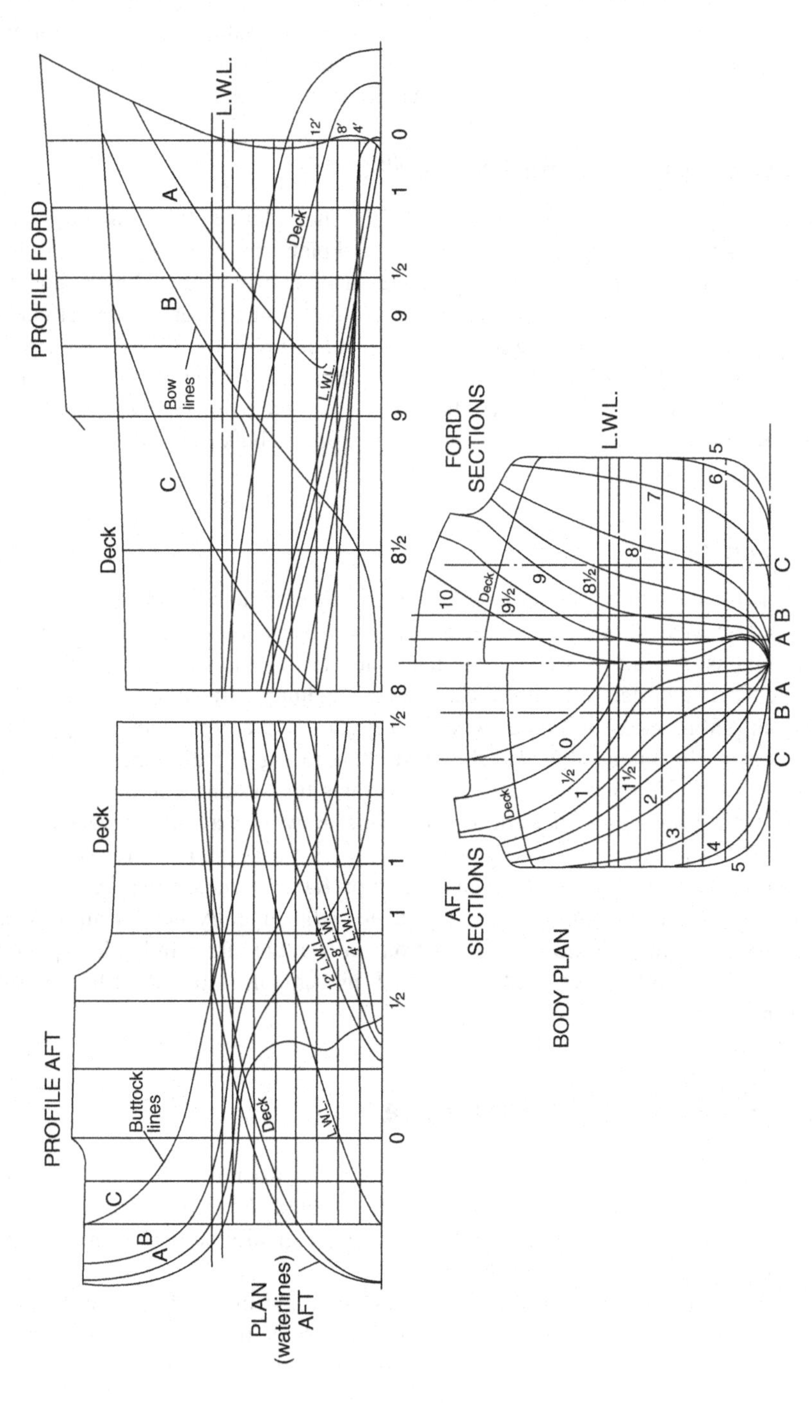

Figure 12.1 Lines plan.

loftsmen for full-size or 10:1 scale fairing. Since the original lines plan was of necessity to a small scale, which varied with the size of ship, the offsets tabulated from widely spaced stations and the fairing were not satisfactory for building purposes. The offsets used for building the ship would subsequently be lifted by the loftsman from the full size or 10:1 scale lines for each frame.

Three-dimensional representation of shell plating

When preparing the layout and arrangement of the shell plating at the drawing stage, it was often difficult to judge the line of seams and plate shapes with a conventional two-dimensional drawing. Shipyards therefore made use of a 'half block model', which was in effect a scale model of half the ship's hull from the center line outboard, mounted on a base board. The model was either made up of solid wooden sections with faired wood battens to form the exterior, or of laminated planes of wood faired as a whole. Finished with a white lacquer, the model was used to draw on the frame lines, plate seams and butts, lines of decks, stringers, girders, bulkheads, flats, stem and stern rabbets, openings in shell, bossings, etc.

Shell expansion

The arrangement of the shell plating taken from a three-dimensional model may be represented on a two-dimensional drawing referred to as a shell expansion plan. All vertical dimensions in this drawing are taken around the girth of the vessel rather than there being a direct vertical projection. This technique illustrates both the side and bottom plating as a continuous whole. In Figure 12.2 a typical shell expansion for a tanker is illustrated. This also shows the numbering of plates and lettering of plate strakes for reference purposes, and illustrates the system where strakes 'run out' as the girth decreases forward and aft. This drawing was often subsequently retained by the shipowner to identify plates damaged in service. However, a word of caution is necessary at this point because since prefabrication became the accepted practice any shell expansion drawing produced will generally have a numbering system related to the erection of fabrication units rather than individual plates. Therefore, single plates were often marked in sequence to aid ordering and production identification.

Loftwork following drawing office

The mold loft in a shipyard was a large covered wooden floor area suitable for laying off ship details at full size. The loft was usually located above the major hull steel workshops, hence the term 'loft', and the practice originally derived from wooden shipbuilding.

When the loftsmen received the scale lines plan, and offsets from the drawing office, the lines would be laid off on the mold loft floor full size and then faired. Fairing was achieved by using wooden battens that were bent to follow the offsets,

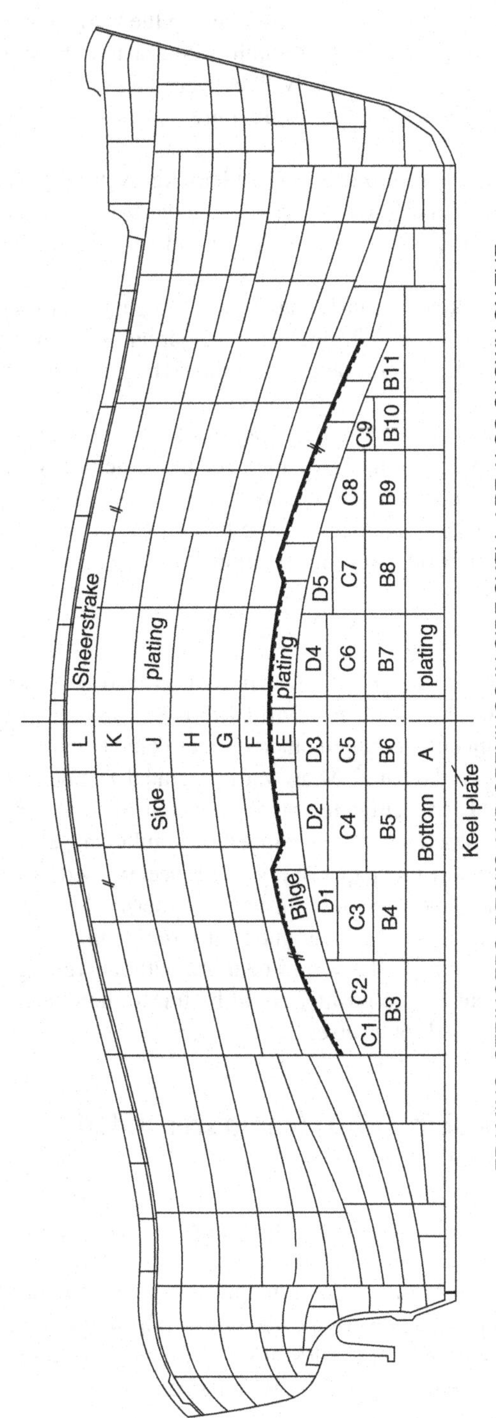

Figure 12.2 Shell expansion.

then held in place while the fair curve was inscribed on the wooden floor. This would mean using a great area of floor, even though a contracted sheer and plan were normally drawn, and aft and forward body lines were laid over one another. Body sections were laid out full size as they were faired to form what was known as a 'scrieve board'.

The scrieve board was used for preparing 'set bars' (curvature to match plate) and bevels (to maintain the web of a ship frame perpendicular to the ship's center line) for bending frames and for making templates and moldings for plates that required cutting and shaping.

Shell plates were developed (translating the curved shape into a flat plate that would later be cut and formed) full size on the loft floor and wooden templates made so that these plates could be marked and cut to the right shape before fitting to the framing on the berth.

The mold loft in effect produced an accurate, full-scale, three-dimensional model of the ship from which production information for the structure could be obtained. However, the process was labor intensive and relatively slow, and therefore expensive. Shipbuilders in Europe, where labor costs were higher, began development of alternative methods to develop production information.

10:1 scale lofting

In the late 1950s the 10:1 lofting system was introduced and was eventually widely adopted. This reduced the mold loft to a virtual drawing office and assisted in the introduction of production engineering methods. Lines could be faired on a 10:1 scale and a 10:1 scale scrieve board created. Many yards operated a flame profiling machine (see Chapter 13) that used 10:1 template drawings to control the cutting operation. In preparing these template drawings the developed or regular shape of the plates was drawn in pencil on to special white paper or plywood sheet painted white, and then the outline was traced in ink on to a special transparent material. The material used was critical, having to remain constant in size under different temperature and humidity conditions, and having a surface that would take ink without 'furring'. Many of the outlines of plates to be cut by the profiler could be traced directly from the scrieve board, for example floors and transverses.

Computer-aided design (CAD)/computer-aided manufacturing (CAM)

The first technical use of computers in the shipbuilding industry occurred in the 1960s and because of the high costs involved was only used by the largest shipbuilders running programs developed in-house on a mainframe computer for naval architecture, including hydrostatics and powering calculations. The hull design was drawn by hand and stored on the computer as tables of offsets for use in the mold loft. Early developments for steel plate cutting and even frame bending were also seen in a few shipyards in the 1960s.

In the late 1970s the graphics terminal and the engineering workstation became readily available and could be linked to a mini computer. These computers cost considerably less than the earlier mainframes, and commercial ship design and construction software became available for them. The larger shipyards quickly adopted these systems. They developed further in the following two decades to run on UNIX workstations and Windows NT machines and have expanded to cover virtually all the computing needs of a large shipyard.

The early 1980s saw the appearance of the personal computer (PC) and several low-cost software packages that performed simple hull design, hydrostatics, and powering estimate tasks. These were popular with small shipyards and also reportedly with some larger shipyards for preliminary design work. They were, however, somewhat limited and incompatible so that it was difficult to build a system that covered all the shipyard's CAD/CAM requirements. During the 1990s the available PC software standardized on hardware, operating systems, programming languages, data interchange file formats and hull geometry, and are now widely used by naval architects and the smaller ship and boat building industry in general.

Ship product model

Software systems for large shipbuilders are based on the concept of the 'ship product model' in which the geometry and the attributes of all elements of the ship derived from the contract design and classification society structural requirements are stored. This model can be visualized at all stages and can be exploited to obtain information for production of the ship (see Figure 12.3).

At the heart of the 'ship product model' is the conceptual creation of the hull form and its subsequent fairing for production purposes, which is accomplished without committing any plan to paper. This faired hull form is generally held in the computer system as a 'wire model', which typically defines the molded lines of all structural items so that any structural section of the ship can be generated automatically from the 'wire model'. The model can be worked on interactively with other stored shipyard standards and practices to produce detailed arrangement and working drawings. The precision of the structural drawings generated enables them to be used with greater confidence than was possible with manual drawings and the materials requisitioning information can be stored on the computer to be interfaced with the shipyard's commercial systems for purchasing and material control. Subassembly, assembly, and block drawings can be created in two- and three-dimensional form and a library of standard production sequences and production facilities can be called up so that the draughtsman can ensure that the structural design uses the shipyard's resources efficiently and follows established and cost-effective practices. Weld lengths and types, steel weights, and detailed parts lists can be processed from the information on the drawing and passed to the production control systems. A three-dimensional steel assembly can be rotated by the draughtsman on screen to assess the best orientation for maximum downhand welding.

The use of three-dimensional drawings is particularly valuable in the area of outfit drawings where items like pipework and ventilation/air-conditioning trunking can be

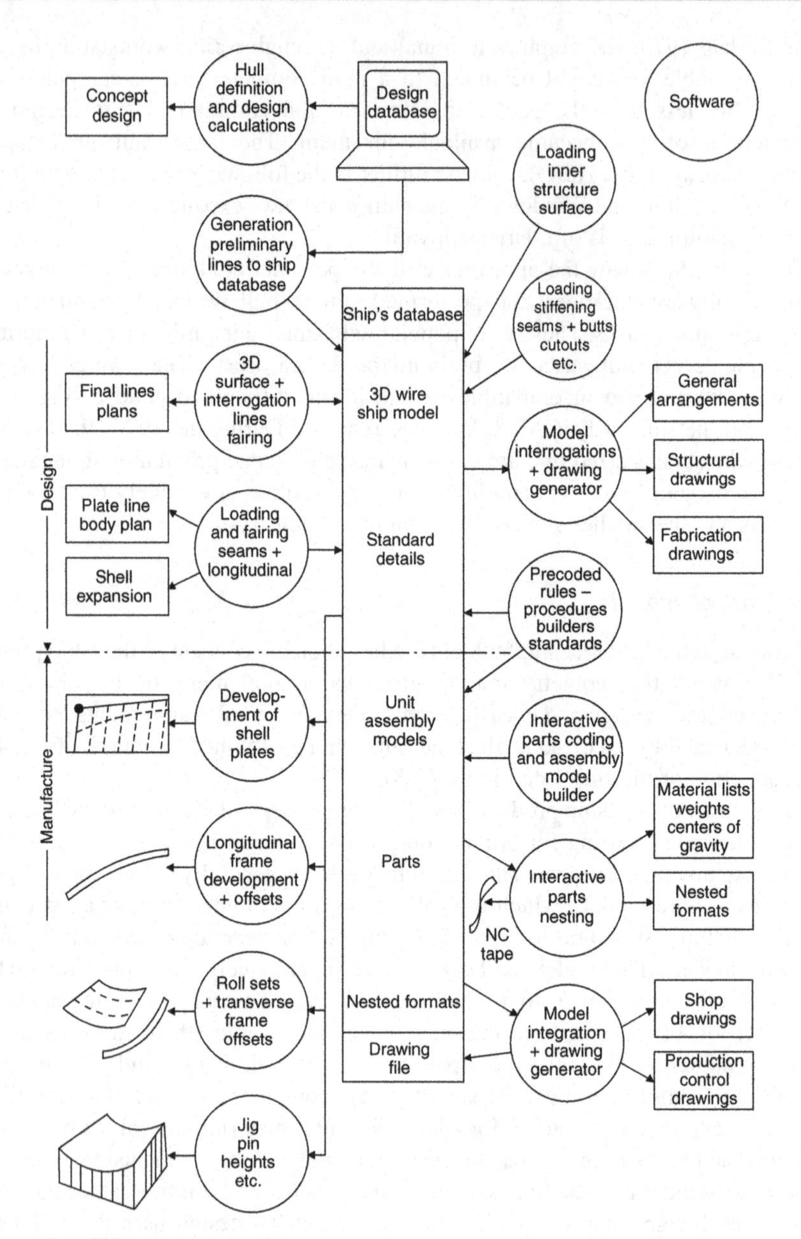

Figure 12.3 Ship product model.

'sighted' in the three-dimensional mode and more accurately measured before being created in the two-dimensional drawing.

Stored information can be accessed so that lofting functions such as preparing information for bending frames and longitudinals, developing shell plates, and

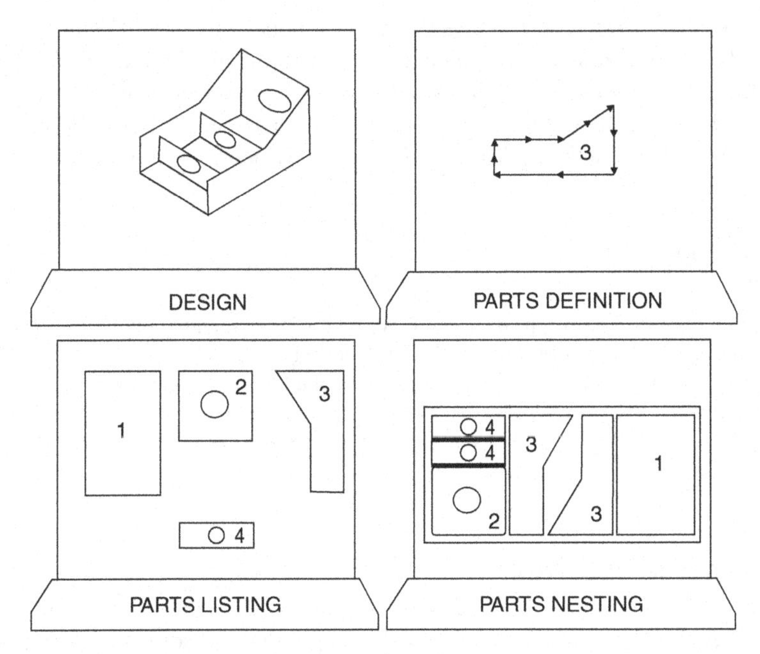

Figure 12.4 Assembly plate parts listing and nesting.

providing shell frame sets and rolling lines or heat-line bending information for plates can be done via the interactive visual display unit.

For a numerically controlled profiling machine the piece parts to be cut can be 'nested', i.e. fitted into the most economic plate, which can be handled by the machine with minimum wastage (see Figure 12.4). This can be done at the drawing stage when individual piece parts are abstracted for steel requisitioning and stored later, being brought back to the screen for interactive nesting. The order in which parts are to be marked and cut can be defined by drawing the tool head around the parts on the graphics screen. When the burning instructions are complete the cutting sequence may be replayed and checked for errors. Instructions for cutting flame planed plates and subsequently joining them into panel assemblies and pin heights of jigs for setting up curved shell plates for welding framing and other members to them at the assembly stage can also be determined (see Figure 12.3).

The basic ship product model also contains software packages for the ship's outfit, including piping, electrical and heating, ventilating and air-conditioning (HVAC) systems for a ship.

More recent developments have seen the hugely increased memory and processing capability of computers harnessed to the ship product model. This now includes all the hull structural information as well as outfitting. All pipes and other systems information is available in the model. There is, as close as possible, a complete representation of the whole ship. This can then be developed into information for part production, cutting and forming, assembly instructions, material lists (bills of materials), welding instructions, and quality assurance checks.

It is also now possible to combine the product model with virtual reality software. This enables visualization of a structure or a compartment within a ship, which can then effectively be entered by the workforce. The work sequence, access requirements, and other details can be developed so that the actual work process is planned completely and production will proceed without any delays or mistakes.

Further reading

CAD/CAM Updates. These appear several times a year in the monthly publication *The Naval Architect*. The Royal Institution of Naval Architects.

Metcalfe, The implementation of CAD in the shipbuilding industry. 79th Andrew Laing Lecture, RINA/IMarEST (May 2011).

Torroja, Alonso, Developments in Computer Aided Ship Design and Production, The Royal Institution of Naval Architects, 1999.

Some useful websites

The following websites are those of a number of providers of software for ship design and construction purposes. They provide much more detail, graphic illustrations of the products, and other information than can be accommodated in a single chapter of this book.

http://www.aveva.com
www.cadcam.autoship.com
www.cadmatic.com
www.foransystem.com
www.napa.fi
www.ShipConstructor.com
www.formsys.com

13 Plate and section preparation and machining

Chapter Outline

Plate and section preparation 135
- Stockyard 136
- Heating 137
- Plate leveling rolls (mangles) 137
- Shot-blasting 137
- Priming paint 137
- Plate handling in machine shops 138

Plate and section part preparation 138
- Plate profile cutting machines 138
- Planing machines 140
- Drilling machines 140
- Guillotines 140
- Presses 140
- Plate rolls 141
- Heat-line bending 142

Frame bending 143
- Cold frame bending 144
- Section profilers 144
- Robotics 144

Further reading 146

Some useful websites 146

This chapter describes the processes that a plate or section undergoes from the time it is received into the shipyard until it is in its final shape ready to be welded into a unit of the ship structure. In recent years many changes have occurred in this sphere of shipyard activity, and substantial economies can be achieved by obtaining an even flow of production at this stage.

Plate and section preparation

For efficient ship production, the steel material must be in an ideal condition when it is ready to be processed, and its marshaling into order for production on arrival is essential to the efficiency of the shipbuilding process.

Ship Construction. DOI: 10.1016/B978-0-08-097239-8.00013-1

Stockyard

On arrival at the shipyard, plates and sections are temporarily stored in the stockyard. In larger shipyards the stockyard is an uncovered space having sufficient area to provide storage for enough plates and sections required for the working of the yard some months ahead. The stockyard is needed because the production schedule of a steelmaker will not conform to the needs of one particular shipyard. Also, the shipyard may not be located near a steelmaker and may choose to purchase steel from another country if prices are lower. The stockyard ensures there is always steel to maintain production. In some cases the quantity of steel in storage might be about three months' stock, but shipyards in countries that are not substantial steel producers may find it necessary to carry much larger stocks of material. In contrast, Far East shipyards closely associated with and/or located near a steel manufacturer may carry as little as one month's stock, and in some cases less.

When the plates and sections are ordered the steel mills are provided with details of the identification code for each item so that they may be marked. On arrival in the stockyard, since the coding is generally in terms of the unit structure for which each item is intended, it is convenient to store the plates and sections in their respective ship and ship unit areas. In other words the material for each ship is allotted an area of the stockyard, this area being subdivided into plots for those items intended for a major structural fabrication block of the ship.

Smaller shipyards may choose to buy steel from a stockist, removing the need for a stockyard, which is an expensive facility. They may also standardize the steel plates used to reduce the need for marshaling. The ideal is 'just in time' delivery of steel, although some stock is kept as a buffer against delivery failures.

Material delivery and storage are controlled in accordance with production engineering practices to suit the ship's construction program.

Steel plates are now stacked horizontally in piles as required. (It was once common practice to store the plates vertically in racks, which was convenient for weathering purposes, but steel is now treated prior to production.) Steel profiles or sections are laid horizontally in convenient batches, usually by size and type.

Where cranes are used for material handling it is important that there be adequate coverage for the full extent of the stockyard, and the material may be delivered to the covered workshop areas. Craneage is generally of the electric overhead gantry type, traversing the stockyard on rails and extending into the workshop area; alternatively, the plates may be placed on a roller conveyor for delivery to the workshops. Lifts are made with magnetic clamps for plates and slings for the sections; the crane capacity can be up to 30 tonnes, to suit the material sizes that the shipyard can process. Plates required for the work being processed may be selected and stored adjacent to a conveyor and then transferred into the preparation workshop in the correct sequence.

The preparation begins with a conveyor system that transports the plates through a sequence of processes to bring it to a suitable condition for cutting.

Heating

Usually, a water wash or gas heating system is provided that heats the plate ready for subsequent paint drying. A wash system will also remove debris, and in some cases snow or ice.

Plate leveling rolls (mangles)

This is a set of plate-straightening rolls through which the plate is passed prior to its being worked. During production at the steelworks, the plates may acquire locked-in stresses, which are released with the use of thermal processes in the shipyard. This can cause distortion, which slows production. Also, during transit to the shipyard, plates may become distorted, and for many of the modern machining processes it is important that the plates should be as flat as practicable. Two sets of rolls may be provided, one for heavy plate and one for light plate, but types are now available that permit a wide range of thicknesses to be straightened.

Shot-blasting

Plates and sections are in almost all cases now shot-blasted to remove rust and millscale. The principles of shot-blasting are dealt with in Chapter 27.

Shot-blasting plant in shipyards is generally of the impeller wheel type, where the abrasive is thrown at high velocity against the steel surface and may be recirculated. The plant is installed so that the plate may pass through in the horizontal position, which allows the use of automated, roller conveyor systems. The only disadvantage of the horizontal plant is the removal of spent abrasive from the top of the plate, which may be relatively easily overcome. The shot is recovered and recycled. As it breaks down into smaller pieces, the mix of whole shot and fragments assists the blasting process. Spent shot and debris is removed for disposal. The whole process is enclosed.

A separate shot-blast plant is often installed for sections.

Priming paint

Following the shot-blasting of plates and sections, the material passes immediately through an airless spray painting plant. In one pass the material is automatically sprayed with a priming paint of controlled coat thickness. A number of suitable priming paints are available; the requirements for these and their formulation may be found in Chapter 27. Following the priming paint stage a drying process may be provided. This is a tunnel with fans to create an airstream, in conjunction with the preheating, so that the plate that emerges is dry and ready for processing. The drying also reduces the potential for damage to the primer coating.

Smaller shipyards that do not use large quantities of steel can purchase the material ready shot-blasted and primed by a supplier, which eliminates the expensive preparation equipment.

Plate handling in machine shops

Throughout the machining shops, overhead electric cranes having capacities from 5 to 30 tonnes, depending on the ship and material sizes, are normally used to transport plates and sections to each machine process. Individual machines may have jib cranes mounted on the frame that can be employed to handle the plate during the machining process.

Distribution of plates in the shop may also be by means of conveyors or electric-powered trollies running on rails in bays between the plate working machines. Highly mechanized plant systems are also available where conveyor systems, flame cutting tables, and aligning and clamping equipment are installed as integrated units.

Plate and section part preparation

A number of the methods in use for forming plates into the required shapes have been in use for many years. This is particularly true of the methods adopted for fitting plates and frames to the curve of the hull, but as we have seen in the previous chapter the information for doing this can now be derived using the CAD/CAM systems available. Cutting flat plates to the required profiles has become highly automated, very sophisticated machine tools having been introduced for this purpose. In the main, cutting is achieved by the use of an oxy-fuel flame or plasma arcs.

Plate profile cutting machines

Where a plate is to be cut into one or more (or a series of) complicated shapes, a profiling machine is employed. Cutting is achieved by the use of an oxy-fuel gas or plasma-arc flame, or plasma-arc profilers are usually numerically controlled. In this case the computer-generated information on the plate geometry is fed directly to the cutting machine. The information is processed so that, in addition to the shape of the plate part, the start and finish of cutting, the speed of cutting, and information to mark the plate are provided. Marking is generally done using an inkjet printing system, although some machines use a powder that is fused onto the plate using the gas heating. Provision can also be made for the shrinkage of the plate part as it cools after cutting. The predicted shrinkage of the plate part is added to the nominal dimensions from the CAD system, so after cooling the part size is correct. The overall arrangement is shown in Figure 13.1.

In some cases, 1:1 template or drawing controlled systems may be used. This can be for small, repetitive parts such as brackets. Such parts are more usually programmed using a basic CAD system.

A full-size template or drawing may be used to control a cutting machine, and can be useful where a single small item is to be cut in large numbers. The size of the item is obviously restricted, and the location from which the item is cut in a large plate is selected by the operator. Where a template is used a mechanical follower may be employed, and where a drawing is used an electronic scanning device may follow the outline. Profilers of this type have a limited application in shipbuilding, but can be

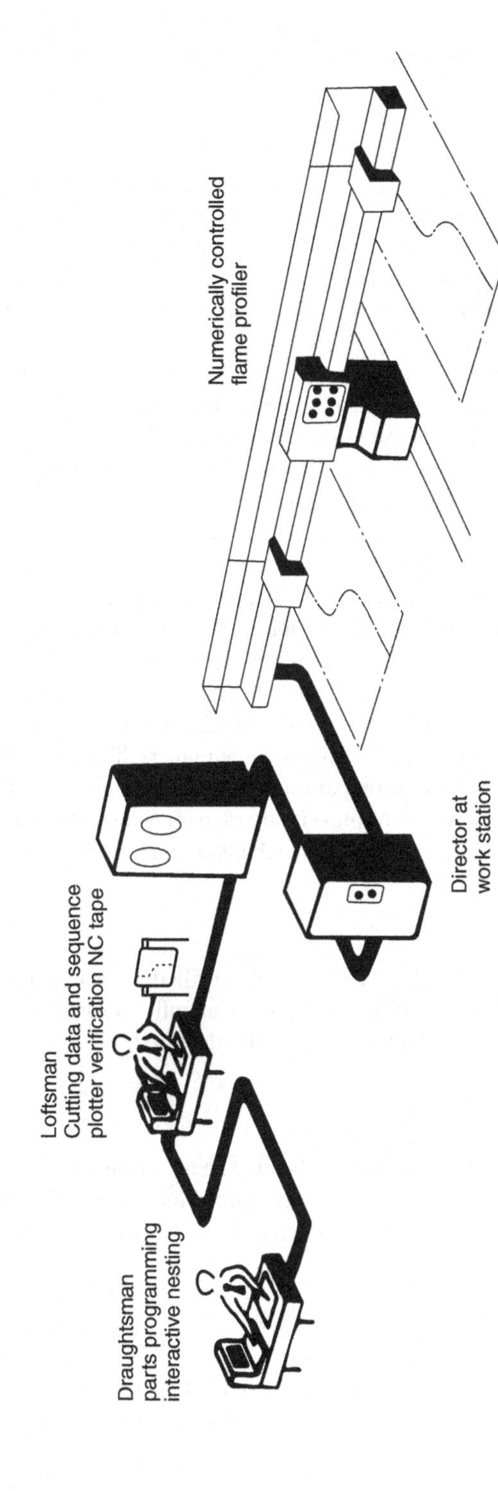

Figure 13.1 Numerical flame cutting control system.

efficient for cutting batches of, say, identical beam knees or stiffener brackets. More basic cutting systems are likely to be found in very small shipyards and in ship repair.

Planing machines

Profiling machines are essentially for use where a plate requires extensive shaping with intricate cuts being made. Many of the plates in the ship's hull, particularly those in straightforward plate panels, decks, tank tops, bulkheads, and side shell, will only require trimming and edge preparation, and perhaps some shallow curves may need to be cut in shell plates. This work may be carried out on a planing machine, usually a flame or plasma-arc planer.

A flame planer consists basically of three beams carrying burning heads and running on two tracks, one either side of the plate working area. One beam carries two burning heads for trimming the plate sides during travel, whilst the other two beams have a single burning head traversing the beam in order to trim the plate ends. Each burning head may be fitted with triple nozzles and is used to give the required edge preparation. Plate, beam, and burning head positioning is manual, but cutting conditions are maintained once set.

Mechanical planing machines were in existence prior to the flame planer, and have an advantage where they are able to cut materials other than steels. Higher cutting speeds than that obtained with flame cutting may be obtained for vertical edges with a rotary shearing wheel on a carriage, where the plate is held by hydraulic clamping. For edge preparations, older machines with a conventional planing tool require a large number of passes and are much slower than flame planers. There are, however, merits in mechanically machined edge preparations where the superior finish is advantageous for critical welds. The use of plasma-arc planers, which also provide a superior and faster cut, tends to make the use of mechanical planers obsolete.

Drilling machines

Some plates or sections may need to have holes drilled in them, for example bolted covers and portable plates. Drilling machines generally consist of a single drilling head mounted on a radial arm that traverses the drill bed.

Guillotines

Smaller 'one-off' plate shapes such as beam knees, various brackets, and flat bar lengths may be cut in a hydraulically operated guillotine. Plate feed to the guillotine is usually assisted by the provision of plate-supporting roller castors, and positioning of the cut edge is by hand.

Presses

Hydraulic presses may be extensively used in the shipyard for a variety of purposes. They are capable of bending, straightening, flanging, dishing, and swaging plates

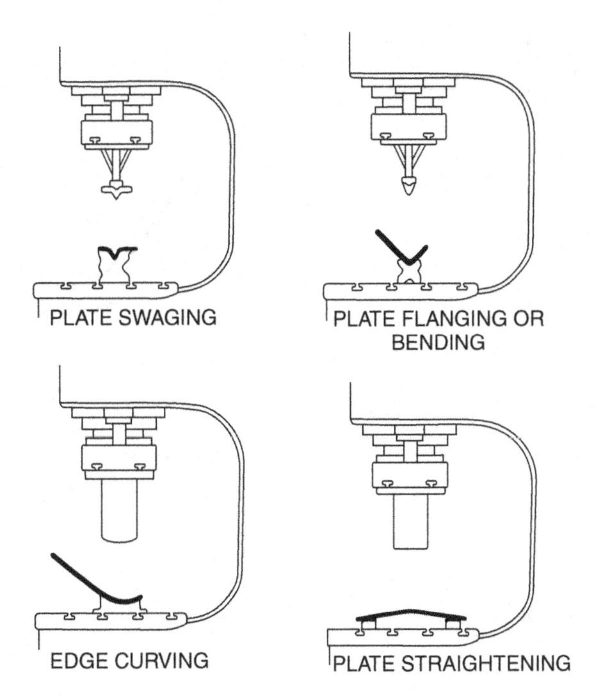

Figure 13.2 Gap presses.

(see Figure 13.2). All of the work is done with the plate cold, and it is possible to carry out most of the work undertaken by a set of rolls. This is done at less capital cost, but the press is slower when used for bending and requires greater skill.

Plate rolls

Heavy-duty bending rolls used for rolling shell plates etc. to the correct curvature are hydraulically operated. Two lower rolls are provided and are made to revolve in the same direction so that the plate is fed between them and a slightly larger diameter top roll that runs idly (see Figure 13.3). Either or both ends of the top roll may be adjusted for height, and the two lower rolls have adjustable centers. With modern bending rolls, plates up to 45 mm thick may be handled and it is possible to roll plates into a half circle. These large rolls are also supplied with accessories to allow them to undertake heavy flanging work with the pressure exerted by the upper beam, for example 'troughing' corrugated bulkhead sections.

Shorter pyramid full-circle rolls are also used in shipyards, these being very useful for rolling plates to a full circle. This may be done to obtain large mast and derrick post sections for example, or bow thruster tunnel. Arrangements are made for removing the rolled full-circle plate by releasing the top roller end bearing. Vertical rolls are also available and may be used to roll plates full circle, but can be much more useful for rolling heavy flats used as facing bars on transverses of large tankers, etc.

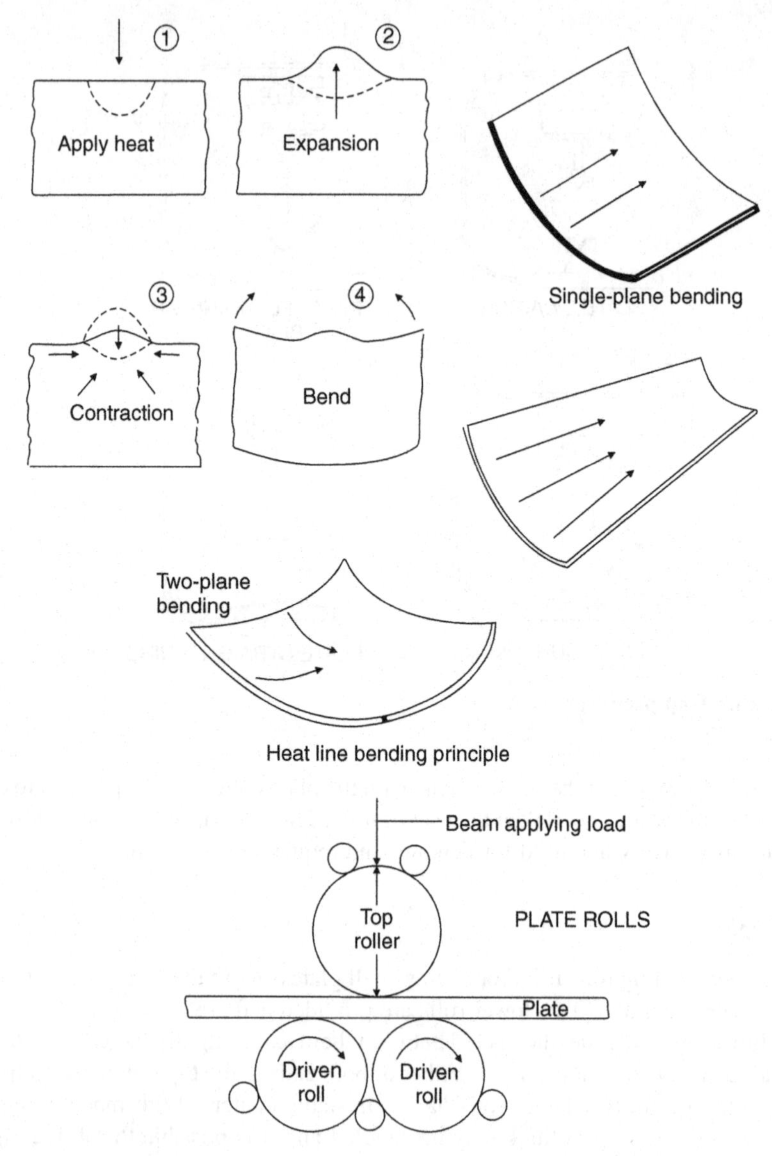

Figure 13.3 Forming shell plates.

Heat-line bending

The 'heat-line' bending procedure is a widely used technique to obtain curvature in steel plates for shipbuilding purposes. It is, however, a process that until recently relied on highly skilled personnel and did not guarantee constant accuracy of shapes formed.

Heat is applied in a line to the surface of a plate by a flame torch, with immediate cooling using air or water. The narrow heated line of material is prevented from expanding in the direction of the plate surface by the large mass of cold plate, and therefore expands outwards perpendicular to the plate surface. On cooling, contraction will take place in the direction of the plate surface, causing the plate to become concave on the side to which heat was applied (see Figure 13.3). An experienced operator is able to make a pattern of such heat lines on a plate, producing controlled distortion to obtain a required shape. Heat-line bending can be more time consuming than using rolls or presses, but it has an advantage in that the plate holds its form more accurately when stiffening and other members are added later in the fabrication process. This is an important consideration since shape inaccuracy can be critical at the erection stage in terms of lost production time. Heat-line bending may be used after cold forming to obtain improved accuracy. It is also used where a double curvature is required, for example on bulbous bow plates.

In recent years fully automated heat-line bending systems have been developed and installed in shipyards. These numerically controlled heat-line bending machines permit highly accurate, reproducible thermal forming of any steel plate using the data originating from the shipyard's CAD system. More than double the productivity of the traditional manual process is achieved.

Frame bending

The traditional system of bending side frames may still be in use for repair work, and is described as follows. A 'set-bar', which is a flat bar of soft iron, is bent to the scrieve line of the frame on the scrieve board and then taken to the frame bending slabs. On these solid cast-iron slabs pierced with holes the line of the frame is marked, and modified to agree with the line of the toe of the frame. As the heated frame on cooling will tend to bend, the set-bar is sprung to allow for this change in curvature before it is fixed down on the bending slabs by means of 'dogs' and pins inserted in the slab holes. Whilst the set-bar is being fixed the frame section is heated in a long oil-fired furnace adjacent to the bending slabs. When at the right temperature it is pulled out onto the slabs and fixed with dogs and pins against the set-bar, as quickly as possible. Tools are available for forcing the frame round against the set-bar, including a portable hydraulic ram, and the toe of the web may require constant hammering to avoid buckling under compression.

As the frames fitted have their webs perpendicular to the ship's center line, all except those immediately amidships will require beveling. A beveling machine is available that is placed in front of the furnace door, and as the frame is removed it passes between its rollers, which are controlled so that the flange is bent at every point to an angle indicated by a bevel board prepared by the mold loft.

Once bent, the bar is put aside to cool, but is fastened down to prevent its warping in the vertical direction. When cold it is checked against the frame line drawn on the slabs. Meanwhile, the set-bar is turned over and used to bend the corresponding frame for the opposite side of the ship.

Cold frame bending

It is now almost universal practice to cold-bend ship frames using commercially available machines for this purpose. The frames are progressively bent by application of a horizontal ram whilst the frame is held by gripping levers (Figure 13.4). Any type of rolled section can be bent in some machines with a limitation on the size of section. Obtaining the correct frame curvature can be achieved by the 'inverse curve' method or numerical control. With the 'inverse curve' method the inverse curve information can be determined for each frame using a CAD/CAM system. The inverse curve is marked onto the straight frame and the frame bent until the inverse curve becomes straight on the curved frame (see Figure 13.4). A hydraulic cold-frame-bending machine can be controlled by numerical control tapes prepared in a similar manner to the numerically controlled flame profilers, the frame line being initially defined from the computer-stored faired hull.

Section profilers

Plate profilers produce very accurate plate components for assembly into ship units and it is important that the supporting stiffening members of the structure are produced with the same accuracy. Section profilers accurately cut, scallop, and bevel, as necessary, the profile edge to be welded to the plate. Cutting is by oxy-fuel gas or plasma and in modern systems is undertaken by a compact robot operating in a viewable enclosed workstation with integral fume extraction. Prior to cutting, the stiffening member can be moved through a marking cell where part identification and, where required, bending lines for inverse curve frame bending (see Figure 13.4) can be marked.

Robotics

Robots have in recent years been provided with improved control features that have made them more adaptable to the workshop floor situation. For example, most robots are now available with some form of 'adaptive control', which provides feedback from the environment permitting, say, automatic adjustment of the robots path and/or its functions. Also, the provision of 'off-line' programming and simulation packages makes it possible to develop and test programs for the robot remotely. Thus, the robot carries on working whilst new programs are produced for it.

Many of the world's shipyards and shipyard systems suppliers have been developing and implementing robots in shipbuilding. A large proportion of these developments have been in fully automating the machine welding processes described in Chapter 9, but other areas of adoption and trial have involved flame and plasma-arc cutting, local shot-blasting and painting, and marking. Robots developed to date for shipyard usage are either associated with large gantry structures or are small portable units. The former often have the movable robot mounted on the traveling gantry with sensors providing the adaptive control, and are employed for cutting and welding processes. The latter can be manually transported or self-propelled, even climbing vertically, or for robotic transportation, and have been used for local welding in difficult situations and cleaning and painting.

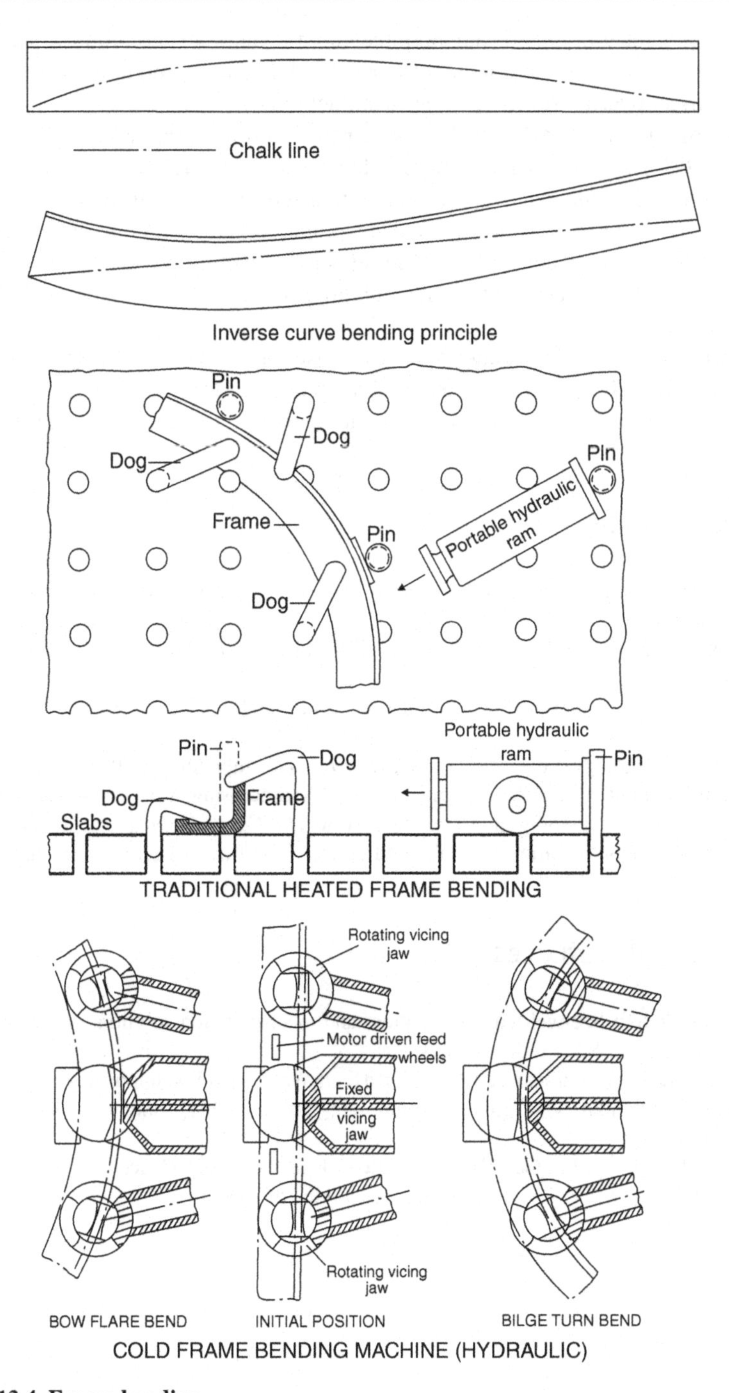

Figure 13.4 Frame bending.

Both the gantry-mounted and portable robotic units are commonly featured in automated shipyard production units for panel, subassembly, and unit fabrication now available to shipyards from specialist manufacturers.

Many of the early and existing robot programming systems are linked to the shipyard's CAD system so that programs developed for the robot can be run 'off line' with the three-dimensional graphics simulating the robot's performance before it is put to work. A more recent patented development has seen the introduction of a robotic welding unit controlled by an advanced machine vision-guided programming system. This is claimed to reduce preprogramming times to zero and no CAD models are required.

Whilst robots have advantages in their use in difficult and unpleasant work conditions and tedious repetitive work situations, their development and adoption is increasingly seen as a means to higher productivity and reduced manufacturing cost.

Further reading

Boekholdt R: *Welding Mechanization and Automation in Shipbuilding Worldwide*. Abington Publishing, 1996.

Kalogerakis: The use of robots in the shipbuilding industry, *The Naval Architect*, July/August 1986.

New advances in efficient stiffener production from Dutch specialist, *The Naval Architect*, July 2004.

Shipbuilding technology, Feature in *The Naval Architect*, July/August 2003.

Speeding throughput with automated plate and profile processing. Van der Giessen-de Noord: creating innovative solutions, A supplement to *The Naval Architect*, September 2002.

Vision-guided robots 'open up new shipbuilding era', *The Naval Architect*, February 2004.

Some useful websites

www.pemamek.com See welding, automation, and mechanization solutions for shipyards.

www.tts-marine.com See production systems.

http://www.intechopen.com/source/pdfs/279/InTech-Welding_robot_applications_in_shipbuilding_industry_off_line_programming_virtual_reality_simulation_and_open_architecture.pdf

http://www.dtic.mil/cgi-bin/GetTRDoc?Location=U2&doc=GetTRDoc.pdf&AD=ADA454642 Considers the economics of robots.

14 Assembly of ship structure

Chapter Outline

Assembly 149
Subassemblies 151
Unit assembly 151
Block assembly 152
Outfit modules 152
Unit erection 154
Joining ship sections afloat 158
Further reading 159
Some useful websites 159

Historically, ships were constructed on a slipway piece by piece. The transition from wood, to iron and then steel made little initial difference to the construction process, and the result was the typical shipyard layout as described in Chapter 11. Exceptions were for military and large passenger ships.

During the Second World War a large number of merchant and war ships were required to be built in a short period of time. These requirements speeded the adoption of welding in shipyards, and often led to the application of mass production techniques in shipbuilding. Prefabrication of ship units, i.e. the construction of individual sections of the ship's structure prior to erection, became a highly developed science. Often, the units were manufactured at a location remote from the shipyard, and erection in the shipyards was carried out with schedules that still look very impressive today. Many of the more spectacular achievements in this field were obtained in the USA, where much of the tonnage required during the war period was constructed using these new advances. Unfortunately, the results of this crash building program were not always entirely satisfactory; the reputation of welded structures, for example, suffered for quite a time as a result of wholesale application without a background of experience.

Also, the adoption of the prefabricated units was dictated by speed and large-scale production, rather than economics. However, the potential benefits of units were identified and shipyards began to make the necessary investment, in additional assembly workshops and larger cranes, to use prefabricated units. The initial low cost of labor made traditional methods still reasonably economic. During the 1950s there was an increase in ship size and the greater quantity of steel required for the structures of these ships was a major driver for the adoption of units. Also, the costs of labor in traditional shipbuilding countries increased.

Ship Construction. DOI: 10.1016/B978-0-08-097239-8.00014-3

From the mid 1950s, prefabrication was gradually being applied to merchant ship construction as shipyards benefitting from a high demand for ships were able to invest in the additional workshops and cranes. Today, all vessels are to varying degrees prefabricated.

With riveted construction unit fabrication was rarely employed; the introduction of welding lent itself more favorably to this form of construction. With welding, simpler unit shapes (no staggered butts are required in the side shell, for example) and at first less critical tolerances could be applied. In the early days of unit construction, the units were assembled over size, so they could be trimmed to fit on the building berth (or dock). What is termed 'green' material was left on the unit edges, typically around 20 mm. The unit was then lifted into place and cut to fit the structure, which had already been erected. The pressures of cost and required speed of construction have led to much greater accuracy of assembly, so that for parallel body in particular, units and blocks can be produced with sufficient accuracy to allow them to be fitted and welded without any adjustment. This provides a saving in man-hours and time.

Units may be constructed under cover, which is an attractive advantage, for example, in Northern Europe and also in hot climates, not only because of working conditions, but because of the better welding conditions. It is possible to turn units over to allow downhand welding, which is easier to perform and likely to provide better results. There is great advantage in keeping vertical and overhead welding to a minimum. Also, central services are more readily available at the shop, with gases for cutting, air for chipping, and electric current for welding being placed where needed. Production planning techniques may be adopted with prefabrication sequences, the material and labor being planned in unit groups and the whole shipbuilding sequence being controlled to fit the time allowed on the berth, or in the building dock.

Once the unit construction method became established, and ships continued to become larger, the units were also made bigger and then several units were assembled into large blocks. Figure 14.1 shows a typical side block for a single hull bulk carrier. When the first large tankers were built, typical units were around 50 tonnes. By 1970, lifting blocks of 500 tonnes were not uncommon. As the ships grew in size they also grew in complexity and outfitting began to be a bottleneck. In 1970, a 200,000-tonne deadweight tanker could be constructed in two or three months in new or modernized shipyards. It was then necessary to speed the outfitting of ships to avoid having several ships outfitting at the same time and to gain the advantages that faster construction offered in terms of increased income. The traditional approach of completing and launching the hull prior to outfitting in a basin or at a quay had to be modified.

Until unit and then block construction was established, the outfit and equipment were fitted at the berth and after launching at the fitting-out quay. Piping systems, ventilation, and the machinery units were fitted into the erected hull as independent items. Likewise, the accommodation spaces were lined out and fitted with the furniture and other fittings in situ. Whilst the production methods applied to the steelwork fabrication and erection were time saving and cost-effective, the haphazard and largely uncontrolled methods of outfitting detracted from the overall production engineering of the ship. Rather than plan outfit installation on a total ship system basis, it is today standard practice to plan this work for zone installation, the zone corresponding to

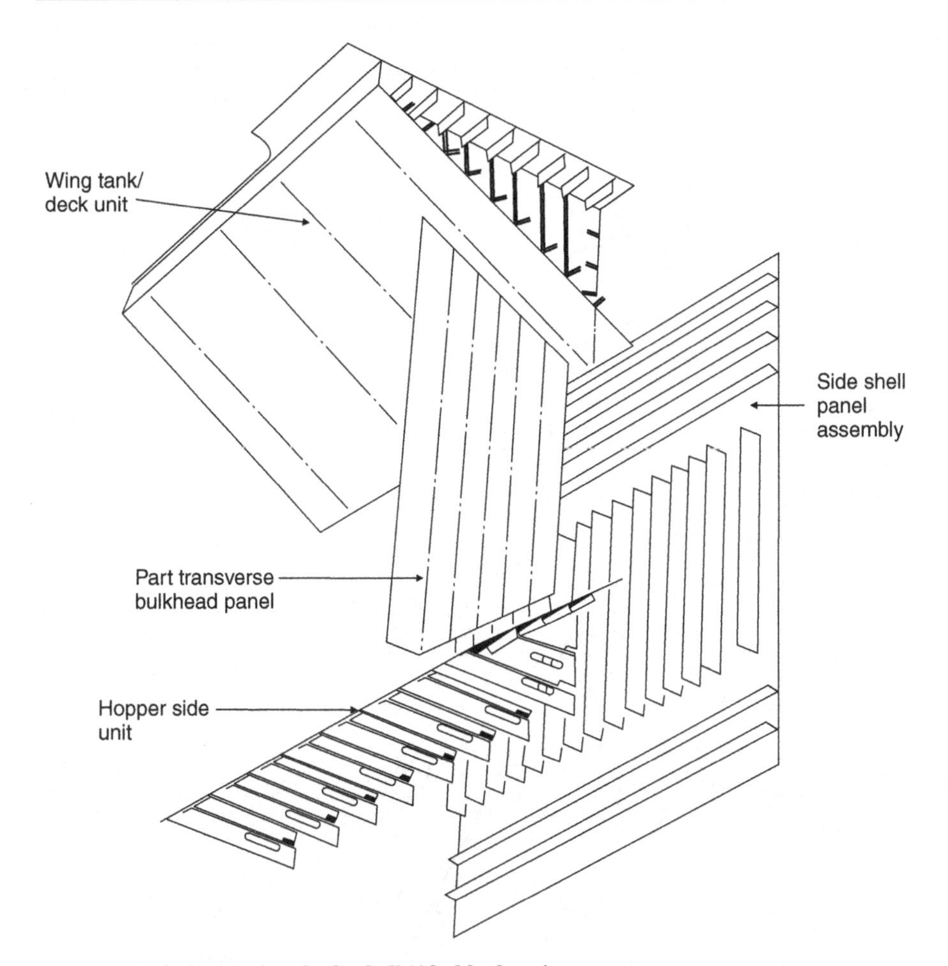

Figure 14.1 Bulk carrier single shell side block unit.

a main compartment area that may be broken down into blocks or smaller assemblies. Pre-outfitting of each assembly or block may be of the order of 85–90%. Both steelwork and outfit are highly planned for each assembly and block unit, and fabrication and outfit installation are undertaken at a workstation where the facilities and material are supplied to the workforce. Figure 14.2 shows a pipe module which could be fitted onto a steel block, or lifted directly into the ship during construction.

Assembly

Assembly is any process that takes a number of steel piece parts, or larger structures, and combines them into a larger structure. For very large ships, assembly consists of a number of stages, putting together increasingly larger elements of the ship. The number of stages varies between shipyards and especially varies with ship size. There

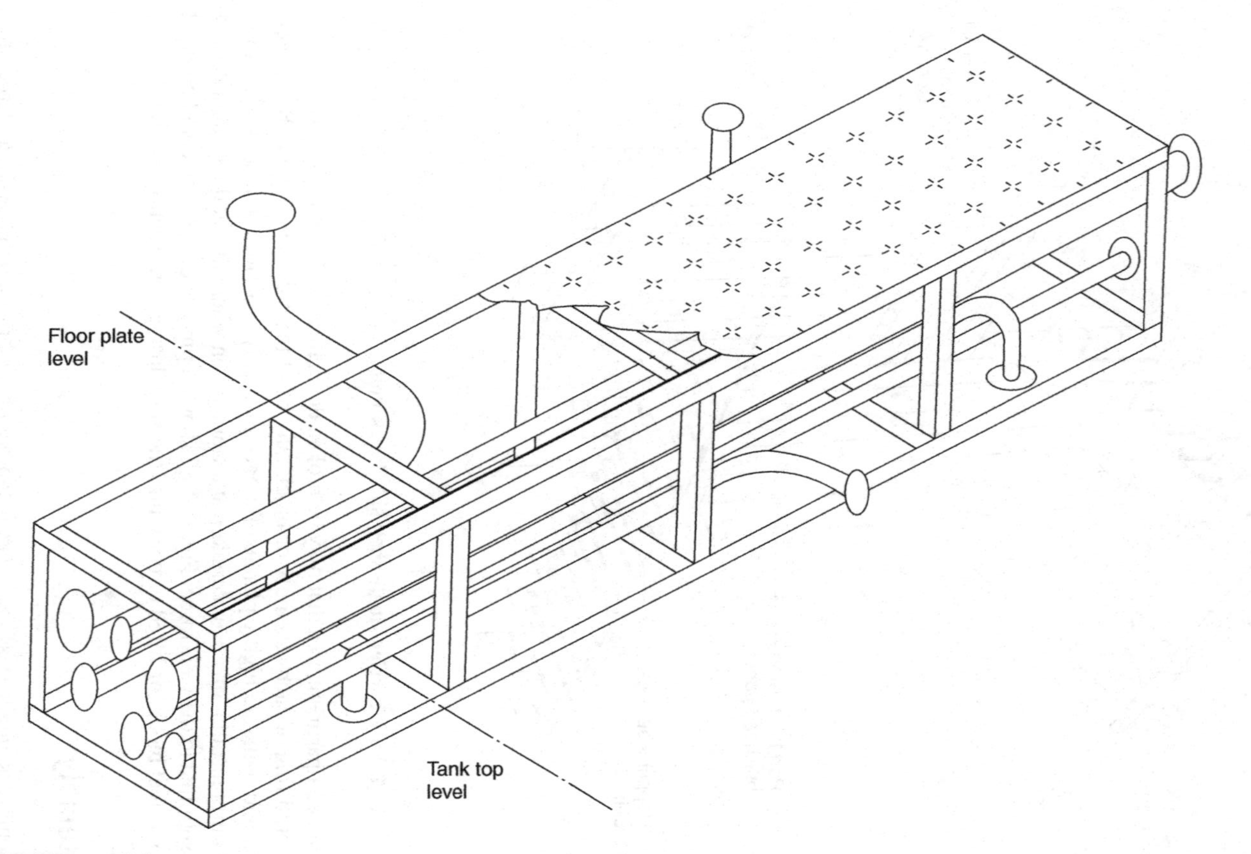

Figure 14.2 Pipe module.

is not really a 'typical' ship, but as an illustration, considering a Panamax bulk carrier with around 10,000 tonnes of steel, a length of around 250 meters and a breadth of 32 meters, then the following assemblies can be described:

- *Minor assembly*: Brackets, intercostal floors or girders, bulwarks, two-dimensional assemblies with five parts and a maximum size of 2 meters by 5 meters, weight below 2 tonnes.
- *Subassembly*: A flat or curved panel, an egg box structure or other internal structure up to around 12 meters by 12 meters, weight up to 20 tonnes.
- *Unit assembly*: A structure that can be erected in the building dock, typically one or two panels and associated internal structure, weight up to 60 tonnes.
- *Block*: An assembly of two or more units into a bottom or side block, weighing up to 200 tonnes.

The above are only illustrative and the dimensions and weights will vary for larger or smaller ships.

Subassemblies

When plates and sections have been machined (cut, bent, beveled, etc.) they are ready for assembly into two-dimensional ship units. Within the fabrication shop there are often arranged a number of bays for different assemblies, for example flat plate panels, curved shell units, matrix or 'egg box' structures and some minor subassemblies. All these may be termed subassemblies if they are subsequently to be built into a larger three-dimensional unit prior to erection. A two-dimensional plate panel assembly may be highly automated (see also 'Welding automation' section in Chapter 10) with prepared plates being placed and tack welded prior to automatic welding of the butts, after which the plates are turned and back welded unless a single-sided weld process has been used. The panel is marked and the stiffeners placed and welded automatically or with semi-automatic process. Marking and welding of two-dimensional panels is often robotically controlled. Minor subassemblies such as deep frames consisting of web and welded face flat are also attached at this stage. Curved shell plates are placed on jigs and welded, and the various stiffening members can be aligned and welded in a similar manner to those on a flat panel assembly. Assembly jigs may also be used for matrix or 'egg box' assemblies, for example structures of solid and bracket plate floors with longitudinal side girders that are to go into double-bottom units.

Unit assembly

In most instances the two-dimensional subassemblies will be built into three-dimensional unit assemblies. The size of the unit assembly will have been decided at an early stage of the planning process, ideally at the structural design stage. Constraints such as lifting capacities and dimensions that can be handled are taken into consideration, as is the provision of breaks at natural features ensuring the blocks are self-supporting and easily accessed, etc. Panel assemblies used in the block may well have dimensions restricted by the plate length that can be handled at the machining stage and this can subsequently influence block length. In the machinery area the size and arrangements can be decided by zone outfit considerations.

Each unit should be designed for maximum downhand welding but may have to be turned for this purpose.

Block assembly

Several units can be combined into a block. Where a typical unit is a double-hull section, a block can combine several of these to form a complete double-bottom block over a plate length, or a side block. Units will have generally been outfitted during assembly, for example the piping in a double-bottom unit, installed when the unit is in the most convenient orientation and access is good. The block is sufficiently large, especially in zones such as machinery spaces or pump rooms, for substantial outfitting before transport to the building dock.

Units can be easily turned to effect outfit installation, particularly those containing machinery flats in the aft engine room areas where pipework etc. can be fitted on the underside of the flat in the inverted position and then it is turned to install equipment above the flat (see Figure 14.3). These already outfitted units can then be assembled into a larger block and additional outfitting installed. Also, outfit modules can be installed into the blocks. A block's center of gravity is calculated and lifting lugs so provided that these operations can be undertaken, and finally the block can be suspended for erection at the building dock or berth and put into place in the correct plane.

Outfit modules

Units of machinery, pipework, and other outfit systems required for a specific zone can be planned, built up into modules and installed as such into a block fabrication. Pipework in particular lends itself to this form of assembly and can, with careful planning at the CAD stage, be arranged in groupings so that pipe bank modules can be arranged for a particular zone. Figure 14.2 is an example of a pipe module. Modules can range from a small pipe bank supported by light framing of pipe hangers, or a complete auxiliary machinery unit on its seating, which has even been test run prior to installation, to a large modular unit that, together with several similar units, constitutes the bulk of a complete engine room. The latter have been developed in one European shipyard, where macro-modules of the order of 10 m × 10 m × 4 m made up of square rolled hollow sections (which function as pillars when installed) and horizontal parts of the ship's structure such as flats are completely outfitted. A number of these macro-modules erected around the main engine are indistinguishable from a conventional engine room. Subcontractors are encouraged by the shipyards to supply their equipment in module form.

Not all outfits can be incorporated into modules and a large number of piece parts have to be provided for fitting in any given zone at a particular time within the

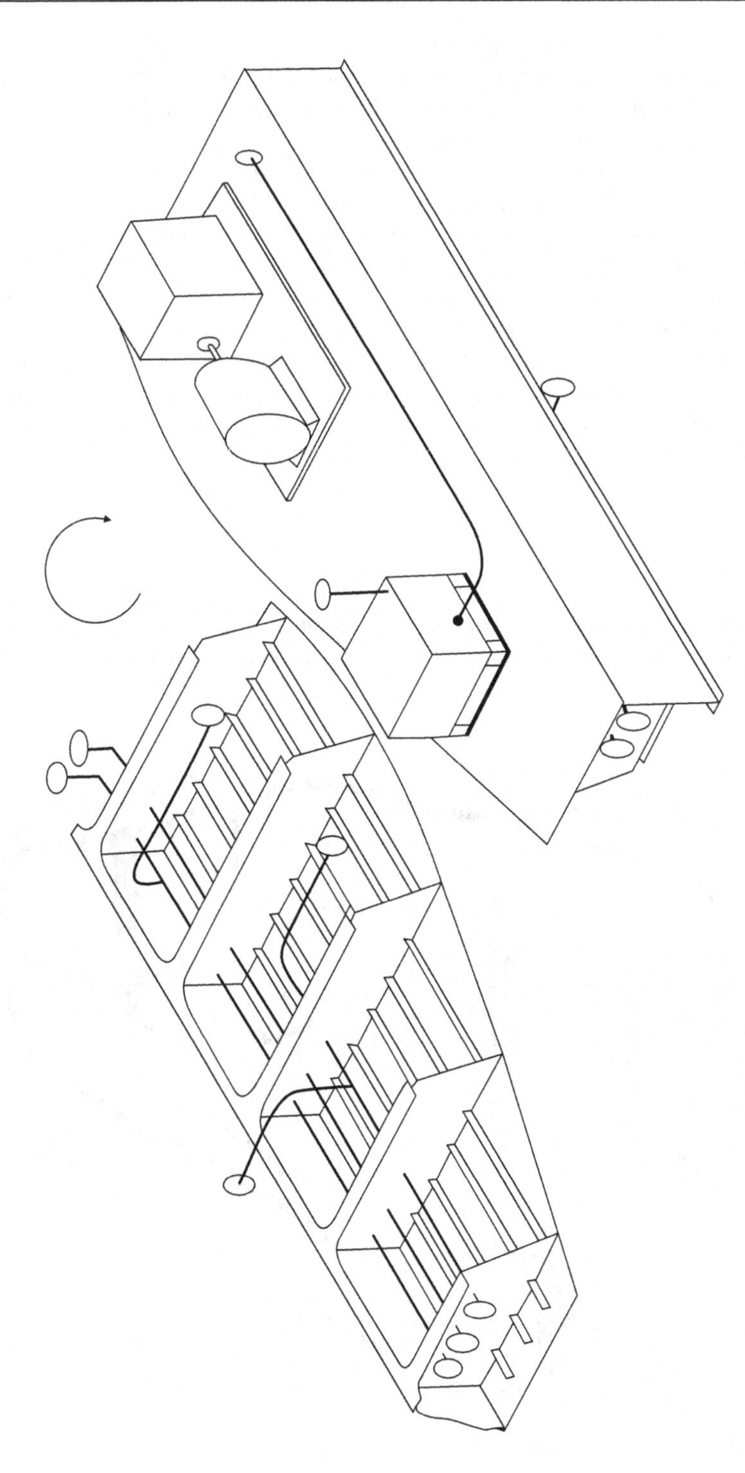

Figure 14.3 Assembly unit outfitted on both sides.

assembly shops. To maintain production engineering standards, a concept of 'palletization' has been developed whereby the piece parts for that zone are generated at the CAD/CAM stage, bought in and/or fabricated, etc., and made available at the workstation when the particular assembly is ready to receive them.

An 'open top' arrangement for blocks or smaller ships being outfitted under cover can facilitate installation of the items and modules.

Superstructure blocks are frequently fabricated separately and pre-outfitted with accommodation before erection as a complete unit. Modular cabin units are a common feature of modern shipbuilding, some companies specializing in their production. Figure 14.4 shows a typical self-supporting cabin/toilet module complete with pipework, ventilation, electrical fittings and wiring, all of which can plug into the ship's systems, and all built-in furniture. An accommodation block must be specifically designed for such modules and the sequence of module access and placement in the block carefully planned.

Unit erection

When any panel and the block assemblies are complete there will be some time buffer before their erection at the building berth, building dock, or within the building hall to

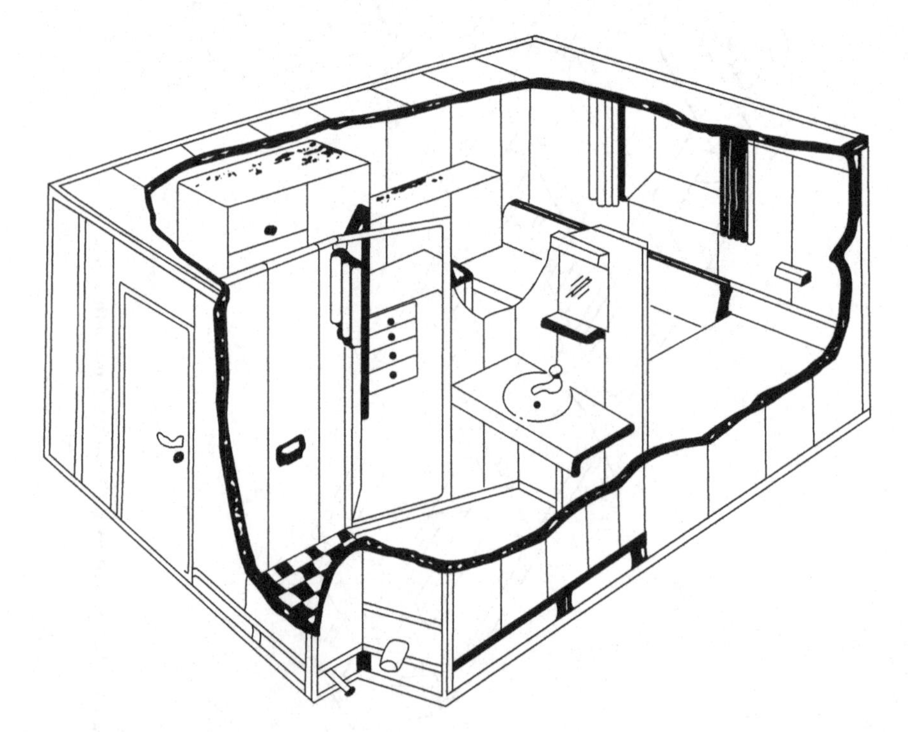

Figure 14.4 Cabin/toilet module.

allow for any mishaps in the production schedule. Stowage is generally adjacent to the berth, dock or building hall, and will vary in size according to the yard's practice, some yards storing a large number of units before transferring them to the berth or dock for erection in order to cut the berth/dock time to a minimum.

Sequences of erection for any particular ship vary from shipyard to shipyard and depend on a number of factors. Experience of previous ship erection schedules and difficulties given the yard's physical and equipment constraints leads to standard practices being established. These are taken into consideration at the structural design stage, as are the desirability of minimizing positional welding and fairing. In general, it is common practice to make a start in the region of the machinery spaces aft, obviously working from the bottom upwards, and also forward and aft. In earlier times this was done to give the engineers and other outfit workers early access to these spaces, but with the amount of pre-outfitting this might not be considered so important. However, this area still requires a larger amount of finishing work. In particular, the boring of the stern for the tailshaft is preferably undertaken when the after sections are fully faired and welded.

For an established medium-sized yard with building berth or hall having modest craneage, typical erection sequences for a general cargo ship, a large double-hull oil tanker, and bulk carrier are shown in Figures 14.5, 14.6, and 14.7 respectively. The block assemblies for the bulk carrier are shown in Figure 14.8.

In erecting the ship units it is important to employ the correct welding sequences. These are arranged to avoid excessive 'locked-in' stresses, and overlapping frames,

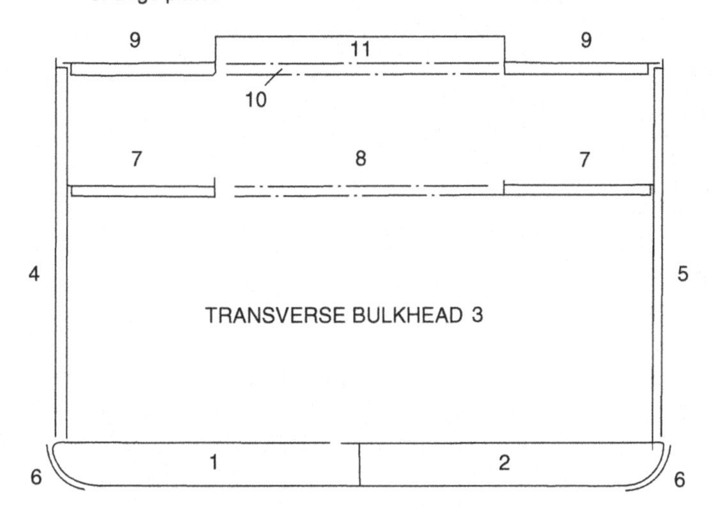

Figure 14.5 An erection sequence for a general cargo ship.

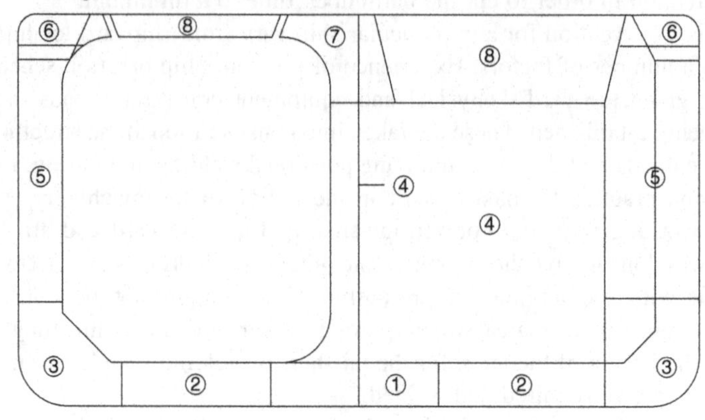

Figure 14.6 An erection sequence for a large double-hull tanker.

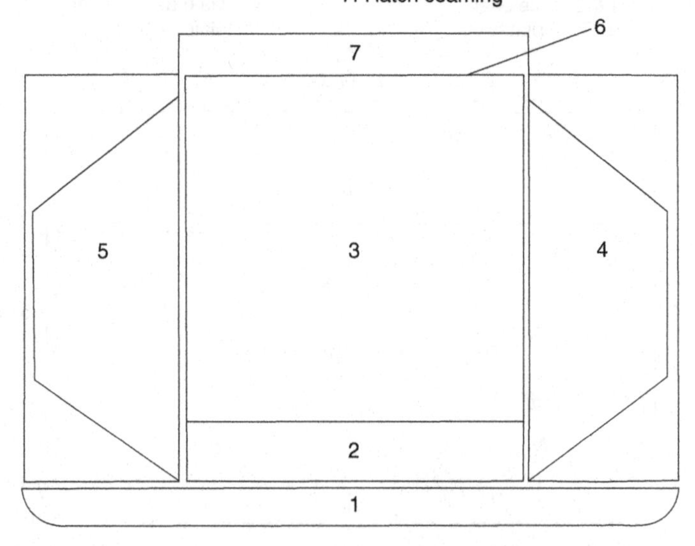

Figure 14.7 An erection sequence for a dry cargo bulk carrier.

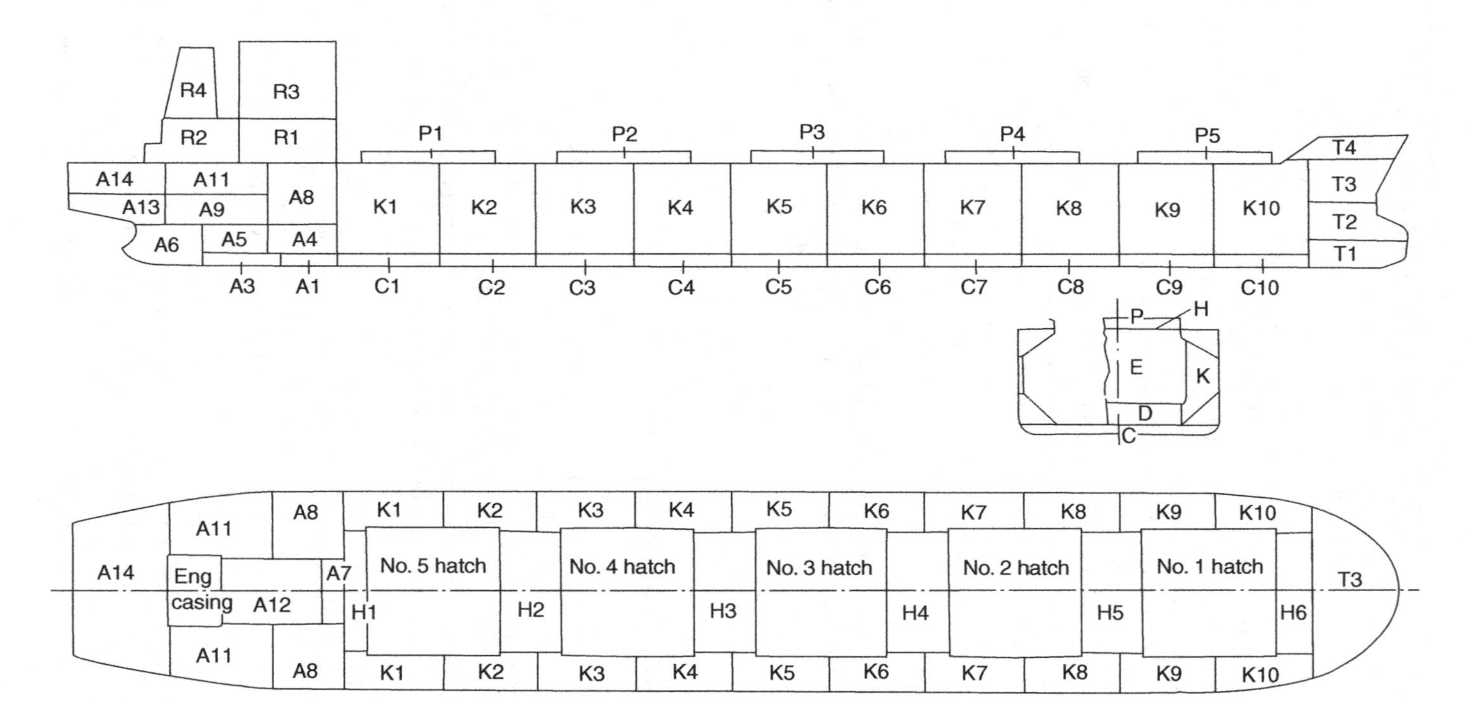

Figure 14.8 Typical block erection for a single-hull bulk carrier.

longitudinals, stiffeners, etc. may be left unwelded across unit seams and butts until these are completed in a similar manner to that described in Chapter 10.

In erecting units tolerances are a problem, more so on three-dimensional units than with two-dimensional units, and particularly at the shaped ends of the ship where 'green' material is still often left. Quality control procedures in the manufacturing shops to ensure correct dimensioning and alignment are very necessary if time-consuming, expensive, and arduous work at the berth is to be avoided. Improvements in this area have been made with the use of accurate jigs for curved shell units, planned weld sequences, and the use of lower heat input welding equipment, dimensional checks on piece parts, and the use of laser alignment tools for setting up data and checking interfaces.

More recently, the use of digital photogrammetry has been adopted in some shipyards. Photographs of a unit, or of an assembly at an earlier production stage, are taken from several locations and software then converts the image data into an accurate, three-dimensional view of the unit. This is useful in the alignment of units and in ensuring successive units are assembled to fit those already in place.

Tolerance allowance data is built up with experience and can become very accurate when building standard ships.

A number of large shipbuilding firms building the super tankers, large chemical carriers, bulk carriers, etc. may assemble much larger 'mega-blocks' that can be the full width of the ship. These are transported to the building berth or a building dock, lifted into place and joined. Rather than erecting 100–150 units at the berth or dock, they may only have to place and join as few as 10–15.

Joining ship sections afloat

Owing to the enormous increase in size of bulk carriers and tankers, some shipyards with restricted facilities, building berth or dock size in particular, have resorted in the past to building the ship in two halves and joining these afloat. This is not a very efficient process and is adopted as an expedient only.

Where the two sections are to be joined afloat, extremely accurate fit-up of the sections is aided by the possibility of ballasting the two ship halves. The two sections may then be pulled together by tackles, and for the finer adjustments hydraulic cylinders may be used, extremely accurate optical instruments being employed to mark off the sections for alignment. One method adopted is that where a cofferdam is arranged in way of the joint, a caisson is brought up against the ship's hull, and the cofferdam and caisson are pumped dry. To balance any tendency for the vessel to hog during the pumping of the cofferdam, it is necessary to shift ballast in the fore and aft sections. Once the spaces are dried out, welding of the complete joint may be undertaken, the resulting weld being X-rayed to test the soundness of such a critical joint. On completion of the paint scheme in way of the joint the caisson is removed.

A similar method makes use of a rubber 'U' form ring rather than a caisson, which needs modification for each ship size.

If a dry dock is available the sections may be aligned afloat and even welded above the waterline, the rest of the joint or the complete joint being secured by strongbacks. The welding of the rest or the whole joint is carried out in the dock.

Further reading

Boekholt R: *Welding Mechanization and Automation in Shipbuilding Worldwide*. Abington Publishing, 1996.

Samsung's mega-block revolution. *Surveyor*, Fall 2005 edition. American Bureau of Shipping publication.

Some useful websites

http://www.stxons.com/service/eng/main.aspx The site includes a 'virtual tour' of the shipyard.

http://www.meyerwerft.de/ Useful pictures and descriptions of the shipyard.

15 Launching

Chapter Outline

End launches 162
- Building slipway 162
- Launching ways and cradle 164
- Lubricant 165
- Releasing arrangements 165
- Launching sequence 168
- Arresting arrangements 168

Side launches 170
Building docks 170
Ship lifts 171
Floating docks 171
Marine railways 172
Further reading 172
Some useful websites 172

This chapter identifies the various means of transferring a ship from the land on which it is constructed to the water. The traditional end launching process is described as this is the most complex method. End launching is now less common as most newer shipyards have a building dock. Other methods are less common and include side launching, a ship lift, a floating dock, or a pontoon.

Whilst many modern shipyards now construct ships in building docks and float them out and some utilize ship lifts, a good number of long-established shipyards are still launching ships in the traditional manner. A significant advance has been provided by computer programs that allow prior assessment of the performance of the ship and the loading it may experience during the launch, both on the ways and in the water.

Launching involves the transference of the weight of the ship from the keel blocks, shores, etc. on which it was supported during construction, to a cradle on which it is allowed to slide into the water. Normally the vessel is launched end on, stern first, but a number of shipyards located on rivers or other narrow channels are obliged to launch the vessel sideways. Vessels have been launched bow first, but this was a rare occurrence as the buoyancy and weight moments, as well as the braking force, are generally more favorable when the vessel is launched stern first. There is also a danger of damage to propellers and rudders as the ship leaves the end of the slipway.

Ship Construction. DOI: 10.1016/B978-0-08-097239-8.00015-5

End launches

On release of a holding mechanism the launching cradle with the ship slides down the ground ways under the action of gravity. When the stern has entered the water the vessel is partly supported by buoyancy and partly by the ground ways. If this buoyancy is inadequate after the center of gravity of the ship has passed the way ends, the ship may tip about the way ends, causing large pressures on the bottom shell and on the ends of the ground ways. To avoid this the greatest depth of water over the way ends should be utilized, and the ground ways extended into the water if necessary. Where this proves impossible it becomes necessary to strengthen the way ends and provide shoring in the bottom shell region, which is likely to be damaged. These remedies are often expensive.

As the vessel travels further into the water the buoyancy becomes sufficient to lift the stern. The vessel then pivots about the forward poppets, i.e. the fore end of the launching cradle. These are designed to take the load thrown on them by the pivoting action, the load being widely distributed in order not to squeeze out the lubricant between the sliding surfaces. Shoring may also be found necessary forward in the ship to prevent structural damage at the time the stern lifts.

Building slipway

Conventional slipways or berths are relatively solid and reinforced with piles to allow them to sustain the weights of ships built upon them. During building the keel blocks take the greater part of the weight, the remainder being carried by shores and, where used, bilge blocks. Foundations under the probable positions of ground ways should also be substantial, since during a launch the ways are subject to large pressures.

Keel blocks are arranged so that the height of keel above the ground is 1.25–1.5 m, giving reasonable access, but not too high so that a large amount of packing is required (see Figure 15.1). At the bow the height of the keel must be sufficient to allow the ship's forefoot to dip the required amount without striking the ground during pivoting when the stern lifts at launch. To suit the declivity of the launching ways determined beforehand, the keel is also inclined to the horizontal at about 1 in 20, or more where the shipyard berths have a larger slope.

To transfer the ship from the building blocks to the launching cradle, the commonest practice is to drive wedges into the launching cradle. This lifts the ship and permits the removal of the keel and bilge blocks together with the shores. In large ships it becomes necessary to split the blocks to remove them, but several types of collapsible blocks have been used to overcome this difficulty. One type is the sand box, which contains sand to a depth of 80–100 mm held in a steel frame located between two of the wooden blocks. This steel frame may be removed and the sand allowed to run out. Another type is a wooden block sawn diagonally, the two halves being bolted so that they collapse on removal of the bolts.

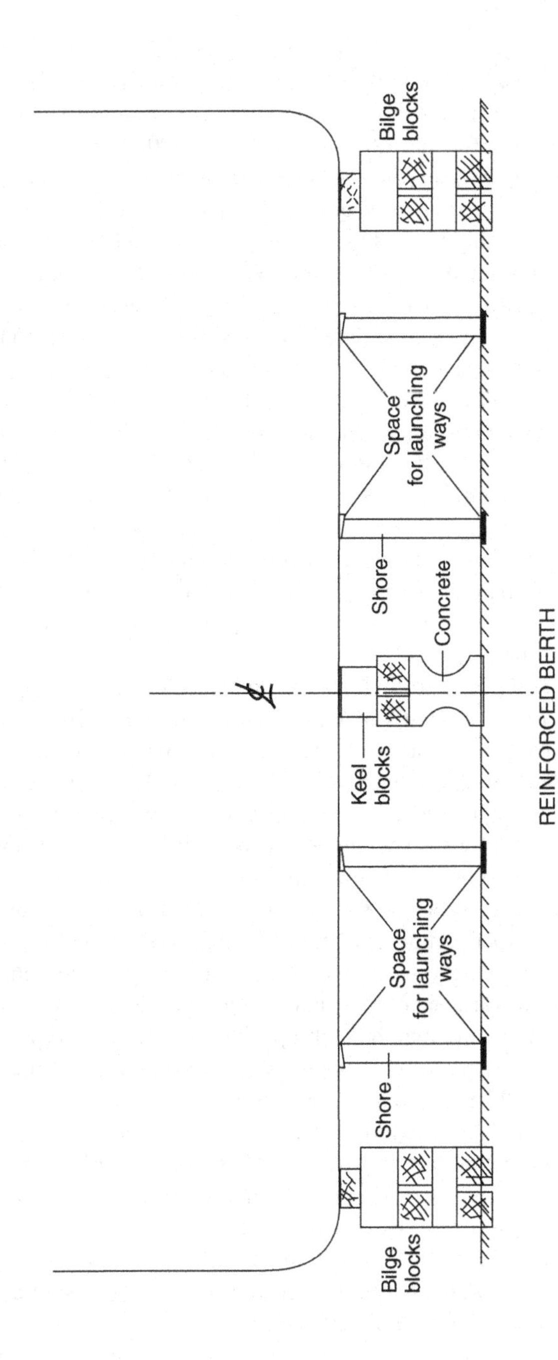

Figure 15.1 Building blocks.

Launching ways and cradle

The fixed ground ways or standing ways on which the cradle and ship slide may be straight or have a fore and aft camber. Transversely the ground ways are normally laid straight but can be canted inwards to suit the ship's rise of floor. Usually, the ground ways have a small uniform fore and aft camber, say 1 in 400, the ways being the arc of a circle of large radius. This means that the lower part of the ways has a greater declivity (say 1 in 16) than the upper part of the ways (say 1 in 25). As a result a greater buoyancy for the same travel of the ship beyond the way ends is obtained, which will reduce the way end pressures. Additional advantages are increased water resistance slowing the vessel and a bow height that is not excessive. The slope of ground ways must be adequate to allow the vessel to start sliding and, if too steep, a large amount of shoring will be required to support the bow; also, the loads on the releasing arrangements will be high. Straight sliding ways have declivities of the order of 1 in 25 to 1 in 16.

Generally two ground ways are fitted, the distance between the ways being about one-third the beam of the ship. It is often desirable that the cradle should be fitted in way of longitudinal structural members and the ground ways over slipway piling, these considerations deciding the exact spacing. Some large ships have been launched on as many as four ways, and in Germany and the Netherlands it is common practice to launch vessels on a single center-line ground way. The width of the ways should be such that the launching weight of the ship does not produce pressures exceeding about 20 tonnes per square meter.

Ground ways are laid on supporting blocks and extend down to the low water mark so that they are covered by at least 1 meter of water at high tide. To guide the sliding ways as they move over the ground ways, a ribband may be fitted to the outer edge of the ground ways. This could be fitted to the inner edge of the sliding ways, but when fitted to the ground ways has the advantage that it aids retention of the lubricating grease. Finally, the ground ways are shored transversely to prevent sideways movement and longitudinally to prevent them from moving down with the ship.

The sliding ways, covering about 80% of the length of the vessel, form the lower part of the cradle, the upper part consisting of packing, wedges, and baulks of timber with some packing fitted neatly to the line of the hull in way of the framing. In very fine-lined vessels the forward end of the cradle, referred to as the forward poppet, will need to be relatively high, and may be built up of vertical timber props tied together by stringers or ribbands. This forward poppet will experience a maximum load that may be as much as 20–25% of the ship's weight when the stern lifts. It is therefore designed to carry a load of this magnitude, but there is a danger in the fine-lined vessel of the forward poppets being forced outwards by the downward force, i.e. the bow might break through the poppets. To prevent this, cross ties or spreaders may be passed below the forefoot of the vessel and brackets may be temporarily fastened to the shell plating at the heads of the poppets. In addition, saddle plates taken under the forefoot of the ship with packing between them and the shell may be fitted to transmit the load to the fore poppets and hence ground ways.

In many modern ships the bow sections are relatively full and little support is required above the fore end of the sliding ways. Here short plate brackets may be temporarily welded between the shell plating and heavy plate wedge rider, as illustrated in Figure 15.2. The design of the forward poppets is based on greater pressures than the lubricant between the sliding ways and ground ways could withstand if applied for any length of time. However, as the duration of pivoting is small, and the vessel has sufficient momentum to prevent sticking at this stage, these high pressures are permissible.

At the after end of the cradle considerable packing may also be required, and again vertical timber props or plate brackets may be fitted to form the after poppet.

Lubricant

For the ship to start sliding on release of the holding arrangements, it is necessary for the ship to overcome the coefficient of friction of the launching lubricant. To do this the slope of the ways under the vessel's center of gravity must exceed the lubricant's coefficient of friction. An estimate of the frictional resistance of the grease must be made before building the ship, since the declivity of the keel is dependent to a large extent on the slope of the ways.

Formerly, melted tallow was applied to the ways, allowed to harden, and then covered with a coat of soft soap. Since the mid 1900s patent mineral-based greases have been applied to the ground ways, these greases being virtually unaffected by temperature changes and insoluble in water whilst adhering firmly to the ways. A commercially marketed petroleum-based launching grease is applied over the mineral grease base coat. This has a coefficient of friction that is low enough to allow initial starting and the maintenance of sliding until the initial resistance of the base coat is overcome by frictional heat. To prevent the petroleum-based grease from soaking into the sliding ways, a base coat may be applied to them. Standing ways that extend into the water may be dried out at low tide prior to the launch and the base coat and grease applied.

Releasing arrangements

Small ships may be released by knocking away a diagonal dog-shore (see Figure 15.3) fitted between the sliding and standing ways.

In most cases, however, triggers are used to release the ship. There are several types available, hydraulic, mechanical, and electrical-mechanical triggers having been used. Electrical-mechanical triggers are commonly used for rapid simultaneous release in modern practice. The hydraulic trigger is less easily installed and less safe. The electrical-mechanical trigger illustrated in Figure 15.3 is generally located near amidships and a small pit is provided in the berth to accommodate the falling levers. A number of triggers will be fitted depending on the size of the vessel to be launched; in the case of the 75,000-tonne bulk carrier for which a launching sequence is given below, six triggers were fitted for the launch.

These triggers are in effect a simple system of levers that allow the large loads acting down the ways to be balanced by a small load on the releasing gear. The principle is often compared with that of a simple mechanical reduction gearing.

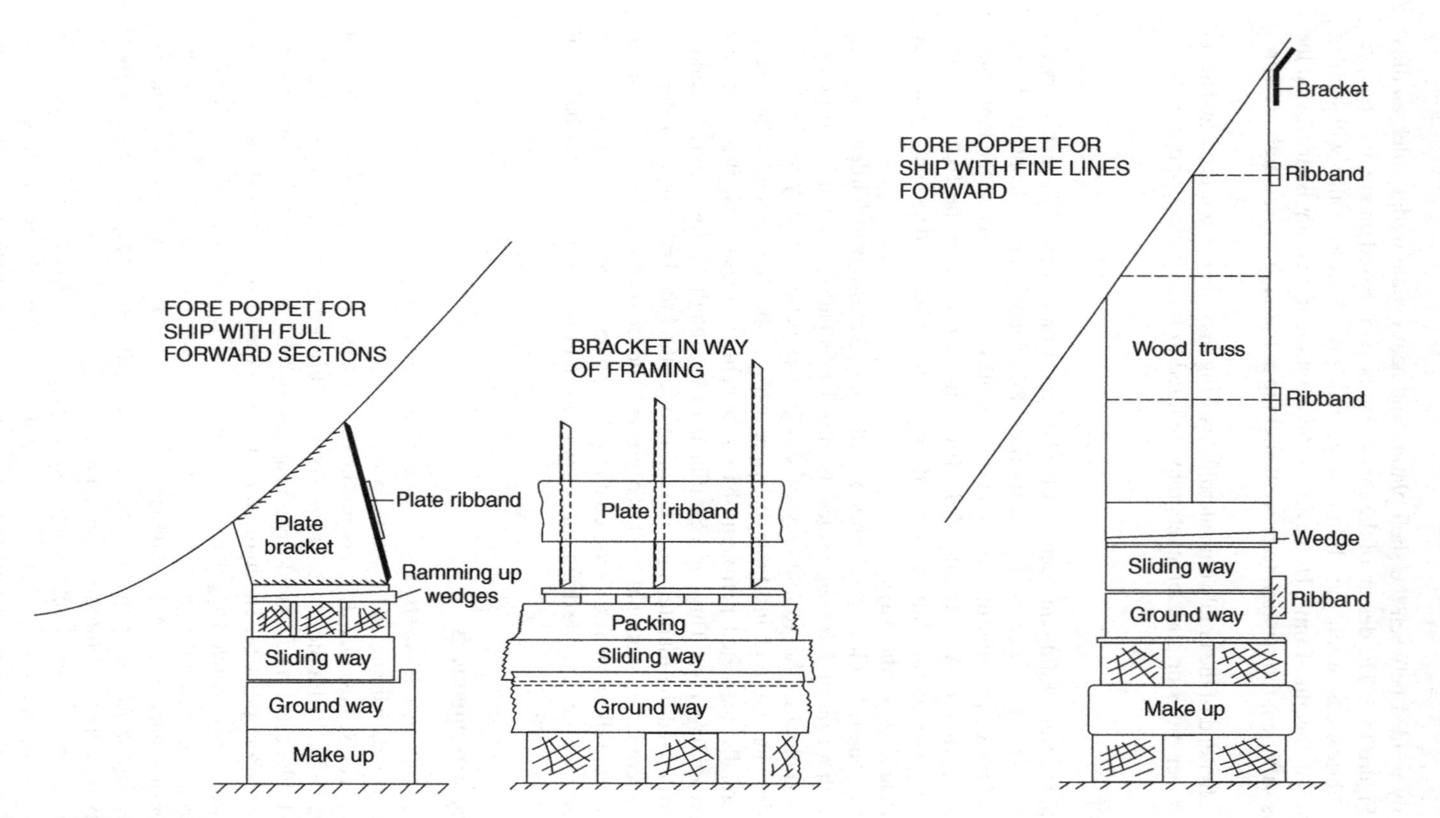

Figure 15.2 Launching ways—fore poppet.

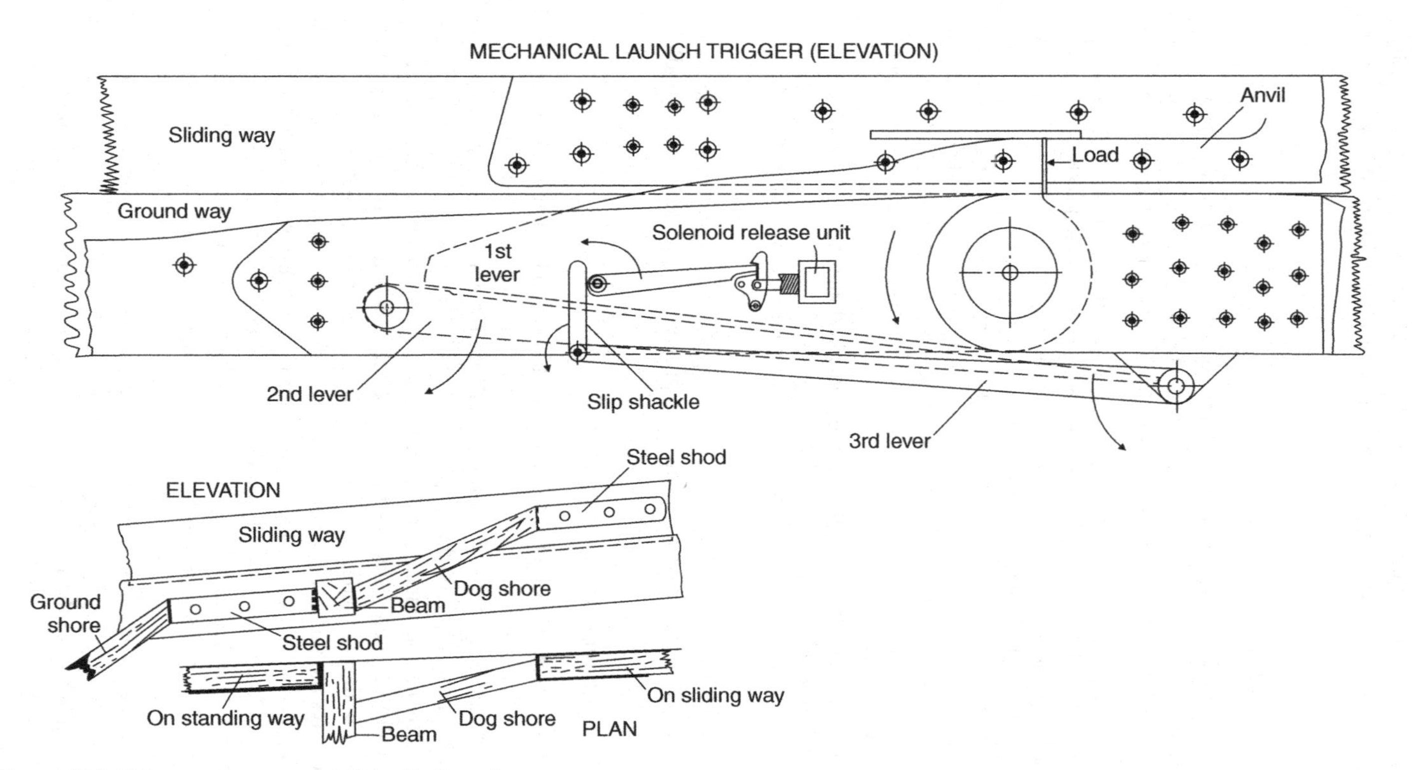

Figure 15.3 Release arrangements for ship launch.

Simultaneous release of the triggers is achieved by means of catches held by solenoids wired in a common circuit. These are released immediately the circuit current is reversed.

Launching sequence

As a guide to the procedure leading up to the launch, the following example is given for the launch of a 75,000-tonne bulk carrier. The launch ways have been built up as the ship is erected from aft; the ways have been greased and the cradle erected.

1. Four days before launch a start is made on ramming up the launch blocks, i.e. driving in the wedges (Figure 15.2) to raise ship off the building blocks. This is done by a dozen or so men using a long ramming pole, a gang working either side of the ship.
2. Two days before the launch a start is made on removing the shores.
3. On the morning of the launch everything is removed up to the high water mark and tumbler shores are put in aft. These are inclined shores that fall away as the ship starts to move.
4. Every second keel block is then removed and the vessel is allowed to settle.
5. An evenly distributed number of keel blocks are then taken out so that only about 20 keel blocks are left supporting the ship.
6. Half an hour before the launch the last remaining keel blocks are removed.
7. The bilge blocks are then removed.
8. The full ship's weight comes on the triggers at the time planned for the launch.
9. Release of triggers on launch by sponsor.

If the vessel fails to start under the action of gravity, the initial movement may be aided by hydraulic starting rams, which are provided at the head of the cradle.

Arresting arrangements

In many cases the extent of the water into which the ship is launched is restricted. It is then necessary to provide means of arresting the motion of the ship once it is in the water. There are a number of methods available for doing this, one or more being employed at most ship launches.

The commonest arrangement is to use chain drags, which are generally arranged symmetrically on either side of the ship. Each chain drag is laid in the form of a horseshoe with its rounded portion away from the water, so that as the ship moves down the ways the forward portion of the drag is pulled through the remainder of the pile. This prevents any excessive shock load in the chain, which would occur if the pile of chain were to be suddenly accelerated to the speed of the ship. The wire rope drag lines are attached to temporary pads on the side of the ship, and supported by rope tricing lines as they are led slightly forward and then aft along the ship's sides. Each drag line is then led forward and shackled to the chain drag (see Figure 15.4). As the ship is released and moves aft the tricing lines are broken in turn, the work done absorbing some of the ship's energy.

To further increase the resistance to motion of the ship, wooden masks may be fitted at the stern of the ship. The mask is made as large as possible but located low down to present a flat surface to the water in the direction of motion. Masks are often

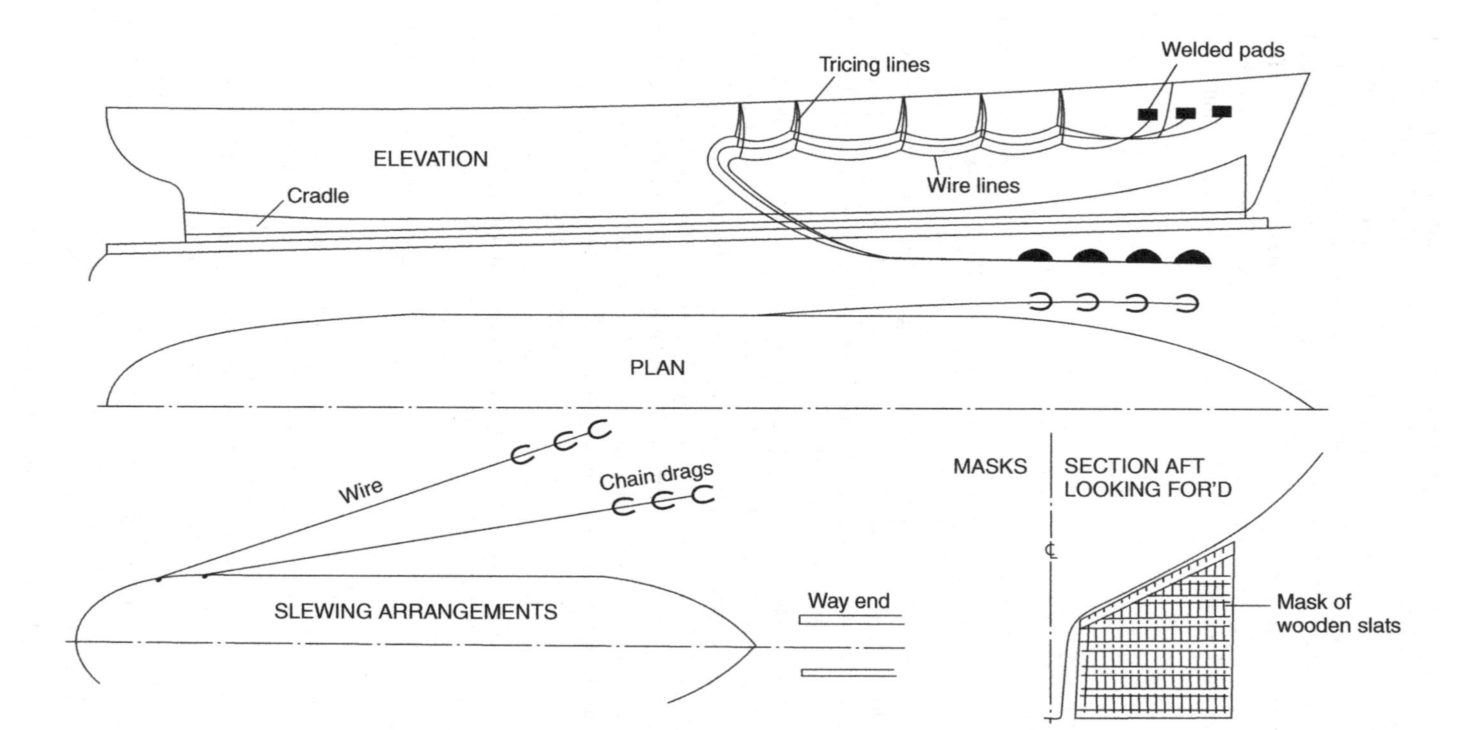

Figure 15.4 Arresting arrangements at launch.

constructed of horizontal pieces of wood with spaces left between each piece to increase the resistance.

One or two shipyards are forced to provide arrangements for slewing the vessel once it has left the ways, as the river into which the ships are launched is very narrow in relation to the ship's length. Chain drags, weights, or anchors may be placed in the water to one side of the building berth for this purpose. These are then made fast to the stern of the ship with drag lines of a predetermined length. Once the vessel is clear of the ways the slack of the lines is taken up and the stern swings so that the ship is pointing up river.

Side launches

Side launching is often used where the width of water available is considerably restricted. There are in fact some advantages to this method, for example the absence of keel declivity, and the relatively simple cradle and short ground ways that do not extend into the water. However, it means that a large area of waterfront is taken up by a single building berth, and the ship is only reasonably accessible from one side during construction. Side launching is usually restricted to small ships, typically no more that around 100 meters in length.

The ground ways are arranged transversely, i.e. at right angles to the line of keel. Sliding ways can also be placed transversely with the packing above them forming the cradle, but they are generally arranged longitudinally. In this case where they are parallel to the keel, the sliding ways are in groups covering two or three ground ways. Packing again forms the cradle with tie pieces between the groups of sliding ways.

One of the features of side launching is the drop where the ground ways are not extended into the water; consequently, large angles of heel occur when the vessel strikes the water. As a result it is necessary to carry out careful stability calculations and close any openings before side launching a vessel. It is true, of course, that stability calculations are also required for a conventional end launch.

Building docks

Perhaps the greatest advantage of the building dock is the relative simplicity of the task of getting the vessel waterborne (see Chapter 11). When it is convenient the dock may be flooded and the vessel floated out. In a building dock it is also possible to construct two ships at the same time. The second of two is usually at the landward end of the dock, and can be separated from the seaward end by an intermediate dock gate. The first ship can then be floated out of the dock while the second remains in the dry. Once the dock is dewatered the second ship can be moved to the seaward end and a third ship started. This allows better utilization of the large and expensive cranes, and also allows double the time to complete the more outfit-intensive zones of the ship, usually the aft machinery spaces and accommodation.

The dock also has a horizontal base, which simplifies alignment of structural units or blocks and management of the ship dimensions.

Calculations are needed to check the stability and loads exerted by the blocks during flooding, the whole problem being similar to that of undocking a vessel that has been dry-docked for survey or other reasons.

In some shipyards conventional berths are fitted at the river or sea end with what is virtually a pair of dock gates. This can be of advantage when working the aft end of the ship and installing the ways. In many instances it also permits higher tides over the way ends when the gates are opened for a launch. A dock gate is ideal in tideless conditions, for example in the Baltic or Mediterranean Seas. The ship can be built with the gate closed, maintaining a dry working environment. When the gate is opened the lower part of the slipway floods and provides sufficient water depth for the launch to be carried out successfully.

Ship lifts

Whilst large ships may be extruded out of building halls on to a slipway (Chapter 11) and large sections transferred by similar means to the head of the slipway and raised onto it for joining and launching, smaller complete ships may be transferred to a ship lift for launching. Rail systems are incorporated into the building hall and lead out to the open ship lift. The best known of these ship lift systems is the patented 'Syncrolift', originally used for slipping ships for repairs and surveys but now also used by some shipyards for launching new ships. Ship lifts basically consist of a series of transverse beams, each supported by wires or chains suspended from a winch on each side. The beams have the transfer rails running longitudinally across them. A platform is also supported by the beams to provide access when the ship is lifted clear of the water. To launch, the ship is positioned on the structure, which can be lowered into the water and the ship floated off the platform. The lift can also be used to recover a ship from the water for repair. The platform is raised and lowered mechanically or hydraulically and is usually provided with transfer arrangements so that the vessel can be moved on or off the platform either laterally or in line with the platform.

Originally, ship lifts were used for fleets of small ships, for example in fishing ports, but have increased in size for large commercial repairs and some construction. The largest lift is able to lift over 20,000 tonnes.

Floating docks

Occasionally, and usually as a temporary expedient, a ship can be built on the land and transferred to a floating dock for launching. The arrangement is similar to a ship lift, with rails to allow the completed ship to be transferred onto the dock. The dock is then sunk in the location or may be towed to deeper water so the ship can be floated off.

As an alternative a ship may be transferred to a pontoon, in an operation similar to an offshore structure loadout. The pontoon will require some reserve buoyancy to maintain stability during the operation and this is usually provided by 'towers' at each corner. Such a specialized pontoon is less expensive than a conventional floating dock.

Marine railways

A few small shipyards use a slipway with a set or rails that extend into the water. The ship is constructed on, or transferred to, a cradle that can run on the rails. For launch, the cradle is lowered down the slipway until the ship floats off. The operation is under control, unlike a dynamic end launch where, once started, the ship cannot be stopped. As with the ship lift, a marine railway can also be used to recover ships for repair.

Further reading

Dunn, Kennedy, and Tibbs, Launching in the 21st century. The Royal Institution of Naval Architects, Drydocks, Launching and Shiplift Conference, 2003.

Pattison, Dixon, and Hodder, Launching and docking: Experiences at VT shipbuilding. The Royal Institution of Naval Architects, Drydocks, Launching and Shiplift Conference, 2003.

Salisbury and Dobb, Recent dynamic launches at Barrow in Furness: Prediction and reality. The Royal Institution of Naval Architects, Drydocks, Launching and Shiplift Conference, 2003.

The vertical shiplift's role in modern shipyards, *The Naval Architect*, February 1995.

Some useful websites

http://www.royalhaskoning.co.uk The company is a designer of dry docks worldwide.
http://www.fsg-ship.de/ The Flensberg shipyard site, includes webcams and video.

Part Five

Ship Structure

Introduction

Ships are the largest, mobile man-made structures in the world. The largest ships are comparable with major buildings, but have additionally to withstand enormous forces from wind and waves. Their structural arrangements are therefore of great importance. Because in most cases ships have to earn by carrying cargo, their structure not only has to be strong, it must also be as light as possible.

The structure of a ship is based on flat and curved steel plate panels, with primary and secondary stiffening. The complete hull structure has to fulfill several important requirements. These include:

- Maintaining watertight integrity
- Separating internal spaces, for example cargo, fuel oil, ballast water
- Resisting forces on the ship, in particular from waves.

The structure also has to conform to the hull shape dictated by hydrodynamics, i.e. the performance in terms of speed and seakeeping specified by the hull designer.

Both the forces on the ship and the capability of the structure have some uncertainty, which must be allowed for in the design. The structure may suffer from changes in material properties over the life of the ship. Some causes of structural failure are hard to predict, for example fatigue and corrosion. The quality of workmanship can also vary considerably.

Basic structural arrangements

The structure of a ship is based primarily on panels, both flat and curved. These are the building blocks of the structure and also provide a convenient breakdown of the structure into units during ship construction (see Chapter 14).

The basic components of the panels are steel plates and steel profiles. The plates form the panels and the profiles are used to provide stiffening to prevent the plates deforming when subjected to forces.

There is a specific terminology for the structural components.

Plating forms the main structural elements of the ship, which are:

- Hull bottom plating, inner and outer bottom
- Hull side plating

- Hull deck plating
- Internal bulkheads
- Internal decks.

Stiffening for the various panels includes:

- Longitudinals that run along the bottom plating and provide significant strength. There may be a keel in the form of a longitudinal center-line girder.
- Deck girders, also longitudinal, which again contribute to the longitudinal strength of the ship.
- Stringers, running longitudinally on the side plating, and where required along the inner side plating for a double-hull ship.
- Frames, which run vertically up the side plating.
- Floors, which are deep transverse frames on the bottom plating.
- Deck beams, which support the deck transversely.

Note

Throughout this part of the book a number of requirements relating to the spacing and scantlings of various structural members are given. These are intended only as examples, which were taken from Lloyd's Register rules at the time of writing and were introduced only to give the student an idea of the variation in dimensions and scantlings found within the ship structure. Other classification society requirements may differ, but basically the overall structure would have the same characteristics. It should be borne in mind that owner's additional requirements and the nature of a particular trade can result in ships having greater scantlings and additional strengthening to the minimum indicated in the following chapters.

16 Bottom structure

Chapter Outline

Keels 175
Single-bottom structure 177
Double-bottom structure 177
Inner bottom plating 180
Floors 180
Transversely framed double bottom 180
Longitudinally framed double bottom 183
Additional stiffening in the pounding region 183
Bottom structure of bulk carriers 184
Testing double-bottom compartments 184
Machinery seats 184

Originally, ships were constructed with single bottoms, liquid fuels and fresh water being contained within separately constructed tanks. The double-bottom structure, which provides increased safety in the event of bottom shell damage, and also provides liquid tank space low down in the ship, evolved during the early part of the twentieth century. Today, only small vessels such as tugs, ferries, and cargo ships of less than 500 gross tonnage have a single-bottom construction. Larger ocean-going vessels are fitted with some form of double bottom.

Keels

At the center line of the bottom structure is located the keel, which is often said to form the backbone of the ship. This contributes substantially to the longitudinal strength and effectively distributes local loading caused when docking the ship. The commonest form of keel is that known as the 'flat plate' keel, and this is fitted in the majority of ocean-going and other vessels (see Figure 16.1a). A form of keel found on smaller vessels is the bar keel (Figure 16.1b). The bar keel may be fitted in trawlers, tugs, etc. and is also found in smaller ferries.

Where grounding is possible this type of keel is suitable with its massive scantlings, but there is always a problem of the increased draft with no additional cargo capacity. If a double bottom is fitted the keel is almost inevitably of the flat plate type, and bar keels are more often associated with open floors, but a flat plate keel can be fitted in way of open floors.

Ship Construction. DOI: 10.1016/B978-0-08-097239-8.00016-7

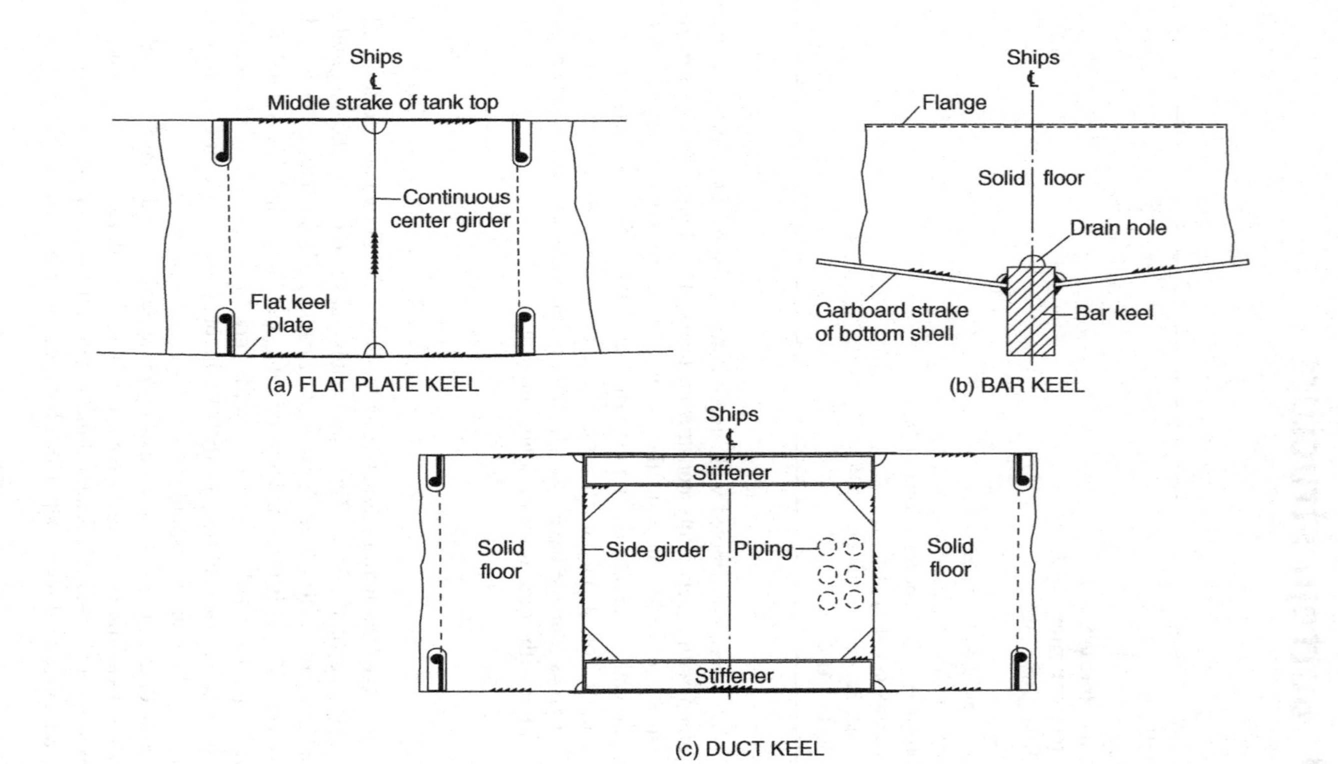

Figure 16.1 Keels.

Duct keels (Figure 16.1c) are provided in the double bottoms of some vessels. These run from the forward engine room bulkhead to the collision bulkhead and are utilized to carry the double-bottom piping. The piping is then accessible when cargo is loaded, an entrance to the duct being provided at the forward end of the engine room. No duct is required aft of the engine room as the piping may be carried in the shaft tunnel. A width of not more than 2.0 m is allowed for the duct, and strengthening is provided at the tank top and keel plate to maintain continuity of strength of the transverse floors.

Single-bottom structure

In smaller ships having single bottoms the vertical plate open floors are fitted at every frame space and are stiffened at their upper edge. A center-line girder is fitted and one side girder is fitted each side of the center line where the beam is less than 10 m. Where the beam is between 10 and 17 m, two side girders are fitted and if any bottom shell panel has a width-to-length ratio greater than 4 additional continuous or intercostal stiffeners are fitted. The continuous center and intercostal side girders are stiffened at their upper edge and extend as far forward and aft as possible.

The single-bottom structure is shown in Figure 16.2 and for clarity a three-dimensional representation of the structure is also provided to illustrate those members that are continuous or intercostal. Both single and double bottoms have continuous and intercostal material, and there is often some confusion in the student's mind as to what is implied by these terms.

A wood ceiling may be fitted across the top of the floors if cargoes are to be carried, but this does not constitute an inner bottom offering any protection if the outer bottom shell is damaged.

Double-bottom structure

An inner bottom (or tank top) may be provided at a minimum height above the bottom shell, and maintained watertight to the bilges. This provides a considerable margin of safety, since in the event of bottom shell damage only the double-bottom space may be flooded. The space is not wasted but utilized to carry oil fuel and fresh water required for the ship, as well as providing ballast capacity.

The minimum depth of the double bottom in a ship will depend on the classification society's requirement for the depth of center girder. It may be deeper to give the required capacities of oil fuel, fresh water, and water ballast to be carried in the bottom. Water-ballast bottom tanks are commonly provided right forward and aft for trimming purposes and if necessary the depth of the double bottom may be increased in these regions. In way of the machinery spaces the double-bottom depth is also increased to provide appreciable capacities of lubricating oil and fuel oil. The increase in height of the inner bottom is always by a gradual taper in the longitudinal direction, no sudden discontinuities in the structure being tolerated.

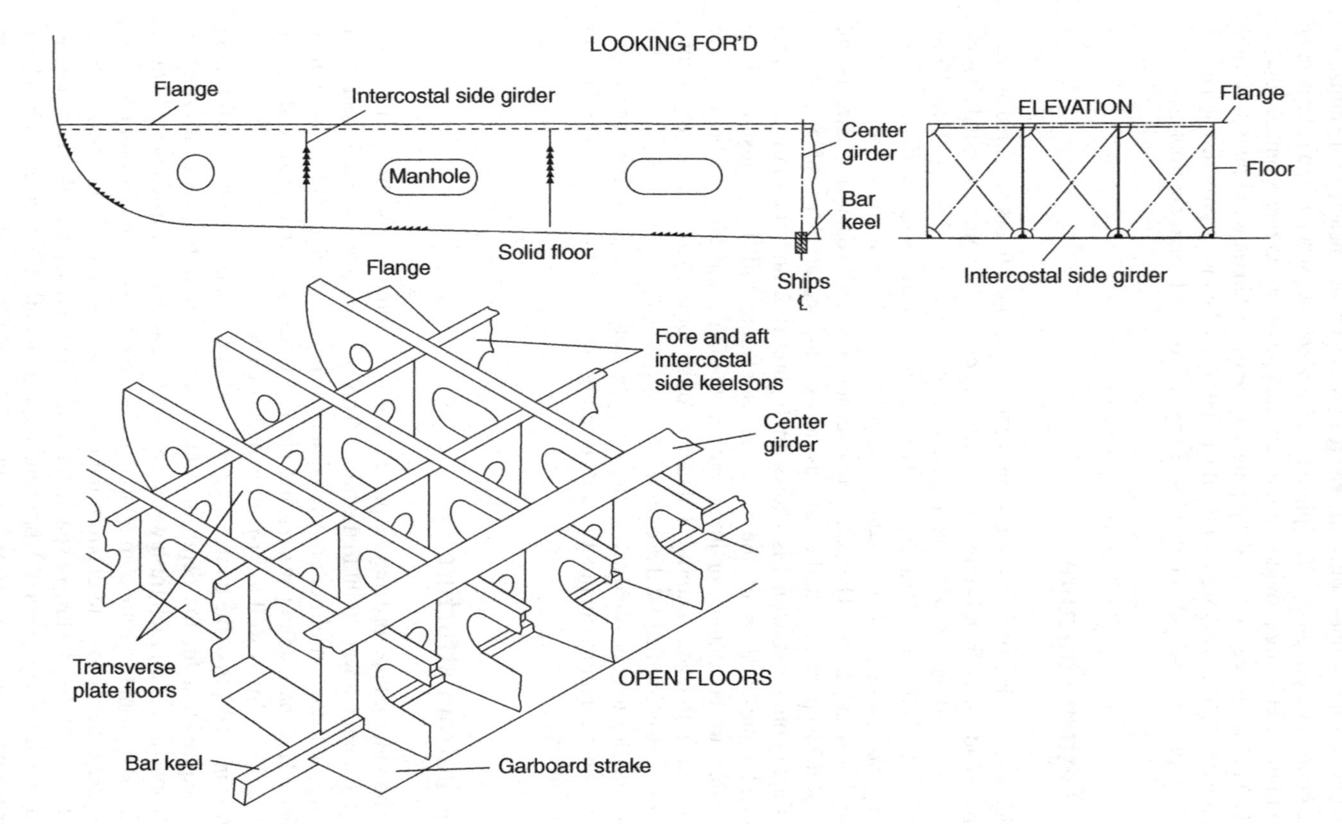

Figure 16.2 Single-bottom construction.

Double bottoms may be framed longitudinally or transversely (see Figure 16.3), but where the ship's length exceeds 120 m it is considered desirable to adopt longitudinal framing. The explanation for this is that on longer ships tests and experience have shown that there is a tendency for the inner bottom and bottom shell to buckle if welded transverse framing is adopted. This buckling occurs as a result of the longitudinal bending of the hull, and may be avoided by having the plating longitudinally stiffened.

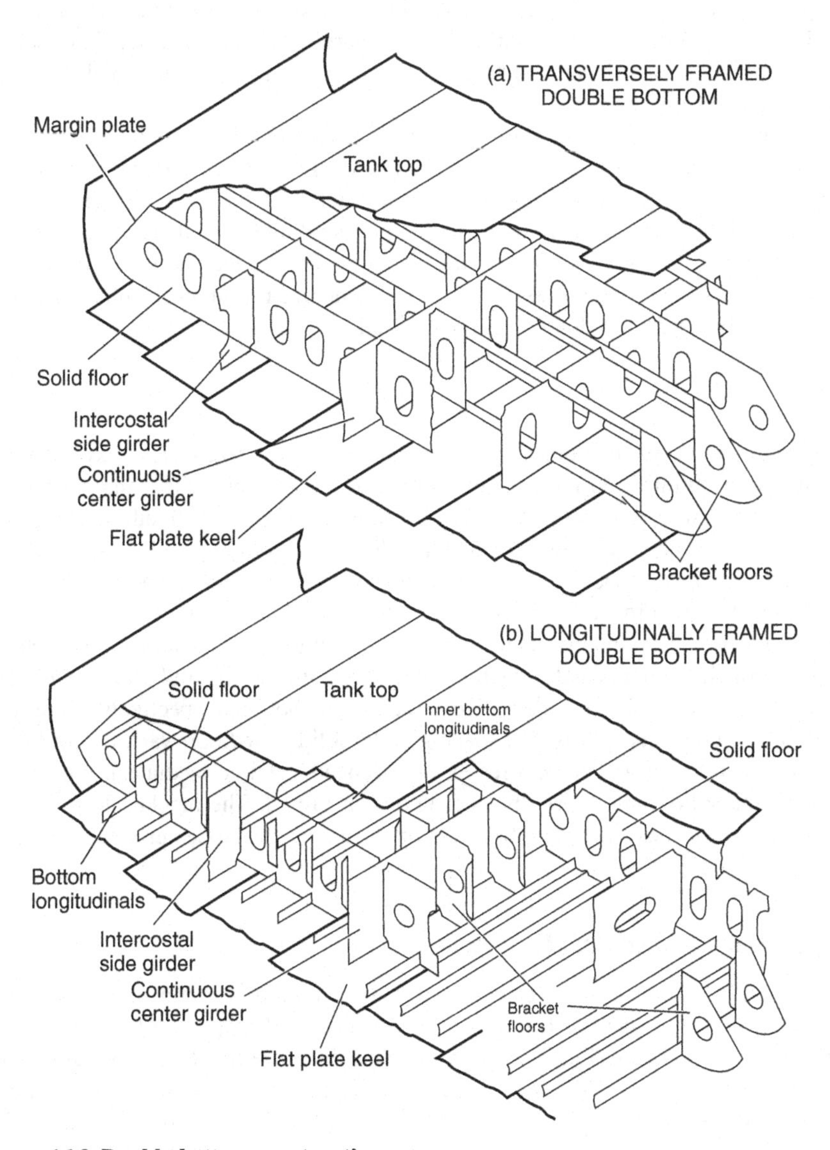

Figure 16.3 Double-bottom construction.

Double bottoms in the way of machinery spaces that are adjacent to the after peak are required to be transversely framed.

Inner bottom plating

The inner bottom plating may, in a general cargo ship, be sloped at the side to form a bilge for drainage purposes. It is not uncommon, however, for it to be extended to the ship's side, and individual bilge wells are then provided for drainage purposes (see Chapter 26). In vessels requiring a passenger certificate it is a statutory requirement for the tank top to extend to the ship's side. This provides a greater degree of safety, since there is a substantial area of bilge that may be damaged without flooding spaces above the inner bottom.

At the center line of the ship the middle strake of the inner bottom may be considered as the upper flange of the center-line docking girder, formed by the center girder and keel plate. It may therefore be heavier than the other strakes of inner bottom plating. Normally, a wood ceiling is provided under a hatchway in a general cargo ship, but the inner bottom plating thickness can be increased and the ceiling omitted. If grabs are used for discharging from general cargo ships the plate thickness is further increased, or a double ceiling is fitted.

Floors

Vertical transverse plate floors are provided both where the bottom is transversely and longitudinally framed. At the ends of bottom tank spaces and under the main bulkheads, watertight or oiltight plate floors are provided. These are made watertight or oiltight by closing any holes in the plate floor and welding collars around any members that pass through the floors. Elsewhere, 'solid plate floors' are fitted to strengthen the bottom transversely and support the inner bottom. These run transversely from the continuous center girder to the bilge, and manholes provided for access through the tanks and lightening holes are cut in each solid plate floor. Also, small air and drain holes may be drilled at the top and bottom respectively of the solid plate floors in the tank spaces. The spacing of the solid plate floors varies according to the loads supported and local stresses experienced. At intermediate frame spaces between the solid plate floors, 'bracket floors' are fitted. The bracket floor consists simply of short transverse plate brackets fitted in way of the center girder and tank sides (see Figures 16.4 and 16.5).

Transversely framed double bottom

If the double bottom is transversely framed, then transverse solid plate floors, and bracket floors with transverse frames, provide the principal support for the inner bottom and bottom shell plating (Figure 16.4). Solid plate floors are fitted at every frame space in the engine room and in the pounding region (see the end of this chapter). Also, they are introduced in way of boiler seats, transverse bulkheads, toes of brackets supporting stiffeners on deep tank bulkheads, and in way of any change in depth of the double

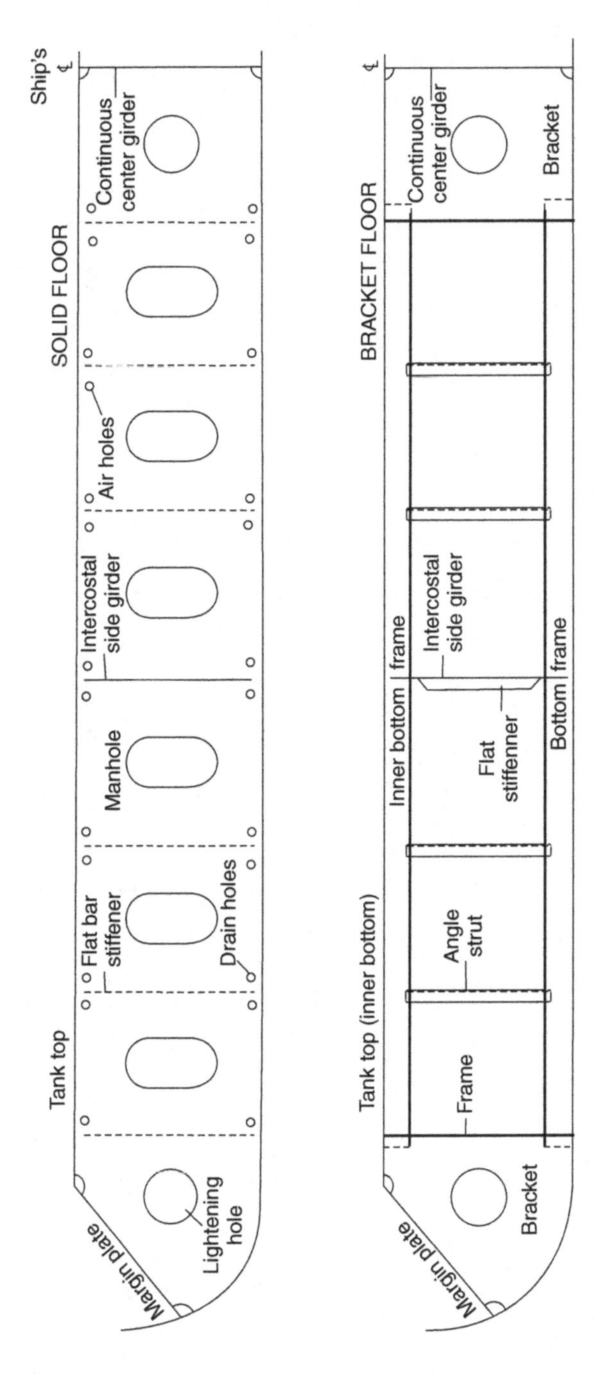

Figure 16.4 Transversely framed double-bottom construction.

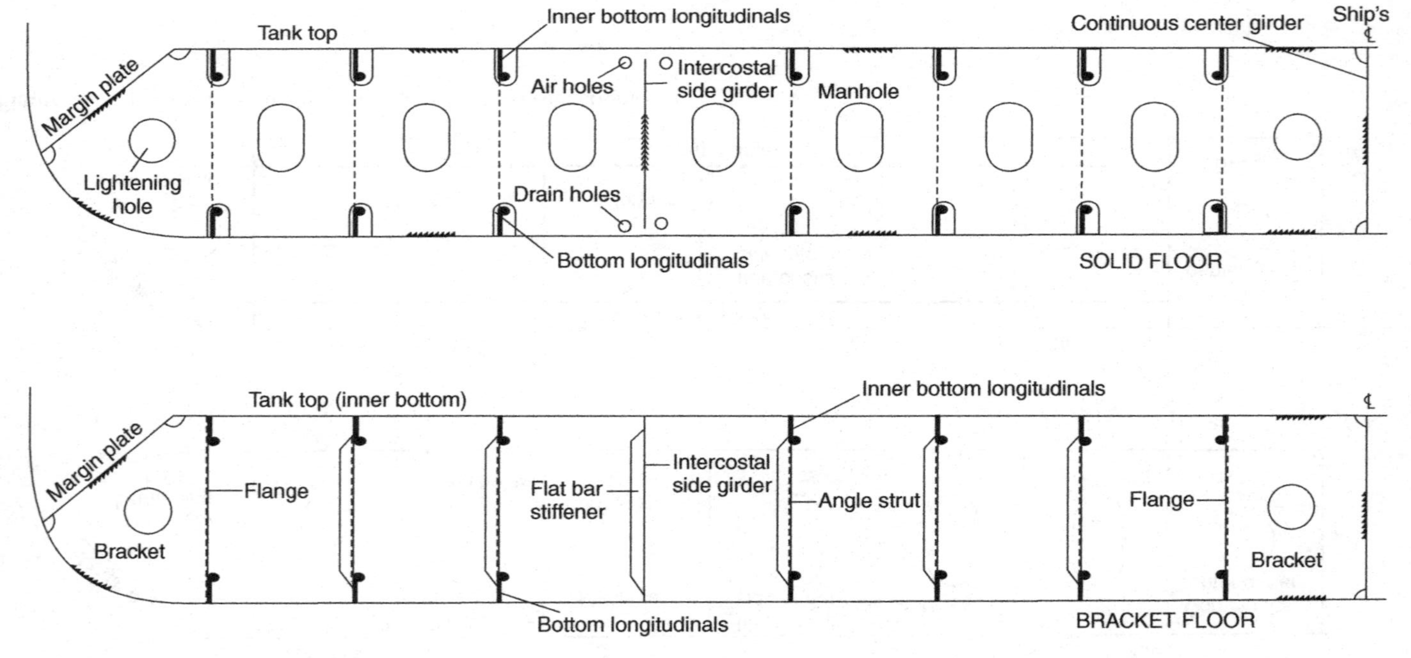

Figure 16.5 Longitudinally framed double-bottom construction.

bottom. Where a ship is regularly discharged by grabs, solid plate floors are also fitted at each frame. Elsewhere, the solid plate floors may be spaced up to 3.0 m apart, with bracket floors at frame spaces between the solid floors. The plate brackets of bracket floors are flanged and their breadth is at least 75% of the depth of the center girder at the bracket floors. To reduce the span of the frames, which should not exceed 2.5 meters, at the bracket floor, vertical angle or channel bar struts may be fitted. Vertical stiffeners, usually in the form of welded flats, will be attached to the solid plate floors, which are further strengthened if they form a watertight or oiltight tank boundary.

One intercostal side girder is provided port and starboard where the ship's breadth exceeds 10 m but does not exceed 20 m, and two are fitted port and starboard where the ship's breadth is greater. In way of the bracket floors a vertical welded flat stiffener is attached to the side girder. Additional side girders are provided in the engine room, and also in the pounding region.

Longitudinally framed double bottom

In a longitudinally framed double bottom, solid plate floors are fitted at every frame space under the main engines, and at alternate frames outboard of the engine seating. They are also fitted under boiler seats, transverse bulkheads, and the toes of stiffener brackets on deep tank bulkheads. Elsewhere, the spacing of solid plate floors does not exceed 3.8 m, except in the pounding region, where they are on alternate frame spaces. At intermediate frame spaces brackets are fitted at the tank side and at the center girder, where they may be up to 1.25 m apart. Each bracket is flanged and will extend to the first longitudinal (Figure 16.5).

One intercostal side girder is fitted port and starboard if the ship's breadth exceeds 14 m, and where the breadth exceeds 21 m two are fitted port and starboard. These side girders always extend as far forward and aft as possible. Additional side girders are provided in the engine room and under the main machinery, and they should run the full length of the engine room, extending three frame spaces beyond this space. Forward, the extension tapers into the longitudinal framing system. In the pounding region there will also be additional intercostal side girders.

As the unsupported span of the bottom longitudinals should not exceed 2.5 m, vertical angle or channel bar struts may be provided to support the longitudinals between widely spaced solid floors.

Additional stiffening in the pounding region

If the minimum designed draft forward in any ballast or part-loaded condition is less than 4.5% of the ship's length, then the bottom structure for 30% of the ship's length forward in sea-going ships exceeding 65 m in length is to be additionally strengthened for pounding.

Where the double bottom is transversely framed, solid plate floors are fitted at every frame space in the pounding region. Intercostal side girders are fitted at a maximum spacing of three times the transverse floor spacing, and half-height intercostal side girders are provided midway between the full-height side girders.

If the double bottom is longitudinally framed in the pounding region, where the minimum designed draft forward may be less than 4% of the ship's length, solid plate floors are fitted at alternate frame spaces, and intercostal side girders fitted at a maximum spacing of three times the transverse floor spacing. Where the minimum designed draft forward may be more than 4% but less than 4.5% of the ship's length, solid plate floors may be fitted at every third frame space and intercostal side girders may have a maximum spacing of four times the transverse floor spacing. As longitudinals are stiffening the bottom shell longitudinally, it should be noted that fewer side girders need be provided than where the bottom is transversely framed to resist distortion of the bottom with the slamming forces experienced.

Where the ballast draft forward is less than 1% of the ship's length, the additional strengthening of the pounding region is given special consideration.

Greater slamming forces (i.e. pounding) are experienced when the ship is in the lighter ballast condition, and is long and slender, by reason of the increased submersion of the bow in heavy weather with impact also on the bow flare.

Bottom structure of bulk carriers

Where a ship is classed for the carriage of heavy, or ore, cargoes longitudinal framing is adopted for the double bottom. A closer spacing of solid plate floors is required, the maximum spacing being 2.5 m, and also additional intercostal side girders are provided, the spacing not exceeding 3.7 m (see Figure 16.6).

The double bottom will be somewhat deeper than in a conventional cargo ship, a considerable ballast capacity being required, and often a pipe tunnel is provided through this space. Inner bottom plating, floors, and girders all have substantial scantlings as a result of the heavier cargo weights to be supported.

Testing double-bottom compartments

Each compartment is tested on completion with a head of water representing the maximum pressure head that may be experienced in service, i.e. to the top of the air pipe. Alternatively, air testing is carried out before any protective coatings are applied. The air pressure may be raised to 0.21 kg/cm^2 and then lowered to a test pressure of 0.14 kg/cm^2. Any suspect joints are then subjected to a soapy liquid solution test. Water head structural tests will be carried out on tanks selected by the surveyor in conjunction with the air tests carried out on the majority of tanks.

Machinery seats

It has already been indicated that in the machinery spaces additional transverse floors and longitudinal intercostal side girders are provided to support the machinery effectively and to ensure rigidity of the structure.

The main engine seatings are in general integral with this double-bottom structure, and the inner bottom in way of the engine foundation has a substantially increased

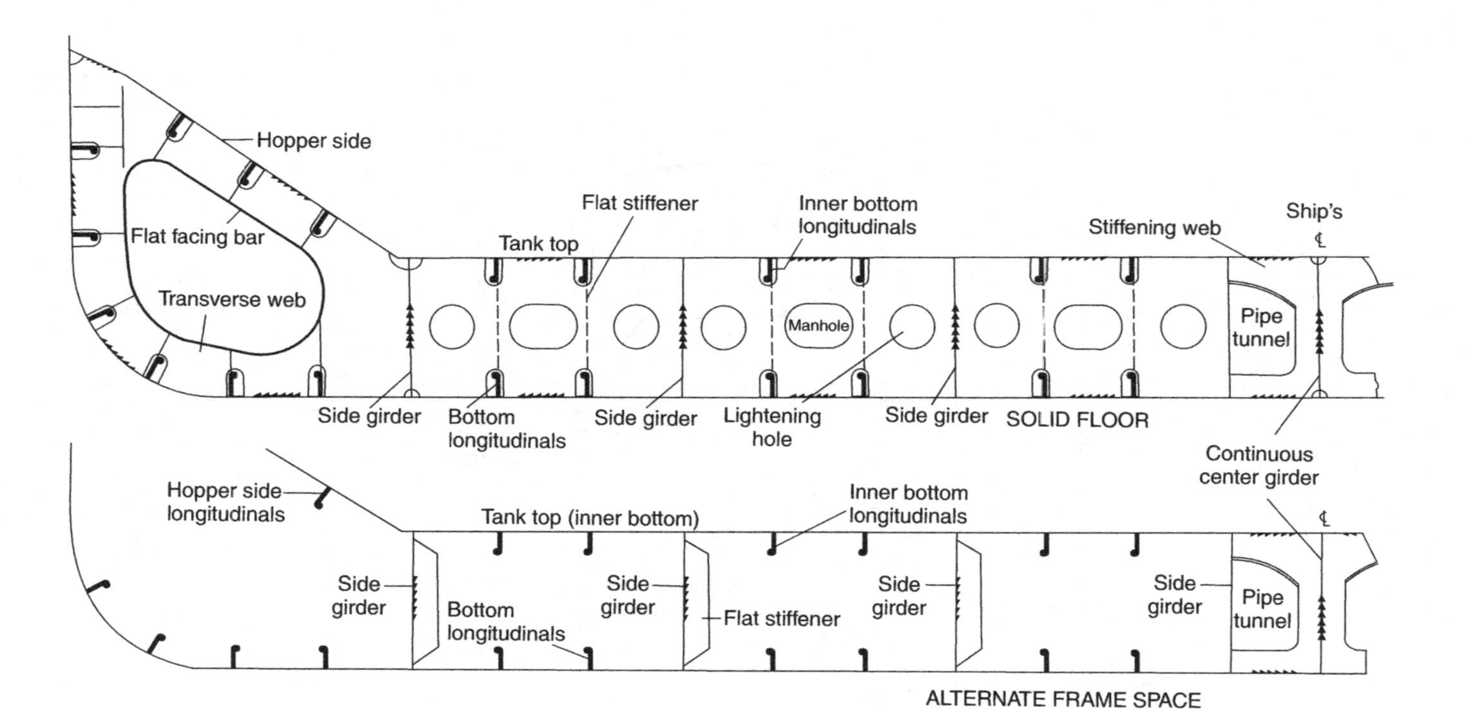

Figure 16.6 Bulk carrier double-bottom construction.

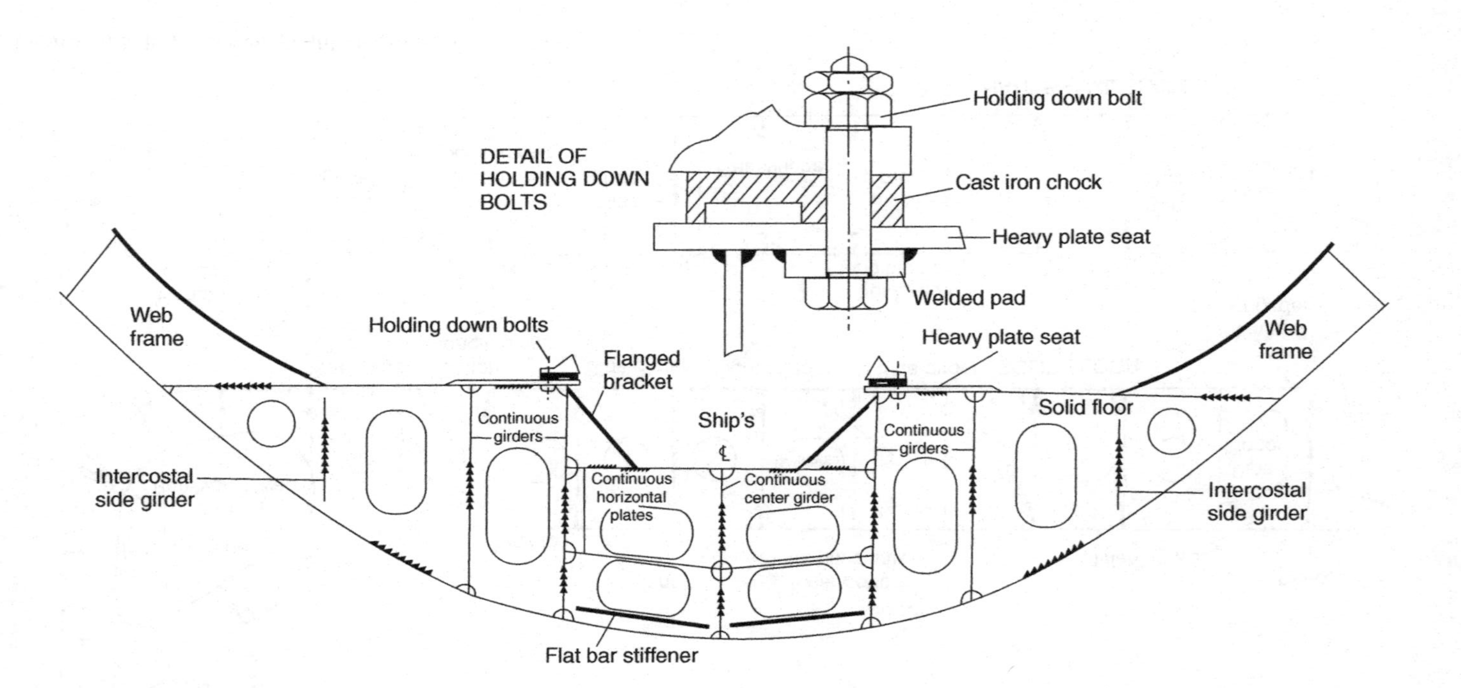

Figure 16.7 Engine seats.

thickness. Often, the machinery is built up on seatings forming longitudinal bearers that are supported transversely by tripping brackets in line with the double-bottom floors, the longitudinal bearers being in line with the double-bottom side girders (see Figure 16.7).

Boiler bearers are similarly fabricated with support from transverse brackets and longitudinal members.

In recent years, particularly for passenger ships, much attention has been given to the reduction of noise and vibration emanating from the machinery installation. To this end, resilient mounting and flexible couplings are now common in such ships.

17 Shell plating and framing

Chapter Outline

Shell plating 189
Bottom shell plating 189
Side shell plating 190
Grades of steel for shell plates 191
Framing 191
Transverse framing 192
Longitudinal framing 193
Tank side brackets 193
Local strengthening of shell plating 195
Additional stiffening for panting 195
Strengthening for navigation in ice 195
Bilge keel 197
Further reading 205
Some useful websites 205

The shell plating forms the watertight skin of the ship and at the same time, in merchant ship construction, contributes to the longitudinal strength and resists vertical shear forces. Internal strengthening of the shell plating may be both transverse and longitudinal, and is designed to prevent collapse of the plating under the various loads to which it is subjected.

Shell plating

The bottom and side shell plating consists of a series of flat and curved steel plates generally of greater length than breadth butt welded together. The vertical welded joints are referred to as 'butts' and the horizontal welded joints as 'seams' (see Figure 17.1). Stiffening members both longitudinal and transverse are generally welded to the shell by continuous fillet welds. Framing is notched in way of welded plate butts and seams.

Bottom shell plating

Throughout the length of the ship the width and thickness of the keel plate remain constant where a flat plate keel is fitted. Its thickness is never less than that of the adjoining bottom plating.

Strakes of bottom plating to the bilges have their greatest thickness over the 40% of the ship's length amidships, where the bending stresses are highest. The bottom

Ship Construction. DOI: 10.1016/B978-0-08-097239-8.00017-9

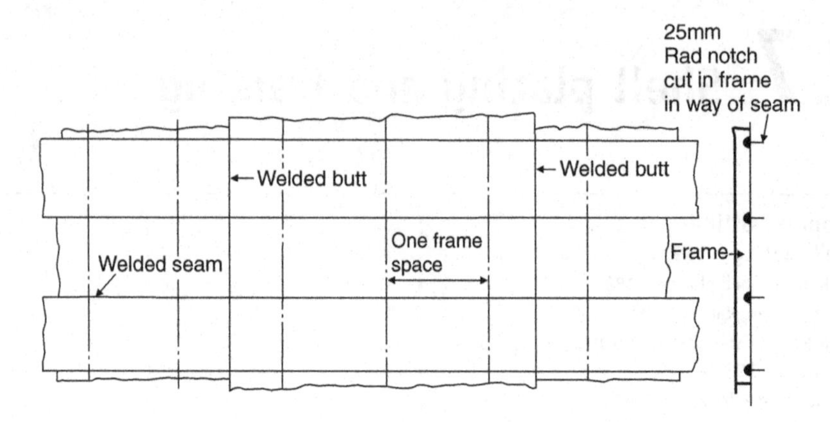

Figure 17.1 Shell plating.

plating then tapers to a lesser thickness at the ends of the ship, apart from increased thickness requirements in way of the pounding region (see Chapter 16).

Side shell plating

As with the bottom shell plating, the greater thickness of the side shell plating is maintained within 40% of the vessel's midship length and then tapers to the rule thickness at the ends. The thickness may be increased in regions where high vertical shear stresses occur, usually in way of transverse bulkheads in a vessel permitted to carry heavy cargoes with some holds empty. There is also a thickness increase at the stern frame connection, at any shaft brackets, and in way of the hawse pipes, where considerable chafing occurs. Further shell plate thickness increases may be found at the panting region, as discussed later in this chapter.

The upper strake of plating adjacent to the strength deck is referred to as the 'sheerstrake'. As the sheerstrake is at a large distance from the neutral axis it has a greater thickness than the other strakes of side shell plating. Also, being in a highly stressed region it is necessary to avoid welded attachments to the sheerstrake, or cutouts that would introduce stress raisers. The upper edge is dressed smooth, and the welding of bulwarks to the edge of the sheerstrake is not permitted within the amidships length of the ship. Scupper openings above the deck over the same length, and at the ends of the superstructure, are also prohibited in larger vessels. The connection between the sheerstrake and strength deck can present a problem, and a rounded gunwale may be adopted to solve this problem where the plating is heavy. This is often the case over the midship portion of large tankers and bulk carriers. Butt welds are then employed to make connections rather than the less satisfactory fillet weld at the perpendicular connection of the vertical sheerstrake and horizontal strength deck stringer plate. The radius of a rounded gunwale must be adequate (not less than 15 times the thickness) and any welded guardrails and fairleads are kept off the radiused plate if possible.

A smooth transition from rounded gunwale to angled sheerstrake/deck stringer connection is necessary at the ends of the ship.

All openings in the side shell have rounded corners, and openings for sea inlets etc. are kept clear of the bilge radius if possible. Where this is not possible openings on or in the vicinity of the bilge are made elliptical.

Grades of steel for shell plates

In large ships it is necessary to arrange strakes of steel with greater notch ductility at the more highly stressed regions. Details of Lloyd's requirements for mild steel and over 40% of the length amidships are given in Table 17.1 as a guide. The rules also require thicker plating for the members referred to in Table 17.1 outside the amidships region to have greater notch ductility.

Framing

The bottom shell may be transversely or longitudinally framed, longitudinal framing being preferred for larger ships, and generally required when the ship's length exceeds 120 meters. The side shell framing may also be transversely or longitudinally framed, transverse framing being adopted in many conventional cargo ships, particularly where the maximum bale capacity is required. Bale capacities are often considerably reduced where deep transverses are fitted to support longitudinal framing. Longitudinal framing is adopted within double-hull spaces and is common within the hopper and topside tanks of bulk carriers. Smaller single-skin bulk carriers

Table 17.1 Lloyd's requirements for mild steel

Requirement	Structural member
1. Grade D where thickness is less than 15 mm, otherwise Grade E.	Sheerstrake or rounded gunwale over 40% of length amidships in ships exceeding 250 m in length.
2. Grade A where thickness is less than 15 mm. Grade B where thickness is 15–20 mm. Grade D where thickness is 20–25 mm. Grade E where thickness is greater than 25 mm.	Sheerstrake and rounded gunwale over 40% of length amidships in ships of 250 m or less in length. Bilge strake (other than for vessels of less than 150 m with double bottom over full breadth).
3. Grade A where thickness is less than 20 mm. Grade B where thickness is 20–25 mm. Grade D where thickness is 25–40 mm. Grade E where thickness is over 40 mm.	Bottom plating including keel. Bilge strake (ships of less than 150 m and with double bottom over full breadth).
4. Grade A where thickness is less than 30 mm. Grade B where thickness is 30–40 mm. Grade D where thickness is greater than 40 mm.	Side plating.

then have transverse frames at the side shell fitted between the hopper and topside tank (see Figure 17.7).

Transverse framing

In a general cargo ship the transverse framing will consist of main and hold frames with brackets top and bottom, and lighter tween deck frames with brackets at the tops only (see Figure 17.2). A typical midship section for a general cargo ship is shown in Figure 17.6.

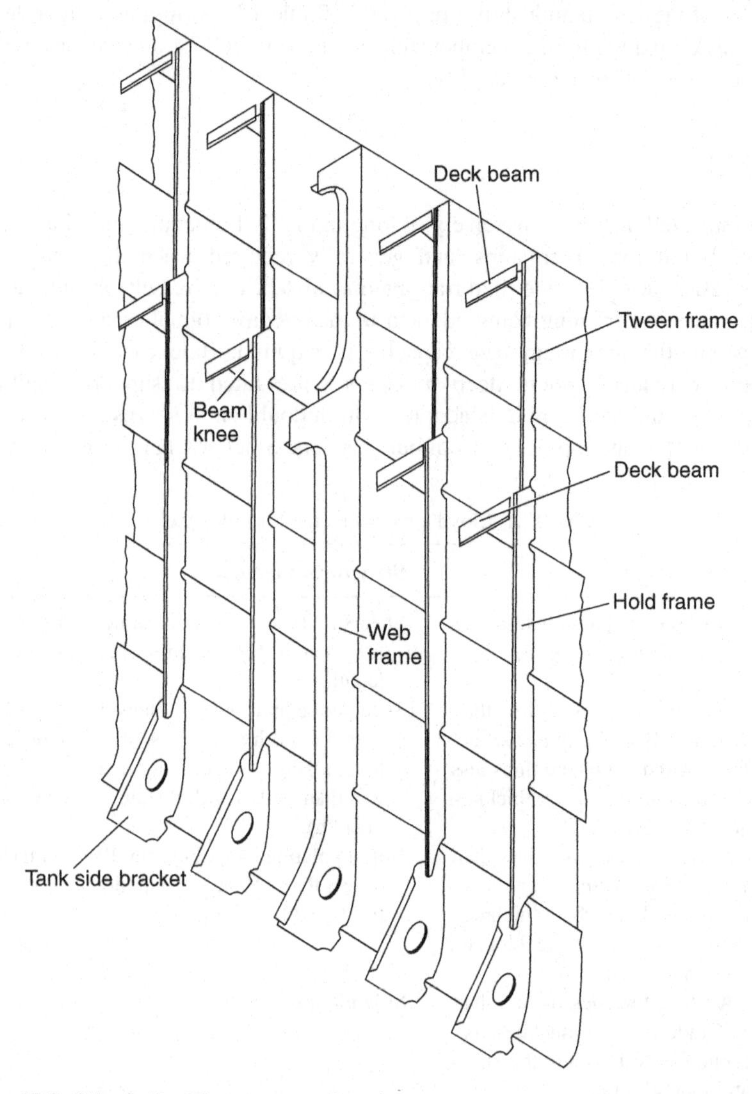

Figure 17.2 Side shell with transverse framing.

Scantlings of the main transverse frames are primarily dependent on their position, spacing and depth, and to some extent on the rigidity of the end connections. In way of tanks such as oil bunkers or cargo deep tanks the side frame size will be increased, except where supporting side stringers are fitted within the tank space. Frames supporting hatch end beams and those in way of deck transverses, where the deck is framed longitudinally, also have increased scantlings.

Web frames, i.e. built-up frames consisting of plate web and face flat, where the web is considerably deeper than the conventional transverse frame, are often introduced along the side shell (see Figure 17.3a). A number are fitted in midship machinery spaces, generally not more than five frame spaces apart, but may be omitted if the size of normal framing is increased. Forward of the collision bulkhead and in any deep tank adjacent to the collision bulkhead, and in tween decks above such tanks, web frames are required at not more than five frame spaces apart. In the tween decks above the after peak tank, web frames are required at every fourth frame space abaft the aft peak bulkhead. In all cases the provision of web frames is intended to increase the rigidity of the transverse ship section at that point.

Longitudinal framing

The longitudinal framing of the bottom shell is dealt with in Chapter 16. If the side shell is longitudinally framed offset bulb sections will often be employed with the greater section scantlings at the lower side shell. Direct continuity of strength is to be maintained, and many of the details are similar to those illustrated for the tanker longitudinals (see Chapter 22). Transverse webs are fitted to support the side longitudinals, these being spaced not more than 3.8 m apart, in ships of 100 m length or less, with increasing spacing being permitted for longer ships. In the peaks the spacing is 2.5 m where the length of ship is less than 100 m, increasing linearly to a spacing of 3.5 m where the length exceeds 300 m.

Larger ships required to have a double hull are longitudinally framed at the sides with transverse webs arranged in line with the floors in the double bottom to ensure continuity of transverse strength. Horizontal perforated flats are fitted between the inner and outer side plating to support the transverse webs (see Figures 17.8 and 17.9).

A roll on roll off (ro-ro) ship midship section is shown in Figure 17.10. This is also longitudinally framed, and the side casing (positioned to maximize the space on the vehicle decks) is similar to the double hull of other types.

A very different ship type is shown in Figure 17.11, that is a catamaran fast ferry. The hull form is radically different and the material for construction is aluminium, to minimize the weight of the structure. The basic longitudinal framing, with transverse webs, is however essentially the same as for the other ship types described in this chapter.

Tank side brackets

The lower end of the frame may be connected to the tank top or hopper side tank by means of a flanged or edge-stiffened tank side bracket, as illustrated in Figure 17.3b.

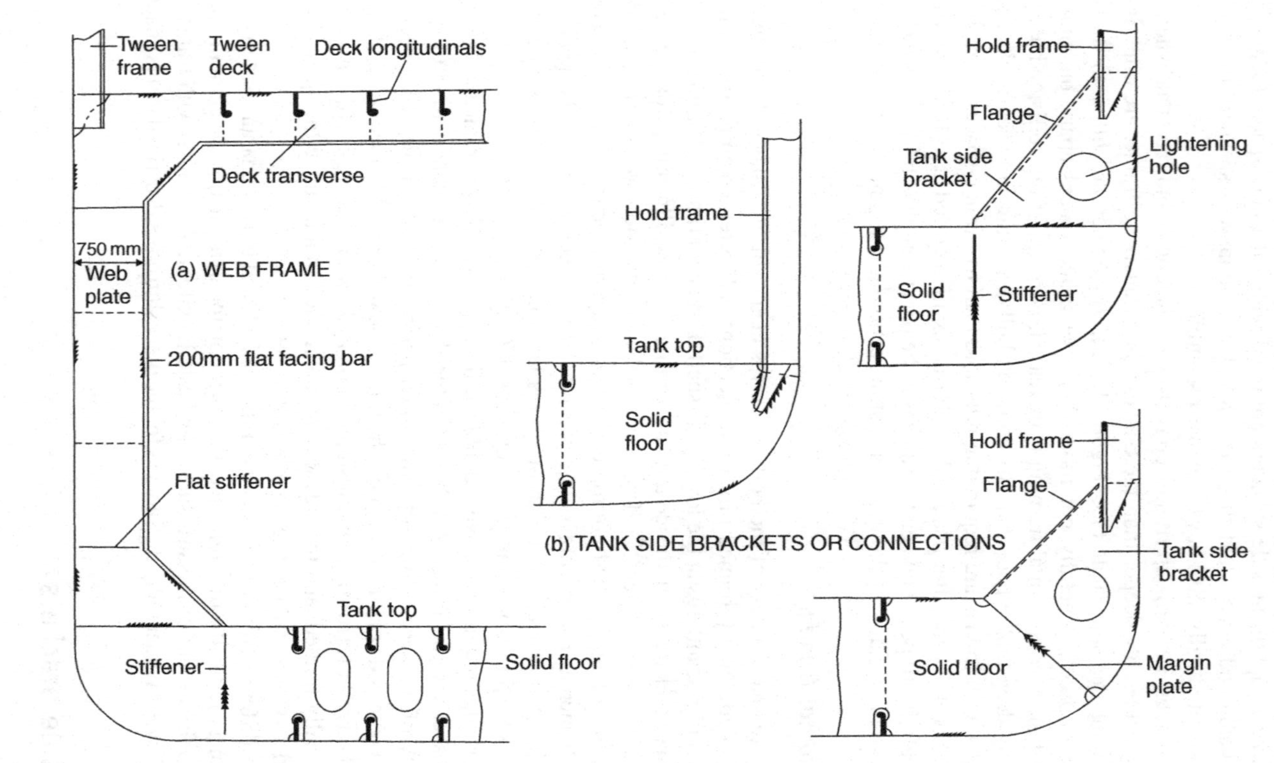

Figure 17.3 Web frame and tank side bracket.

Local strengthening of shell plating

The major region in which the shell plating is subjected to local forces at sea is at the forward end. Strengthening of the forward bottom shell for pounding forces is dealt with in Chapter 16. Panting, which is discussed in Chapter 8, will also influence the requirements for the scantlings and strengthening of the shell forward and to a lesser extent at the aft end. Where a ship is to navigate in ice, a special classification may be assigned depending on the type and severity of ice encountered (see Chapter 4), and this will involve strengthening the shell forward and in the waterline region.

Additional stiffening for panting

Additional stiffening is provided in the fore peak structure, the transverse side framing being supported by any, or a combination of the following arrangements:

1. Side stringers spaced vertically about 2 m apart and supported by struts or beams fitted at alternate frames. These 'panting beams' are connected to the frames by brackets and if long may be supported at the ship's center line by a partial wash bulkhead. Intermediate frames are bracketed to the stringer (see Figure 17.4).
2. Side stringers spaced vertically about 2 m apart and supported by web frames.
3. Perforated flats spaced not more than 2.5 m apart. The area of perforations is not less than 10% of the total area of the flat.

Aft of the forepeak in the lower hold or deep tank spaces panting stringers are fitted in line with each stringer or perforated flat in the fore peak extending back over 15% of the ship length from forward. These stringers may be omitted if the shell plating thickness is increased by 15% for vessels of 150 m length or less, decreasing linearly to a 5% increase for vessels of 215 m length or more. However, where the unsupported length of the main frames exceeds 9 m, panting stringers in line with alternate stringers or flats in the fore peak are to be fitted over 20% of the ship's length from forward whether the shell thickness is increased or not. Stringers usually take the form of a web plate with flat facing bar.

In tween deck spaces in the forward 15% of the ship's length, intermediate panting stringers are fitted where the unsupported length of tween frame exceeds 2.6 m in lower tween decks or 3 m in upper tween decks. Alternatively, the shell thickness may be increased as above.

In the aft peak space and in deep tween decks above the aft peak similar panting arrangements are required for transverse framing, except that the vertical spacing of panting stringers may be up to 2.5 m apart.

If the fore peak has longitudinal framing and the depth of tank exceeds 10 m the transverse webs supporting the longitudinals are to be supported by perforated flats or an arrangement of transverse struts or beams.

Strengthening for navigation in ice

If a vessel is to be assigned a special features notation for navigation in first-year ice (see Chapter 4), the additional strengthening required involves primarily an increase

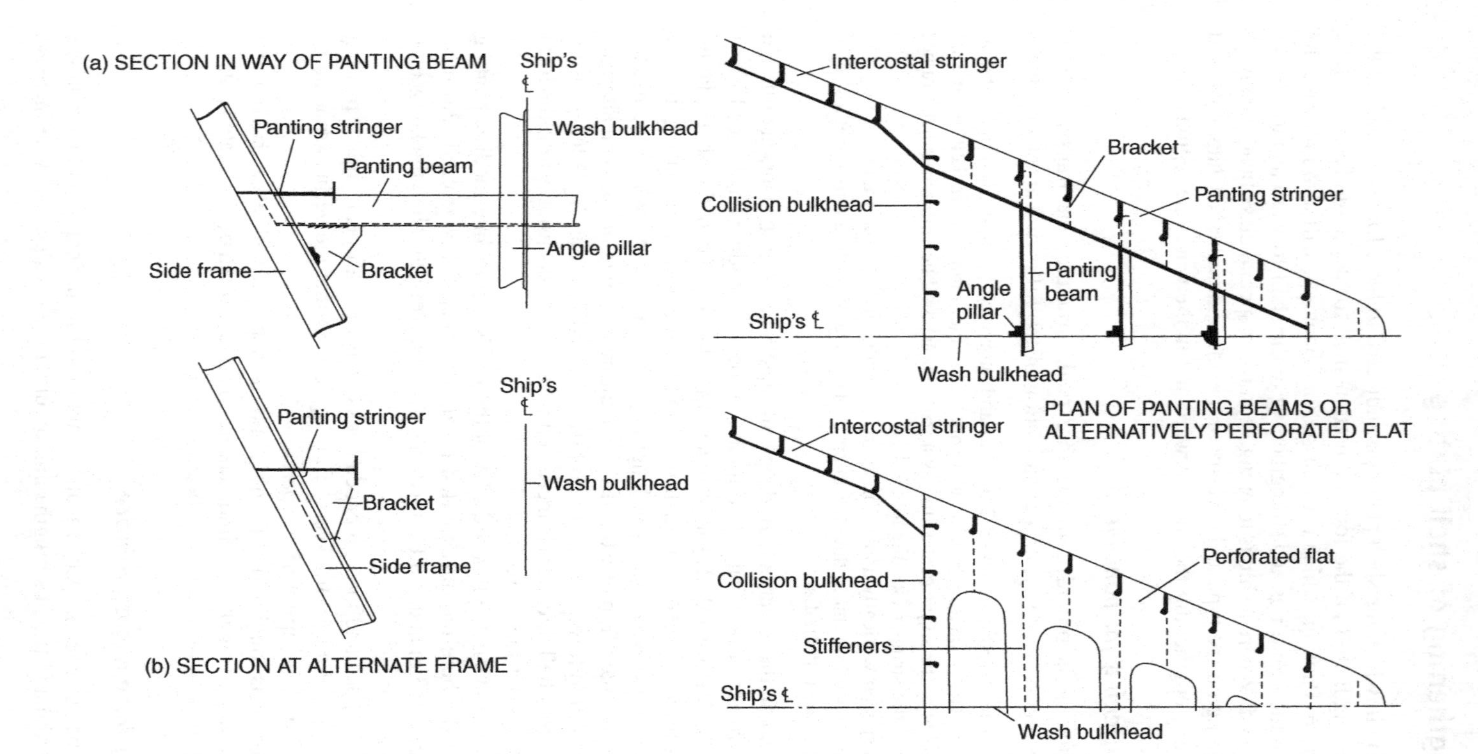

Figure 17.4 Panting arrangements forward.

in plate thickness and frame scantlings in the waterline region and the bottom forward, and may require some modifications and strengthening at the stem, stern, rudder and bossings, etc.

A main ice belt zone is defined that extends above the ice load waterline (i.e. normally the summer load waterline) and below the ice light waterline (i.e. lightest waterline ship navigates ice in). The extent of this zone depends on the ice class assigned.

The shell plating thickness in this zone is greater than on a conventional ship and increases with severity of ice class and with position from aft to forward. For the more severe ice conditions the thickness of the side shell is also increased below the main ice belt zone for at least 40% of the length from forward.

Transverse main and intermediate frames of heavier scantlings are fitted in way of the main ice belt zone. Where the shell is longitudinally framed, longitudinals of increased scantlings are fitted in way of the ice belt zone. Both transverse and longitudinal frame scantling requirements are dependent on the severity of ice class assigned and distance of frame from forward. Transverse ice framing is supported by ice stringers and decks, and longitudinal framing by web frames, the scantlings of which are increased with severity of ice class and distance from forward.

Strengthening for addition of 'Icebreaker' notation to ship-type notation and assignment of special features notation for navigation in multi-year ice and the associated requirements for plating and framing at the bow and stern are too extensive to be covered adequately in this text.

Bilge keel

Most ships are fitted with some form of bilge keel, the prime function of which is to help damp the rolling motion of the vessel. Other relatively minor advantages of the bilge keel are protection for the bilge on grounding, and increased longitudinal strength at the bilge.

The damping action provided by the bilge keel is relatively small but effective, and virtually without cost after the construction of the ship. It is carefully positioned on the ship so as to avoid excessive drag when the ship is under way, and to achieve a minimum drag, various positions of the bilge keel may be tested on the ship model used to predict power requirements. This bilge keel then generally runs over the midship portion of the hull, often extending further aft than forward of amidships and being virtually perpendicular to the turn of the bilge.

There are many forms of bilge keel construction, and some quite elaborate arrangements have been adopted in an attempt to improve the damping performance whilst reducing any drag. Care is required in the design of the bilge keel, for although it would not be considered as a critical strength member of the hull structure, the region of its attachment is fairly highly stressed owing to its distance from the neutral axis. Cracks have originated in the bilge keel and propagated into the bilge plate causing failure of the main structure. In general, bilge keels are attached to

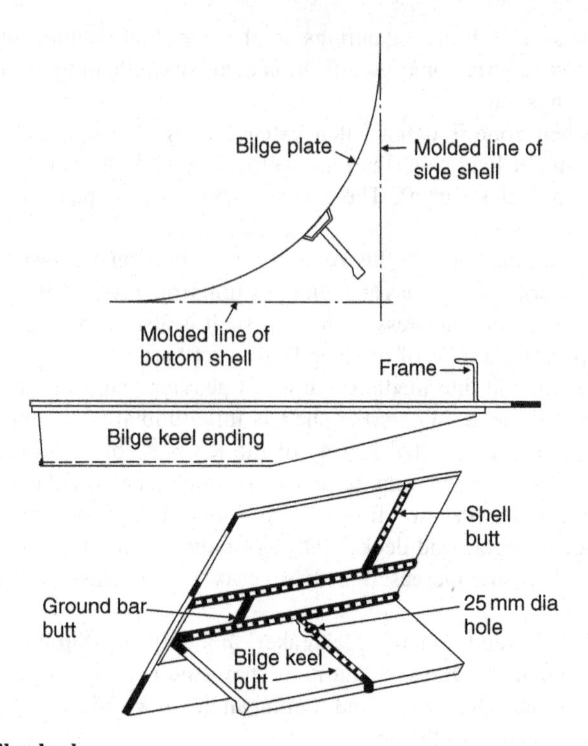

Figure 17.5 Bilge keels.

a continuous ground bar with the butt welds in the shell plating, ground bar, and bilge keel staggered (see Figure 17.5). Direct connection between the ground bar butt welds and the bilge plate and bilge keel butt welds and the ground bar are avoided. In ships over 65 m in length, holes are drilled in the bilge keel butt welds as shown in Figure 17.5.

The ground bar thickness is at least that of the bilge plate or 14 mm, whichever is the lesser, and the material grade is the same as that of the bilge plate. Connection of the ground bar to the shell is by continuous fillet welds and the bilge keel is connected to the ground bar by light continuous or staggered intermittent weld. The latter lighter weld ensures that should the bilge keel be fouled, failure occurs at this joint without the bilge plate being damaged.

Bilge keels are gradually tapered (at least 3 to 1) at their ends and finish in way of an internal stiffening member.

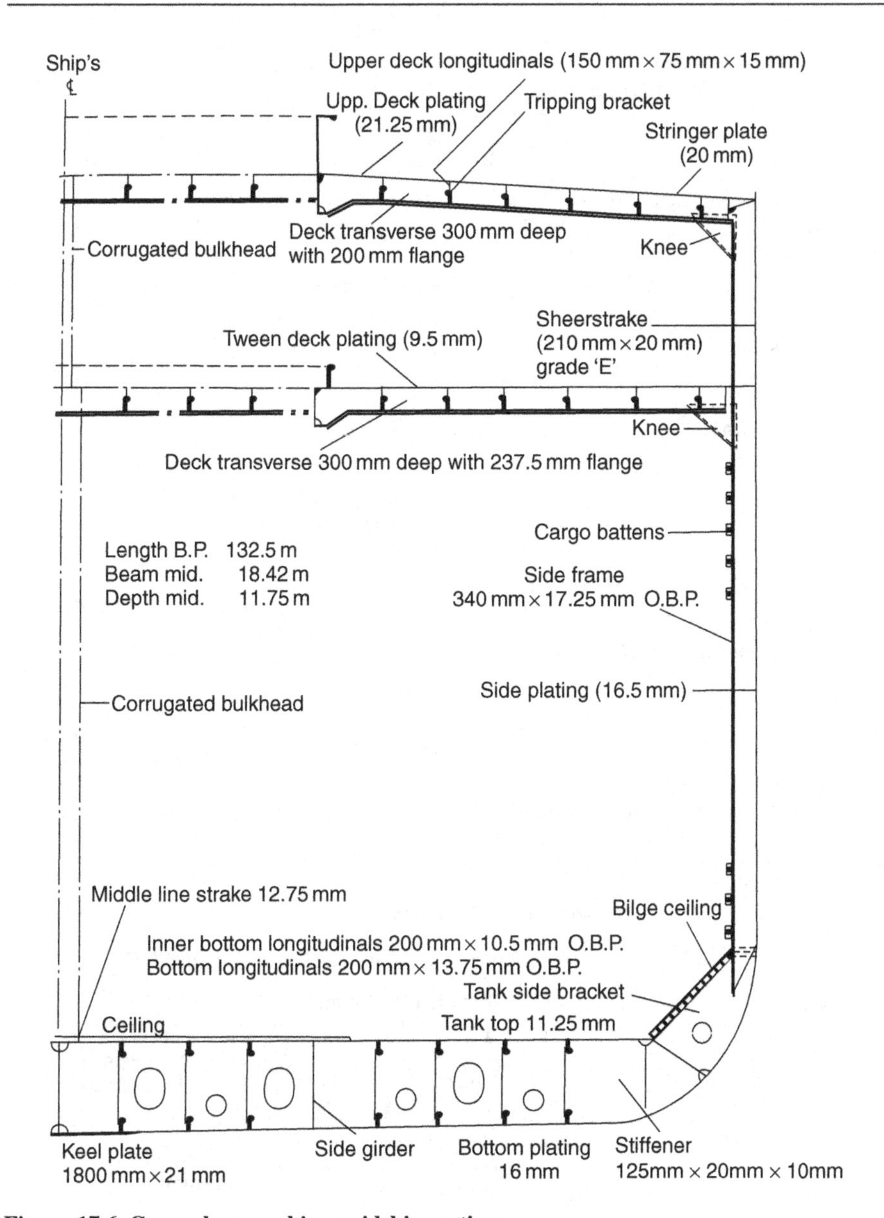

Figure 17.6 General cargo ship—midship section.

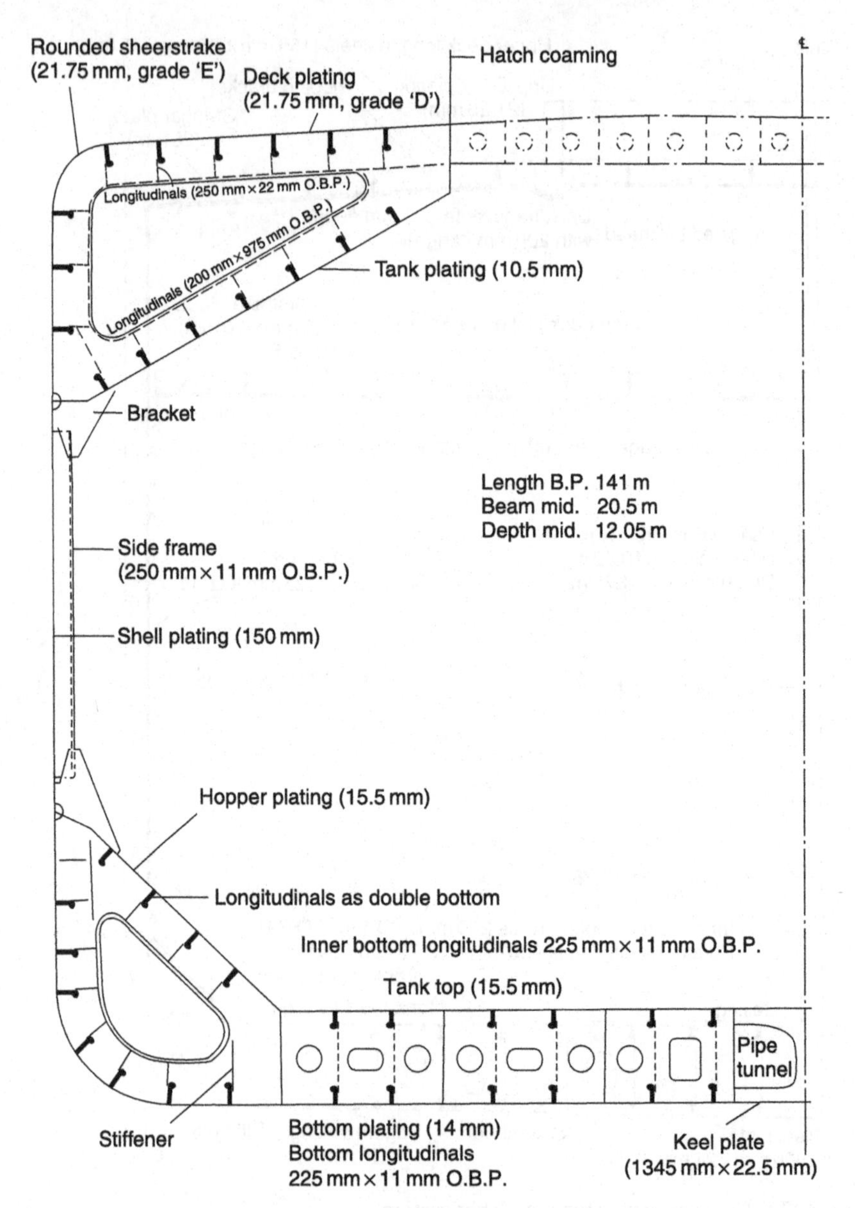

Figure 17.7 Bulk carrier—single-side skin midship section.

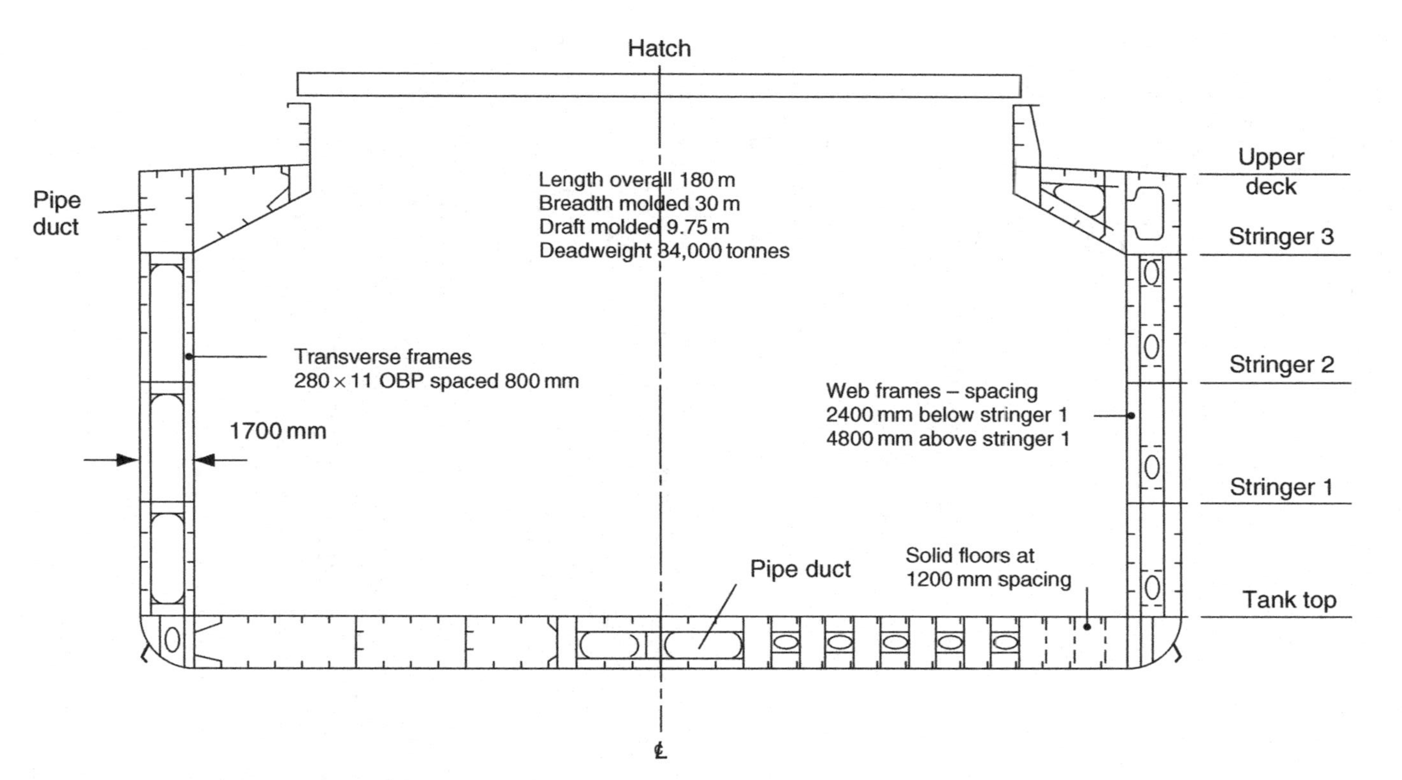

Figure 17.8 Bulk carrier—double-side skin midship section.

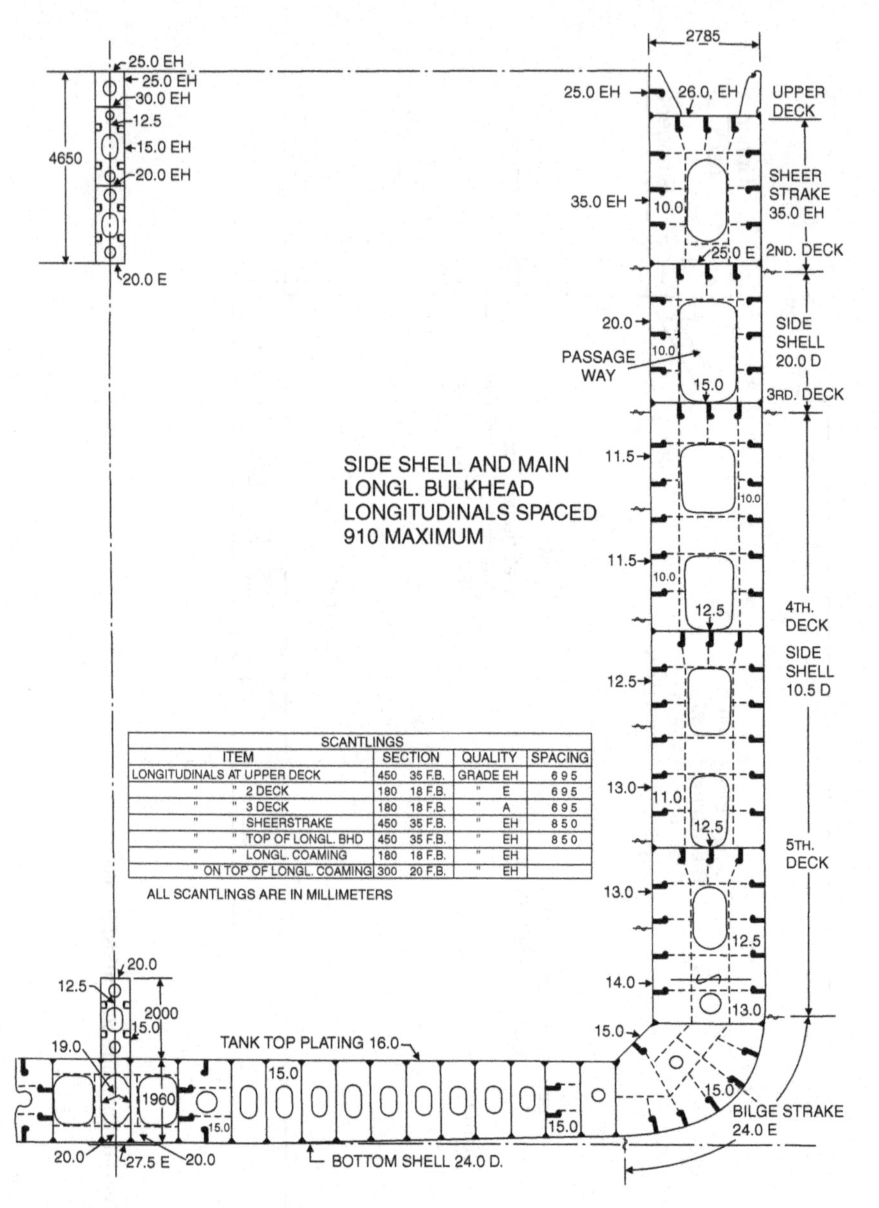

SCANTLINGS			
ITEM	SECTION	QUALITY	SPACING
LONGITUDINALS AT UPPER DECK	450 35 F.B.	GRADE EH	6 9 5
" " 2 DECK	180 18 F.B.	" E	6 9 5
" " 3 DECK	180 18 F.B.	" A	6 9 5
" " SHEERSTRAKE	450 35 F.B.	" EH	8 5 0
" " TOP OF LONGL. BHD	450 35 F.B.	" EH	8 5 0
" " LONGL. COAMING	180 18 F.B.	" EH	
" ON TOP OF LONGL. COAMING	300 20 F.B.	" EH	

ALL SCANTLINGS ARE IN MILLIMETERS

Figure 17.9 Container ship—midship section.

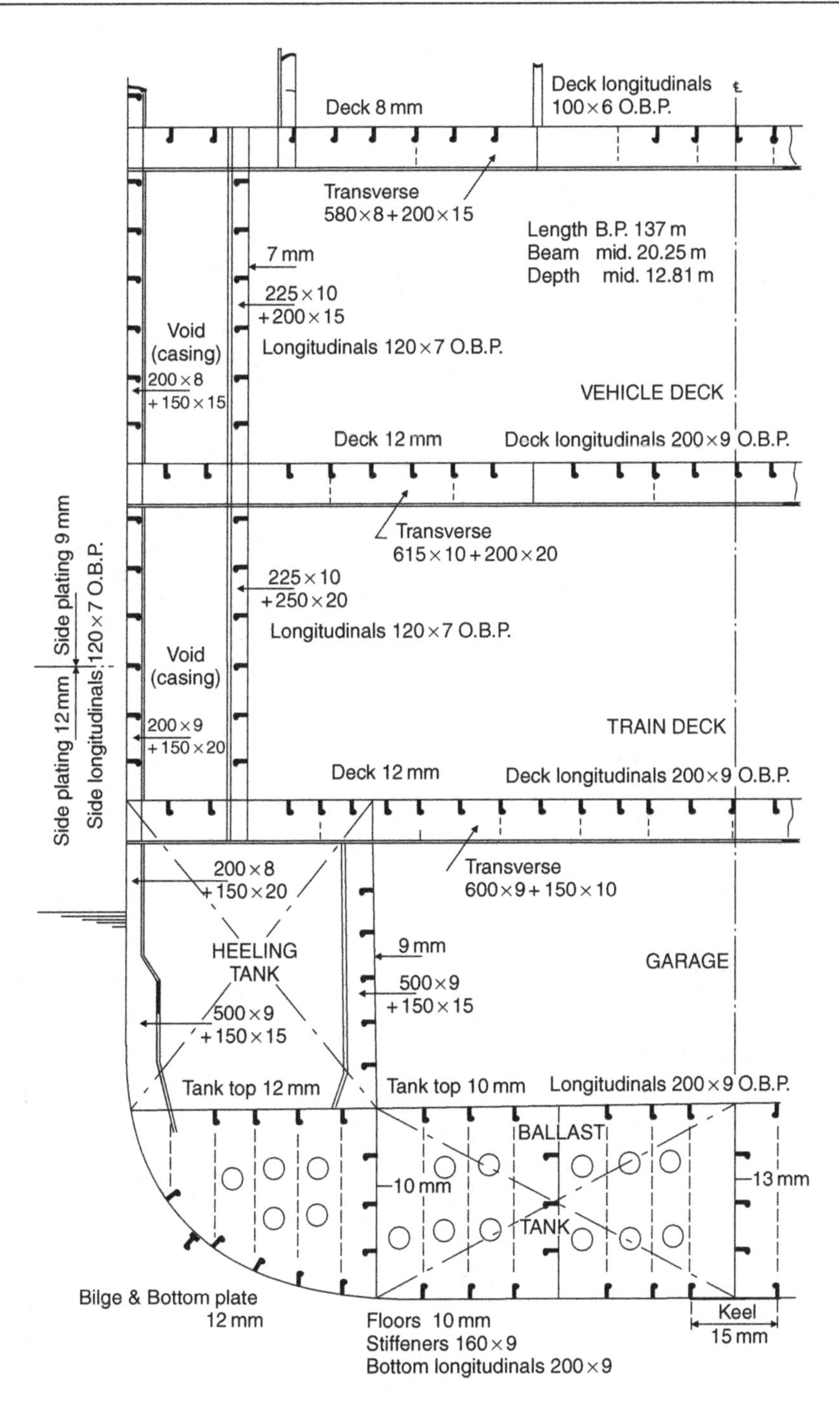

Figure 17.10 Midship section of ro-ro ship.

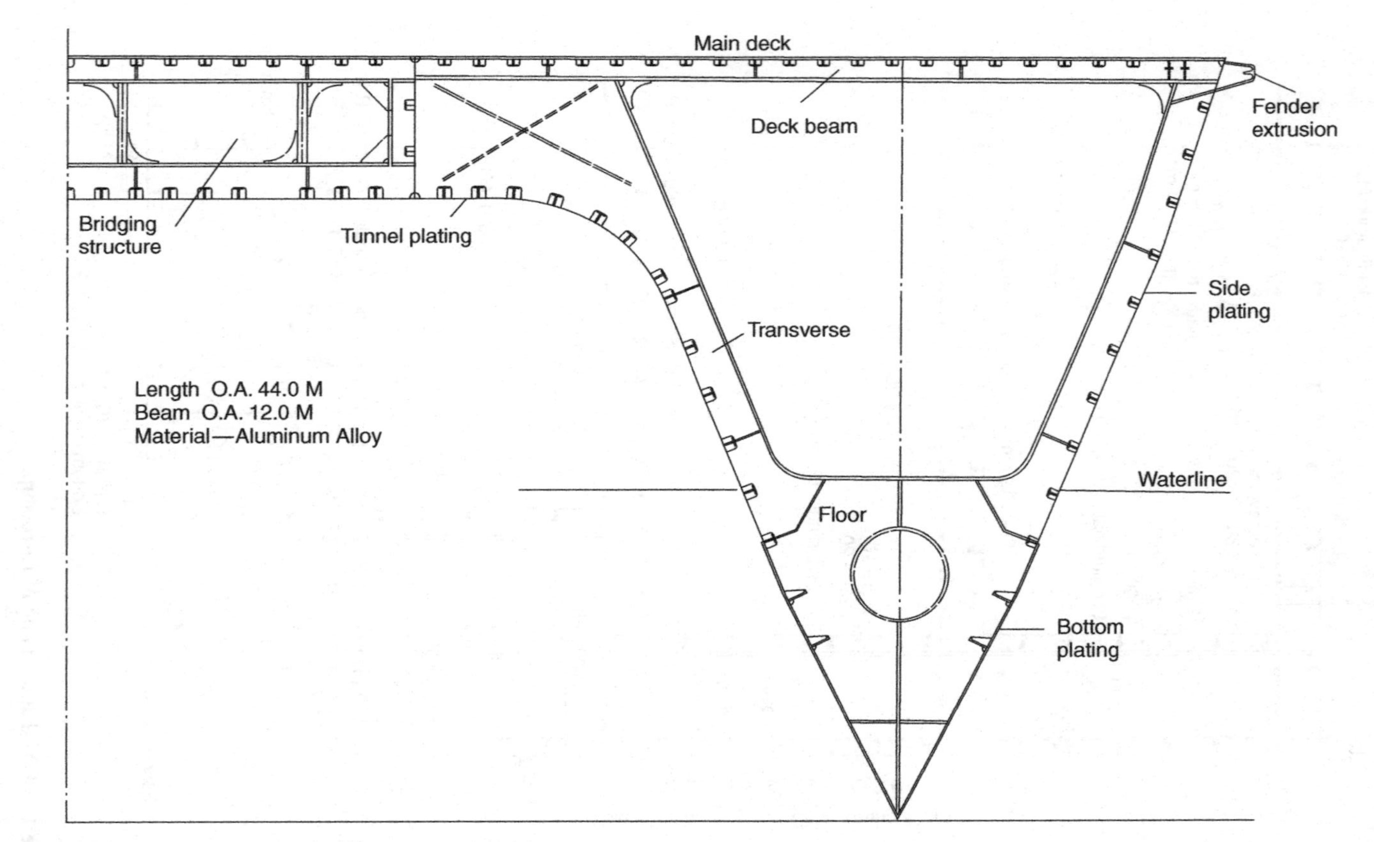

Figure 17.11 High speed craft (catamaran)—section.

Further reading

Common Structural Rules for Bulk Carriers. IACS publication.

Some useful websites

www.iacs.org.uk Copies of the IACS's *Common Structural Rules for Bulk Carriers* are available on this website.

18 Bulkheads and pillars

Chapter Outline

Bulkheads 207
Spacing of watertight bulkheads—cargo ships 208
Spacing of watertight bulkheads—passenger ships 209
Construction of watertight bulkheads 209
Testing watertight bulkheads 213
Watertight doors 213
Deep tanks 214
Construction of deep tanks 216
Testing deep tanks 216
Topside tanks 218
Shaft tunnel 218
Construction of the shaft tunnel 218
Pillars 220
Spacing of hold pillars 220
Pillar construction 220
Small pillars 223
Further reading 223
Some useful websites 223

This chapter deals with the internal vertical structure of the ship. Much of this structure, particularly the pillars and to some extent the transverse strength bulkheads, is responsible for carrying the vertical loading experienced by the ship. The principal bulkheads subdivide the ship hull into a number of large watertight compartments, and their construction and spacing is discussed. Also considered are the boundaries of other smaller compartments, such as deep tanks and the shaft tunnel.

Bulkheads

Vertical partitions in a ship arranged transversely or fore and aft are referred to as 'bulkheads'. Those bulkheads that are of greatest importance are the main hull transverse and longitudinal bulkheads, dividing the ship into a number of watertight compartments. Other lighter bulkheads, named 'minor bulkheads', which act as screens further subdividing compartments into small units of accommodation or stores, are of little structural importance.

Ship Construction. DOI: 10.1016/B978-0-08-097239-8.00018-0

The main hull bulkheads of sufficient strength are made watertight in order that they may contain any flooding in the event of a compartment on one side of the bulkhead being bilged or breached. Further, they serve as a hull strength member, not only carrying some of the ship's vertical loading but also resisting any tendency for transverse deformation of the ship. As a rule the strength of the transverse watertight bulkheads is maintained to the strength deck, which may be above the freeboard deck. Finally, each of the main hull bulkheads has often proved a very effective barrier to the spread of a hold or machinery space fire.

Spacing of watertight bulkheads—cargo ships

The minimum number of transverse watertight bulkheads that must be fitted in a dry cargo ship are stipulated. A collision bulkhead must be fitted forward, an aft peak bulkhead must be fitted, and watertight bulkheads must be provided at either end of the machinery space. This implies that for a vessel with machinery amidships the minimum possible number of watertight bulkheads is four. With the machinery aft this minimum number may be reduced to three, the aft peak bulkhead being at the aft end of the machinery space.

Of these bulkheads, perhaps the most important is the collision bulkhead forward. It is a fact that the bow of at least one out of two ships involved in a collision will be damaged. For this reason a heavy bulkhead is specified and located so that it is not so far forward as to be damaged on impact. Neither should it be too far aft so that the compartment flooded forward causes excessive trim by the bow. Classification societies require the location for ships whose length does not exceed 200 m as not less than 5% and not greater than 8% of the ship's length (Lloyd's length) from the fore end of the load waterline. A minimum distance of 10 meters may also be specified to ensure that the bulkhead is effective. As a rule this bulkhead is fitted at the minimum distance in order to gain the maximum length for cargo stowage. The aft peak bulkhead is intended to enclose the stern tubes in a watertight compartment, preventing any emergency from leakage where the propeller shafts pierce the hull. It is located well aft so that the peak when flooded would not cause excessive trim by the stern. Machinery bulkheads provide a self-contained compartment for engines and boilers, preventing damage to these vital components of the ship by flooding in an adjacent hold. They also localize any fire originating in these spaces.

A minimum number of watertight bulkheads will only be found in smaller cargo ships. As the size increases the classification society will recommend additional bulkheads, partly to provide greater transverse strength, and also to increase the amount of subdivision. Table 18.1 indicates the number of watertight bulkheads recommended by Lloyd's Register for any cargo ship. These should be spaced at uniform intervals, but the shipowner may require for a certain trade a longer hold, which is permitted if additional approved transverse stiffening is provided. It is possible to dispense with one watertight bulkhead altogether, with Lloyd's Register approval, if adequate approved structural compensation is introduced. In container ships the spacing is arranged to suit the standard length of containers carried.

Table 18.1 Bulkheads for cargo ships

Length of ship (meters)		Total number of bulkheads	
Above	**Not exceeding**	**Machinery midships**	**Machinery aft**
	65	4	3
65	85	4	4
85	105	5	5
105	115	6	5
115	125	6	6
125	145	7	6
145	165	8	7
165	190	9	8
190	To be considered individually		

Each of the main watertight hold bulkheads may extend to the uppermost continuous deck, but in the case where the freeboard is measured from the second deck they need only be taken to that deck. The collision bulkhead extends to the uppermost continuous deck and the aft peak bulkhead may terminate at the first deck above the load waterline provided this is made watertight to the stern, or to a watertight transom floor.

In the case of bulk carriers a further consideration may come into the spacing of the watertight bulkheads where a shipowner desires to obtain a reduced freeboard. It is possible with bulk carriers to obtain a reduced freeboard under The International Load Line Convention 1966 (see Chapter 31) if it is possible to flood one or more compartments without loss of the vessel. For obvious reasons many shipowners will wish to obtain the maximum permissible draft for this type of vessel and the bulkhead spacing will be critical. Additionally, SOLAS amendments now require that bulk carriers constructed after 1 July 1999 and of 150 meters or more in length of single-side skin construction and designed to carry solid bulk cargoes of 1000 kg/cubic meter or more when loaded to the summer load line must be able to withstand flooding of any one cargo hold.

Spacing of watertight bulkheads—passenger ships

Where a vessel requires a passenger certificate (carrying more than 12 passengers), it is necessary for that vessel to comply with the requirements of the International Convention on Safety of Life at Sea, 1974 (see Chapter 29). Under this convention the subdivision of the passenger ship is strictly specified, and controlled by the authorities of the maritime countries who are signatories to the convention. In the United Kingdom the controlling authority is the Marine and Coastguard Agency.

Construction of watertight bulkheads

The plating of a flat transverse bulkhead is generally welded in horizontal strakes, and convenient two-dimensional units for prefabrication are formed. Smaller bulkheads may be erected as a single unit; larger bulkheads are in two or more units. It has always been the practice to use horizontal strakes of plating since the plate thickness increases with depth below the top of the bulkhead. The reason for this is that the plate thickness is directly related to the pressure exerted by the head of water when a compartment on one side of the bulkhead is flooded. Apart from the depth, the plate thickness is also influenced by the supporting stiffener spacing.

Vertical stiffeners are fitted to the transverse watertight bulkheads of a ship, the span being less in this direction and the stiffener therefore having less tendency to deflect under load. Stiffening is usually in the form of welded inverted ordinary angle bars, or offset bulb plates, the size of the stiffener being dependent on the unsupported length, stiffener spacing, and rigidity of the end connections. Rigidity of the end connections will depend on the form of end connection, stiffeners in holds being bracketed or simply directly welded to the tank top or underside of deck, whilst upper tween stiffeners need not have any connection at all (see Figure 18.1). Vertical stiffeners may be supported by horizontal stringers, permitting a reduction in the stiffener scantling as a result of the reduced span. Horizontal stringers are mostly found on those bulkheads forming the boundaries of a tank space, and in this context are dealt with later.

It is not uncommon to find in present-day ships swedged and corrugated bulkheads, the swedges like the troughs of a corrugated bulkhead being so designed and spaced as to provide sufficient rigidity to the plate bulkhead in order that conventional stiffeners may be dispensed with (see Figure 18.2). Both swedges and corrugations are arranged in the vertical direction like the stiffeners on transverse and short longitudinal pillar bulkheads. Since the plating is swedged or corrugated prior to its fabrication, the bulkhead will be plated vertically with a uniform thickness equivalent to that required at the base of the bulkhead. This implies that the actual plating will be somewhat heavier than that for a conventional bulkhead, and this will to a large extent offset any saving in weight gained by not fitting stiffeners.

At the lower end of transverse hold bulkheads in bulk carriers a bulkhead stool is generally fitted (see Figure 18.3) at the lower end of the bulkhead. This provides a shedder surface for cargo removal rather than a tight corner at the bulkhead/tank top interface. Inclined shedder plates or gussets are also fitted between the corrugations directly above the stool. The addition of a vertical plate bracket under the inclined shedder plate at the midpoint of the corrugation stiffens and supports the corrugation against collapse under load.

The boundaries of the bulkhead are double continuously fillet welded directly to the shell, decks, and tank top.

A bulkhead may be erected in the vertical position prior to the fitting of decks during prefabrication on the berth or assembly into a building block. At the line of the tween decks a 'shelf plate' is fitted to the bulkhead and when erected the tween decks land on this plate, which extends 300–400 mm from the bulkhead. The deck is lap welded to the shelf plate with an overlap of about 25 mm. In the case of a corrugated bulkhead it becomes necessary to fit filling pieces between the troughs in way of the shelf plate.

If possible the passage of piping and ventilation trunks through watertight bulkheads is avoided. However, in a number of cases this is impossible and to maintain the integrity of the bulkhead the pipe is flanged at the bulkhead. Where a ventilation trunk passes through, a watertight shutter is provided.

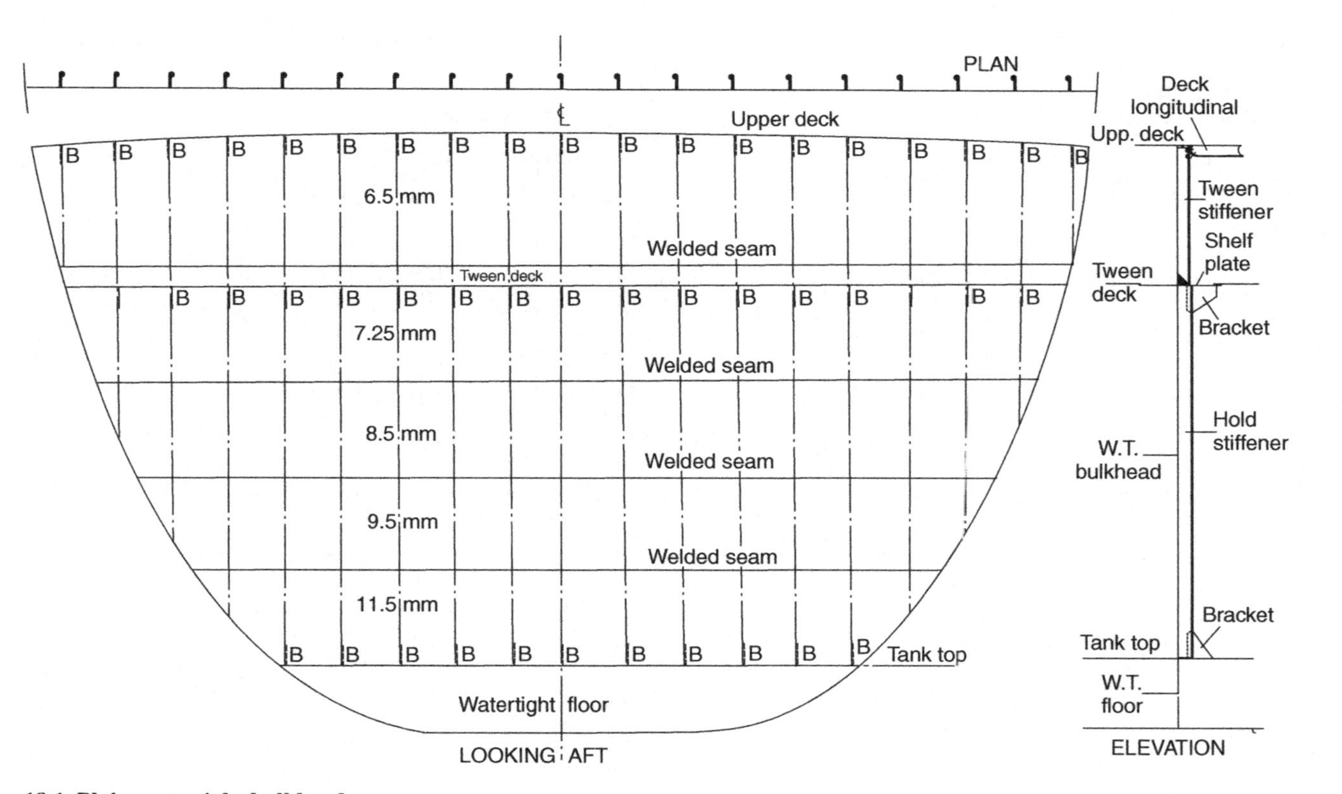

Figure 18.1 Plain watertight bulkhead.

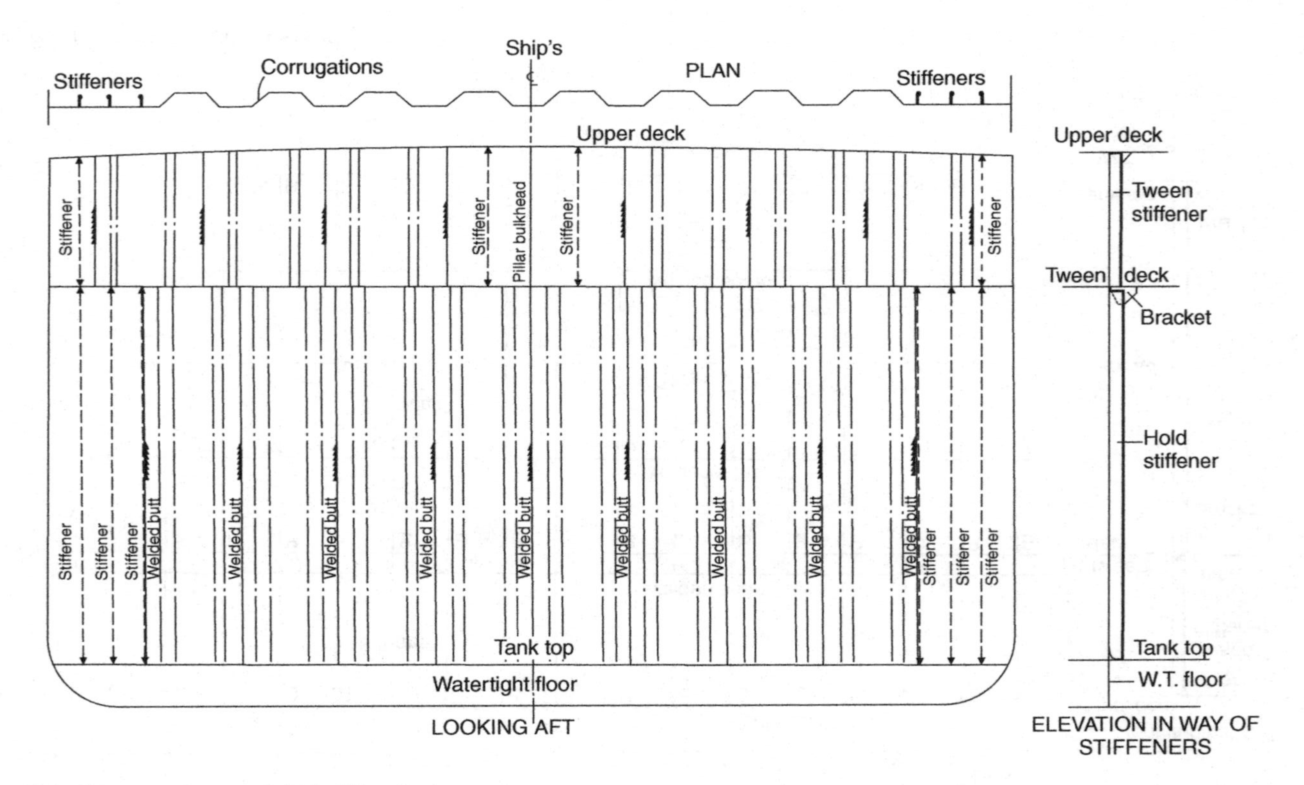

Figure 18.2 Corrugated watertight bulkhead.

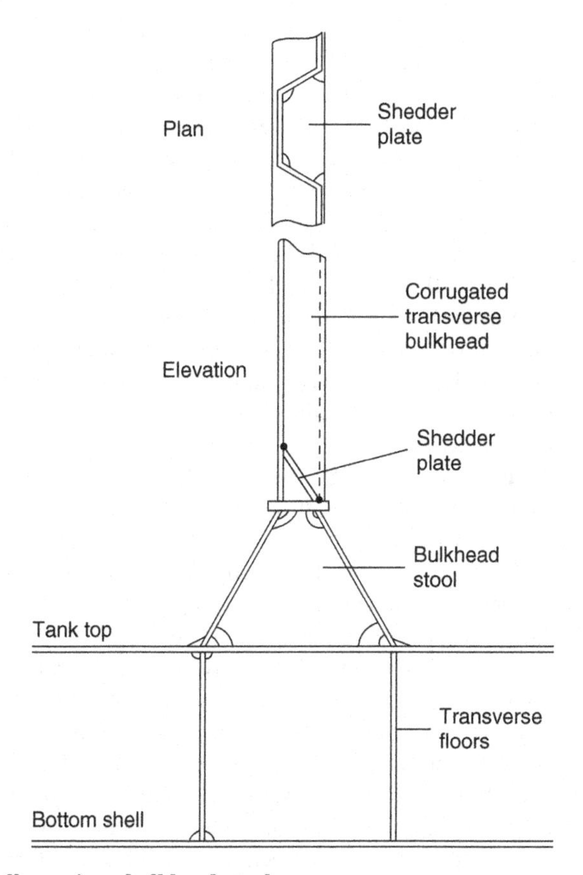

Figure 18.3 Bulk carrier—bulkhead stool.

Testing watertight bulkheads

Both the collision bulkhead, as the fore peak bulkhead, and the aft peak bulkhead, provided they do not form the boundaries of tanks, are to be tested by filling the peaks with water to the level of the load waterline. All bulkheads, unless they form the boundaries of a tank that is regularly subject to a head of liquid, are hose tested. Since it is not considered prudent to test ordinary watertight bulkheads by filling a cargo hold, the hose test is considered satisfactory.

Watertight doors

In order to maintain the efficiency of a watertight bulkhead it is desirable that it remains intact. However, in some instances it becomes necessary to provide access between compartments on either side of a watertight bulkhead and watertight doors are fitted for this purpose. A particular example of this in cargo ships is the direct means of access required between the engine room and the shaft tunnel. In passenger ships watertight doors are more frequently found where they allow passengers to pass between one point of the accommodation and another.

Where a doorway is cut in the lower part of a watertight bulkhead care must be taken to maintain the strength of the bulkhead. The opening is to be framed and reinforced if the vertical stiffeners are cut in way of the opening. If the stiffener spacing is increased to accommodate the opening, the scantlings of the stiffeners on either side of the opening are increased to give an equivalent strength to that of an unpierced bulkhead. The actual opening is kept as small as possible, the access to the shaft tunnel being about 1000–1250 mm high and about 700 mm wide. In passenger accommodation the openings would obviously be somewhat larger.

Mild steel or cast steel watertight doors fitted below the waterline are either of the vertical or horizontal sliding type. A swinging hinged type of door could prove impossible to close in the event of flooding and is not permitted. The sliding door must be capable of operation when the ship is listed 15°, and be opened or closed from the vicinity of the door as well as from a position above the bulkhead deck. At this remote control position an indicator must be provided to show whether the door is open or closed. Vertical sliding doors may be closed by a vertical screw thread, which is turned by a shaft extending above the bulkhead and fitted with a crank handle. This screw thread turns in a gunmetal nut attached to the top of the door, and a crank handle is also provided at the door to allow it to be closed from this position. Often horizontal sliding doors are fitted, and these may have a vertical shaft extending above the bulkhead deck, which may be operated by hand from above the deck or at the door. This can also be power driven by an electric motor and worm gear, the vertical shaft working through bevel wheels, and horizontal screwed shafts turning in bronze nuts on the door. The horizontal sliding door may also be opened and closed by a hydraulic ram with a hydraulic hand pump and with control at the door and above the bulkhead deck (see Figure 18.4). With the larger number of watertight doors fitted in passenger ships the doors may be closed by means of hydraulic power actuated by remote control from a central position above the bulkhead deck.

When in place all watertight doors are given a hose test, but those in a passenger ship are required to be tested under a head of water extending to the bulkhead deck. This may be done before the door is fitted in the ship.

In approved positions in the upper tween decks well above the waterline, hinged watertight doors are permitted. These may be similar to the weathertight doors fitted in superstructures, but are to have gunmetal pins in the hinges.

Deep tanks

Deep tanks were often fitted adjacent to the machinery spaces amidships to provide ballast capacity, improving the draft with little trim, when the ship was light. These tanks were frequently used for carrying general cargoes and also utilized to carry specialist liquid cargoes. In cargo liners where the carriage of certain liquid cargoes is common practice it was often an advantage to have the deep tanks adjacent to the machinery space for cargo heating purposes. However, in modern cargo ships they may need to be

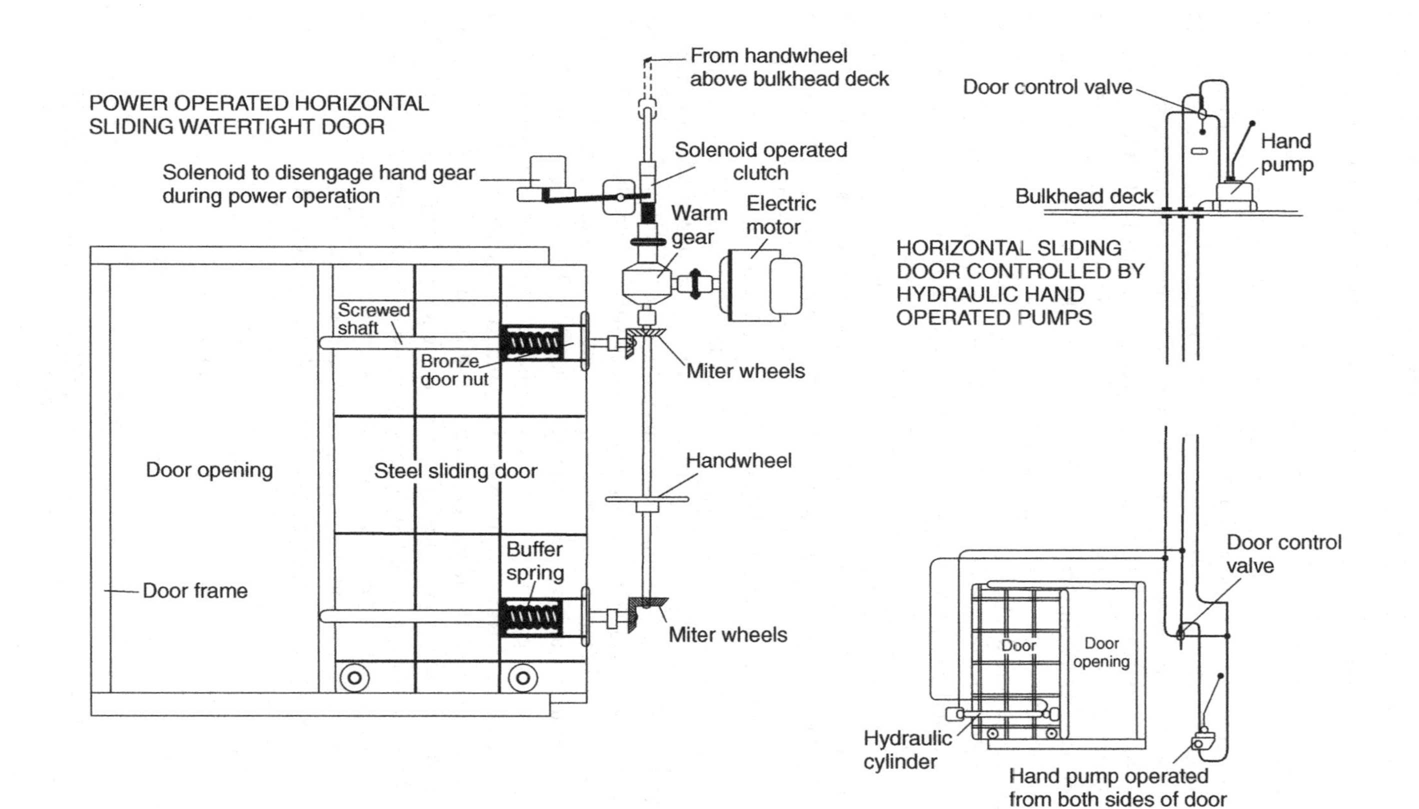

Figure 18.4 Watertight doors.

judiciously placed in order to avoid excessive stresses in different conditions of loading. Most ships now have their machinery arranged aft or occasionally three-quarters aft, and are fitted with deep tanks forward to improve the trim in the light conditions.

Construction of deep tanks

Bulkheads that form the boundaries of a deep tank differ from hold bulkheads in that they are regularly subjected to pressure from a head of liquid. The conventional hold bulkhead may be allowed to deflect and tolerate high stresses on the rare occasions when it has to withstand temporary flooding of a hold, but deep tank bulkheads, which are regularly loaded in this manner, are required to have greater rigidity and be subject to lower stresses. As a result the plate and stiffener scantlings will be larger in way of deep tanks, and additional stiffening may be introduced.

The greater plating thickness of the tank boundary bulkheads increases with tank depth, and with increasing stiffener spacing. To provide the greater rigidity the vertical stiffeners are of heavier scantlings and more closely spaced. They must be bracketed or welded to some other form of stiffening member at their ends. Vertical stiffener sizes may be reduced, however, by fitting horizontal girders that form a continuous line of support on the bulkheads and ship's side. These horizontal girders are connected at their ends by flanged brackets and are supported by tripping brackets at the toes of the end brackets, and at every third stiffener or frame. Intermediate frames and stiffeners are effectively connected to the horizontal girders (see Figure 18.5).

Where deep tanks are intended to carry oil fuel for the ship's use, or oil cargoes, there will be a free surface, and it is necessary to fit a center-line bulkhead where the tanks extend from side to side of the ship. This is to reduce the free surface effect. The center-line bulkhead may be intact or perforated, and where intact the scantlings will be the same as for boundary bulkheads. If perforated, the area of perforations is sufficient to reduce liquid pressures, and the bulkhead stiffeners have considerably reduced scantlings, surging being avoided by limiting the perforation area.

Both swedged and corrugated plating can be used to advantage in the construction of deep tanks since, without the conventional stiffening, tanks are more easily cleaned. However, the use of corrugations does cause problems in ship construction. With conventional welded stiffening it may be convenient to arrange the stiffeners outside the tank so that the boundary bulkhead has a plain inside for ease of cleaning.

In cargo ships where various liquid cargoes are carried, arrangements may be made to fit cofferdams between deep tanks. As these tanks may also be fitted immediately forward of the machine space, a pipe tunnel is generally fitted through them with access from the engine room. This tunnel carries the bilge piping as it is undesirable to pass this through the deep tanks carrying oil cargoes.

Testing deep tanks

Deep tanks are tested by subjecting them to the maximum head of water to which they might be subject in service (i.e. to the top of the air pipe). This should not be less than 2.45 m above the crown of the tank.

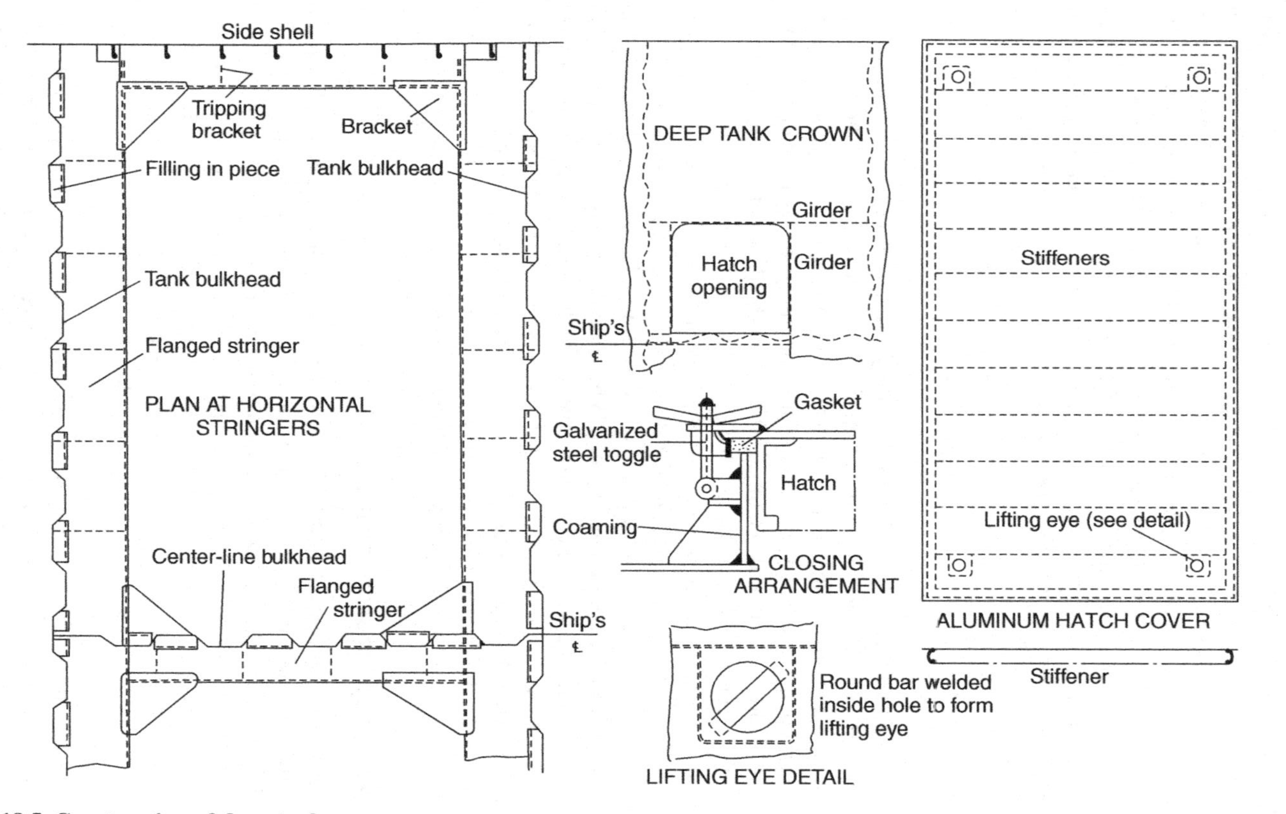
Side shell
Tripping bracket
Bracket
Filling in piece
Tank bulkhead
Tank bulkhead
Flanged stringer
PLAN AT HORIZONTAL STRINGERS
Center-line bulkhead
Flanged stringer
Ship's ℄
DEEP TANK CROWN
Girder
Hatch opening
Girder
Ship's ℄
Gasket
Galvanized steel toggle
Hatch
Coaming
CLOSING ARRANGEMENT
Stiffeners
Lifting eye (see detail)
ALUMINUM HATCH COVER
Stiffener
Round bar welded inside hole to form lifting eye
LIFTING EYE DETAIL

Figure 18.5 Construction of deep tanks.

Topside tanks

Standard general bulk carriers are fitted with topside tanks that may be used for water ballast, and in some instances are used for the carriage of light grains. The thickness of the sloping bulkhead of this tank is determined in a similar manner to that of the deep tank bulkheads. In present practice, as indicated in Chapter 17, the topside tank is generally stiffened internally by longitudinal framing supported by transverses (see Figure 17.7). Transverses are arranged in line with the end of the main cargo hatchways, and in large ships a fore and aft diaphragm may be fitted at half the width of the tank, between the deck and the sloping plating.

Shaft tunnel

When the ship's machinery is not located fully aft it is necessary to enclose the propeller shaft or shafts in a watertight tunnel between the aft end of the machinery space and the aft peak bulkhead. This protects the shaft from the cargo and provides a watertight compartment that will contain any flooding resulting from damage to the watertight gland at the aft peak bulkhead. The tunnel should be large enough to permit access for inspection and repair of the shafting. A sliding watertight door that may be opened from either side is provided at the forward end in the machinery space bulkhead. Two means of escape from the shaft tunnel must be provided, and as a rule there is a ladder in a watertight trunk leading to an escape hatch on the deck above the waterline, at the aft end of the shaft tunnel. Where the ship narrows at its after end the aftermost hold may be completely plated over at the level of the shaft tunnel to form a tunnel flat, as the narrow stowage space either side of the conventional shaft tunnel cannot be utilized. The additional space under this tunnel flat is often used to stow the spare tail shaft. Shaft tunnels also provide a convenient means of carrying piping aft, which is then accessible and protected from cargo damage.

Construction of the shaft tunnel

The thickness of the tunnel plating is determined in the same manner as that for the watertight bulkheads. Where the top of the tunnel is well rounded the thickness of the top plating may be reduced, but where the top is flat it is increased. Under hatchways the top plating must be increased in thickness unless it is covered by wood of a specified thickness. Vertical stiffeners supporting the tunnel plating have similar scantlings to the watertight bulkhead stiffeners and their lower end is welded to the tank top (see Figure 18.6). On completion the shaft tunnel structure is subject to a hose test.

At intervals along the length of the shaft, stools are built that support the shaft bearings. A walkway is installed on one side of the shaft to permit inspection and, as a result, in a single screw ship the shaft tunnel will be offset from the ship's center line. This walkway is formed by gratings laid on angle bearers supported by struts etc.; any piping is then led along underneath the walkway.

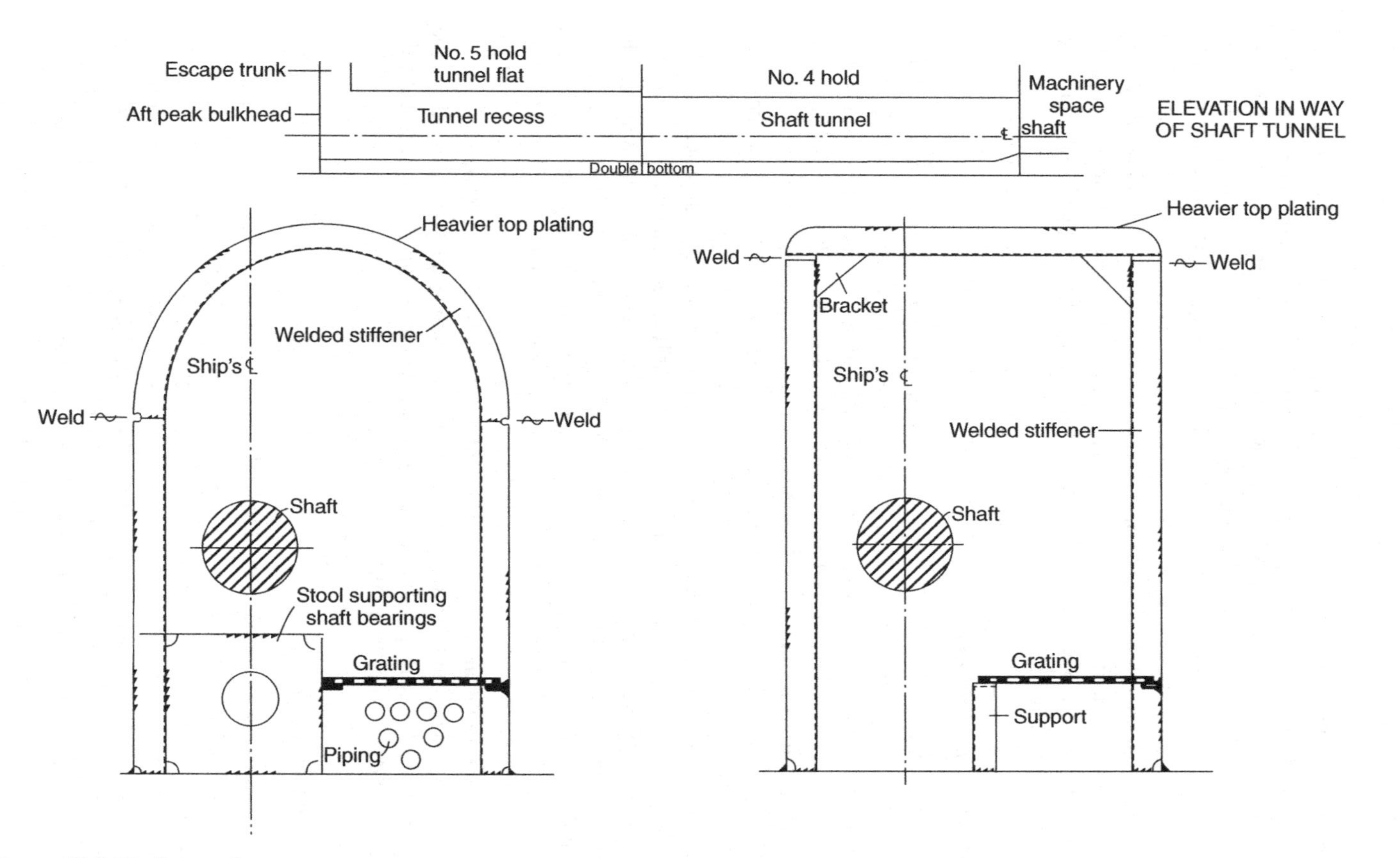

Figure 18.6 Shaft tunnels.

Pillars

The prime function of the pillars is to carry the load of the decks and weights upon the decks vertically down to the ship's bottom structure where these loads are supported by the upward buoyant forces. A secondary function of pillars is to tie together the structure in a vertical direction. Within the main hull of a cargo ship two different forms of pillar may be found, those in the holds invariably fulfilling the first function and those in the machinery spaces fulfilling the latter function. Hold pillars, primarily in compression, are often without bracket connections at their ends, whilst machinery space pillars are heavily bracketed at their ends to permit tensile loadings. This latter type of pillar may also be found in tank spaces where the crown of tank under pressure can put the pillar in tension.

Spacing of hold pillars

Since pillars located in holds will interfere with the stowage arrangements, widely spaced pillars of large fabricated section are used rather than small, solid, closely spaced pillar systems. The arrangement most often found in cargo ships is a two-row pillar system, with pillars at the hatch corners or mid-length of hatch supporting deck girders adjacent to the hatch sides. As the deck girder size is to some extent dependent on the supported span, where only a mid-hatch length pillar is fitted the girder scantlings will be greater than that where two hatch corner pillars are fitted. In fact, pillars may be eliminated altogether where it is important that a clear space should be provided, but the deck girder will then be considerably larger, and may be supported at its ends by webs at the bulkhead. Substantial transverse cantilevers may also be fitted to support the side decks. Pillars may also be fitted in holds on the ship's center line at the hatch end, to support the heavy hatch end beams securely connected to and supporting the hatch side girders. In a similar position it is not unusual to find short corrugated fore and aft pillar bulkheads. These run from the forward or aft side of the hatch opening to the adjacent transverse bulkhead on the ship's center line.

To maintain continuity of loading the tween pillars are arranged directly above the hold pillars. If this is not possible stiffening arrangements should be made to carry the load from the tween pillar to the hold pillar below.

Pillar construction

It has already been seen that the hold pillar is primarily subject to a compressive loading, and if buckling is to be avoided in service the required cross-section must be designed with both the load carried and length of pillar in mind. The ideal section for a compressive strut is the tubular section and this is often adopted for hold pillars, hollow rectangular and octagonal sections also being used. For economic reasons the sections are fabricated in lengths from steel plate, and for the hollow rectangular section welded channels or angles may also be used (see Figure 18.7). A small flat bar or cope bar may be tack welded inside these pillar sections to allow them to be welded externally.

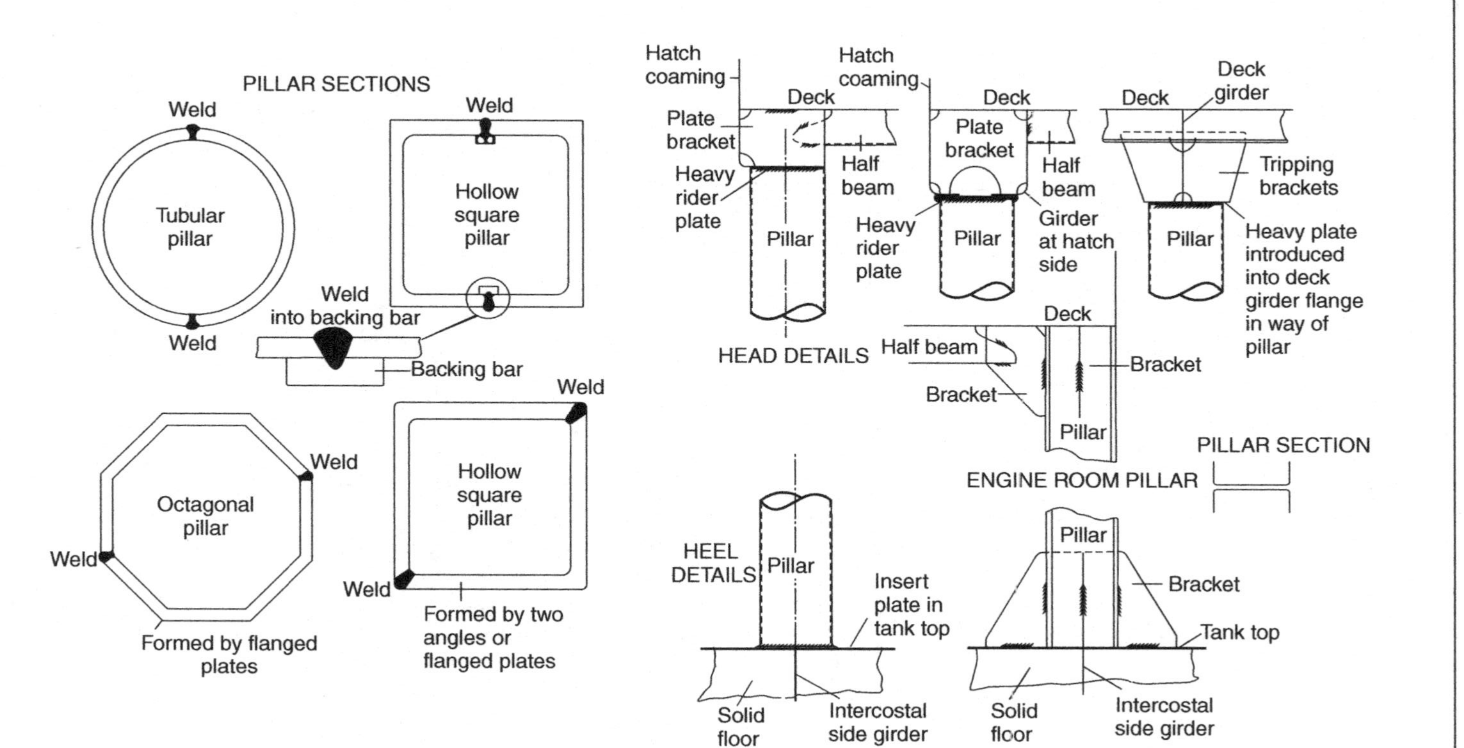

Figure 18.7 Pillars.

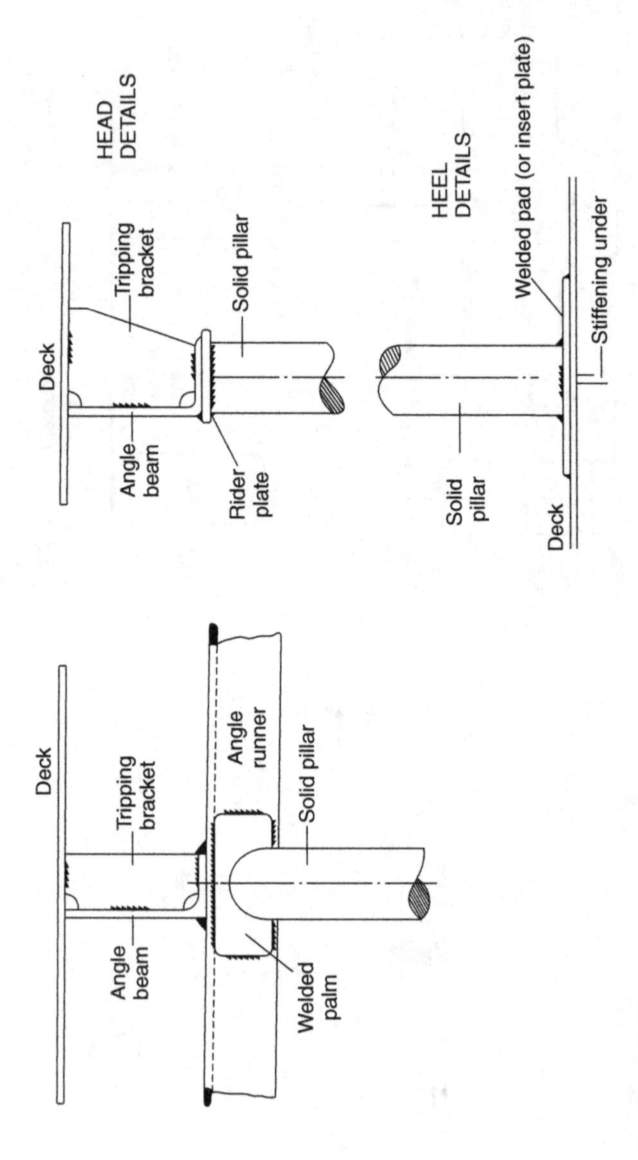

Figure 18.8 Small solid pillars.

Pillars have a bearing fit, and it is important that the loads at the head and heel of the pillar should be well distributed. At the head of the pillar a continuous weld is made to a doubling plate supported by brackets. Details of the head fitting vary from ship to ship and depend very much on the form of hatch side or deck girder that they support. The heel of the hold pillar lands on a heavy doubling or insert plate at the tank top and it is commonly arranged that the point of loading will coincide with a solid floor/side girder intersection in the double bottom below. Where this is not possible partial floors and short intercostal side girders may be fitted to distribute the load.

Machinery space pillars are fabricated from angles, channels, or rolled steel joists, and are heavily bracketed to suitably stiffened members (Figure 18.7).

Small pillars

Within the accommodation and in relatively small vessels, solid round steel pillars having diameters seldom exceeding 150 mm may be fitted. These may have forged palms at their head and heel, the head being welded to a continuous angle fore and aft runner that supports the deck. Alternatively, the pillar head may have a direct continuous weld connection to an inverted angle beam or deck girder, with suitable tripping brackets fitted directly above. The heel is then directly welded to the deck, which is suitably stiffened below (see Figure 18.8).

Rolled hollow steel section pillars of similar size with direct welded head and heel fittings are commonly used today in lieu of small solid pillars.

Further reading

Common Structural Rules for Bulk Carriers. IACS publication.

SOLAS Consolidated Edition, Chapter XII: Additional safety measures for bulk carriers, 2004.

Ultimate strength analysis of bulk carrier corrugated bulkheads, The Naval Architect (August 2002).

Some useful websites

www.iacs.org.uk Copies of the IACS's *Common Structural Rules for Bulk Carriers* are available on this website.

19 Decks, hatches, and superstructures

Chapter Outline

Decks 226
- Deck plating 226
- Deck stiffening 229

Hatches 229
- Hatch coamings 232
- Hatch covers 232

Bulwarks 235
- Construction of bulwarks 235

Superstructures and deckhouses 237
- Forecastle 237
- Bridge structures 238
- Poop structure 238
- Passenger ship superstructures 238
- Weathertight doors 240

Further reading 240
Some useful websites 240

Decks at different levels in a ship serve various functions; they may be either watertight decks, strength decks, or simply cargo and passenger accommodation decks. Watertight decks are fitted to maintain the integrity of the main watertight hull, and the most important is the freeboard deck, which is the uppermost deck having permanent means of closing all openings in the exposed portions of that deck. Although all decks contribute to some extent to the strength of the ship, the most important is that which forms the upper flange of the main hull girder, called the 'strength deck'. Lighter decks that are not watertight may be fitted to provide platforms for passenger accommodation and permit more flexible cargo-loading arrangements. In general cargo ships these lighter decks form tweens that provide spaces in which goods may be stowed without their being crushed by a large amount of other cargo stowed above them.

To permit loading and discharging of cargo, openings must be cut in the decks, and these may be closed by nonwatertight or watertight hatches. Other openings are required for personal access through the decks, and in way of the machinery space casing openings are provided that allow the removal of machinery items when necessary, and also provide light and air to this space. These openings are protected

Ship Construction. DOI: 10.1016/B978-0-08-097239-8.00019-2

by houses or superstructures, which are extended to provide accommodation and navigating space. Forward and aft on the uppermost continuous deck a forecastle and often a poop may be provided to protect the ends of the ship at sea.

Decks

The weather decks of ships are cambered to facilitate the draining of water from the decks in heavy weather. The camber may be parabolic or straight,the latter being preferred in many cases to allow panel lines to be used for the production of the deck panels. There may be advantages in fitting horizontal decks in some ships, particularly if containers are carried and regular cross-sections are desired. Short lengths of internal deck or flats are as a rule horizontal.

Decks are arranged in plate panels with transverse or longitudinal stiffening, and local stiffening in way of any openings. Longitudinal deck girders may support the transverse framing, and deep transverses the longitudinal framing (see Figure 19.1).

Deck plating

The heaviest deck plating will be found abreast the hatch openings of the strength deck. Plating that lies within the line of the hatch openings contributes little to the longitudinal strength of the deck and it is therefore appreciably lighter. As the greatest longitudinal bending stresses will occur over the midship region, the greatest deck plate thickness is maintained over 40% of the length amidships, and it tapers to a minimum thickness permitted at the ends of the ship. Locally, the plating thickness may be increased where higher stresses occur owing to discontinuities in the structure or concentrated loads.

Other thickness increases may occur where large deck loads are carried, where fork-lift trucks or other wheeled vehicles are to be used, and in way of deep tanks. Where the strength deck plating exceeds 30 mm it is to be Grade B steel and if it exceeds 40 mm Grade D over the midships region, at the ends of the superstructure, and in way of the cargo hold region in container ships. The stringer plate (i.e. the strake of deck plating adjacent to the sheerstrake) of the strength deck, over the midship region and container ship cargo hold area, of ships less than 260 meters in length, is to be of Grade B steel if 15–20 mm thick, Grade D if 20–25 mm thick, and Grade E if more than 25 mm thick. Where the steel deck temperatures fall below 0 °C in refrigerated cargo ships the steel will be of Grades B, D, and E depending on thickness.

On decks other than the strength deck, the variation in plate thickness is similar, but lighter scantlings are in use.

Weather decks may be covered with wood sheathing or an approved composition (usually composition for reasons of economy), which not only improves their appearance, but also provides protection from heat in way of any accommodation. Since wood provides some additional strength, reductions in the deck plate thickness are permitted, and on superstructure decks the plating thickness may be further

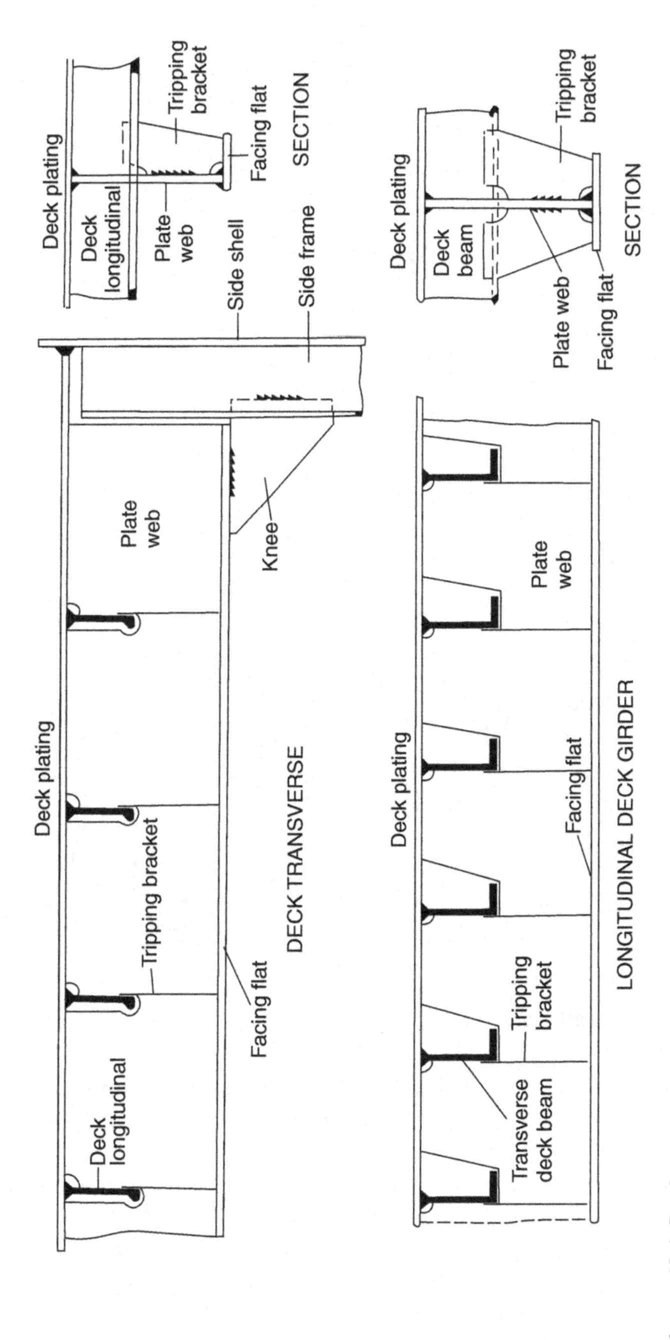

Figure 19.1 Deck supports.

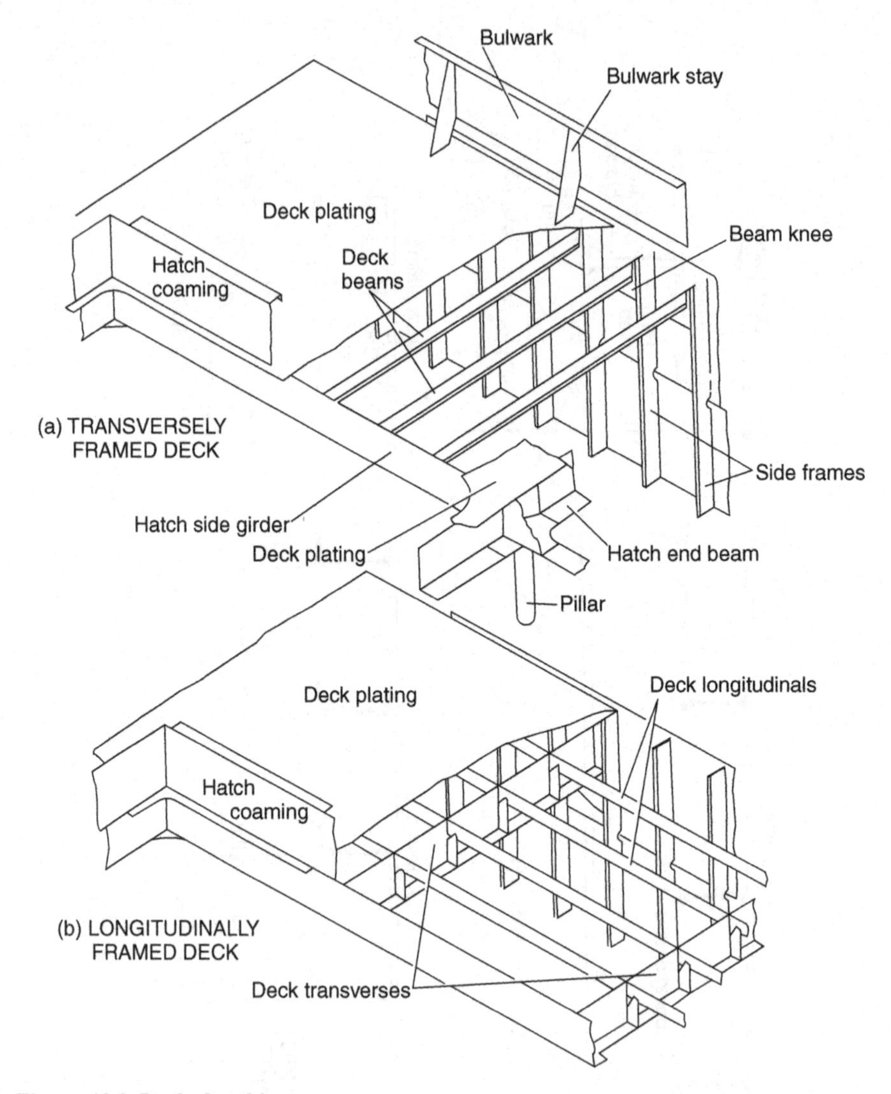

Figure 19.2 Deck sheathing.

decreased within deckhouses, if sheathed. Before fitting any form of sheathing the deck is treated to prevent corrosion between the deck plating and sheathing (see Figure 19.2).

Any openings abreast the hatch openings in a deck are kept to a minimum and clear of the hatch corners. If such openings are cut, compensation is required to restore the sectional area of deck. All large openings in the decks have well-rounded corners, to avoid stresses, with insert plates fitted, unless the corners are parabolic or elliptical with the major axis fore and aft, local stress concentrations being reduced if the latter type of corner is cut (see Figure 19.4).

Deck stiffening

Decks may be framed transversely or longitudinally, but outside the line of openings it is preferred that longitudinal framing should be adopted for the strength deck.

When the decks are longitudinally framed the scantlings of the longitudinals are dependent on their spacing, the length of ship, whether they are inside or outside the line of hatch openings, their span, and the deck loading. Deck transverses support the longitudinals, and these are built from a deep web plate with flange or welded face flat, and are bracketed to the side frame (Figure 19.1). Within the forward 7.5% of the ship's length, the forecastle and weather deck transverses are closely spaced and the longitudinal scantlings increased, the additional transverse and longitudinal stiffening forward being designed to avoid buckling of the deck plating on impact when shipping seas.

Transversely framed decks are fitted with deck beams at every frame, and these have scantlings that are dependent on their span, spacing, and location in the ship. Those fitted right forward on weather decks, like the longitudinal framing forward, have heavier scantlings, and the frame spacing is also decreased in this region so they will be closer together. Beams fitted in way of deep tanks, peak tanks, and oil bunkers may also have increased scantlings as they are required to have the same rigidity as the stiffeners of the tank boundary bulkheads. Deck beams are supported by longitudinal deck girders that have similar scantlings to deck transverses fitted with any longitudinal framing. Within the forward 7.5% of the ship's length these deck girders are more closely spaced on the forecastle and weather decks. Elsewhere the spacing is arranged to suit the deck loads carried and the pillar arrangements adopted. Each beam is connected to the frame by a 'beam knee' and abreast the hatches 'half beams' are fitted with a suitable supporting connection at the hatch side girder (see Figure 19.3).

Both longitudinals and deck beam scantlings are increased in way of cargo decks where fork-lift trucks and other wheeled vehicles, which cause large point loads, are used.

In way of the hatches fore and aft side girders are fitted to support the inboard ends of the half beams and transverses. At the ends of the hatches heavy transverse beams are fitted and these may be connected at the intersection with the hatch side girder by horizontal gusset plates (Figure 19.4). Where the deck plating extends inside the coamings of hatches amidships the side coaming is extended in the form of tapered brackets.

Hatches

The basic regulations covering the construction and means of closing hatches in weathertight decks are contained within the Conditions of Assignment of Freeboard of the Load Line Convention (see Chapter 31). Lloyd's Register provides formulae for determining the minimum scantlings of steel covers, which will be within the requirements of the Load Line Convention. Only the maximum permitted stresses and

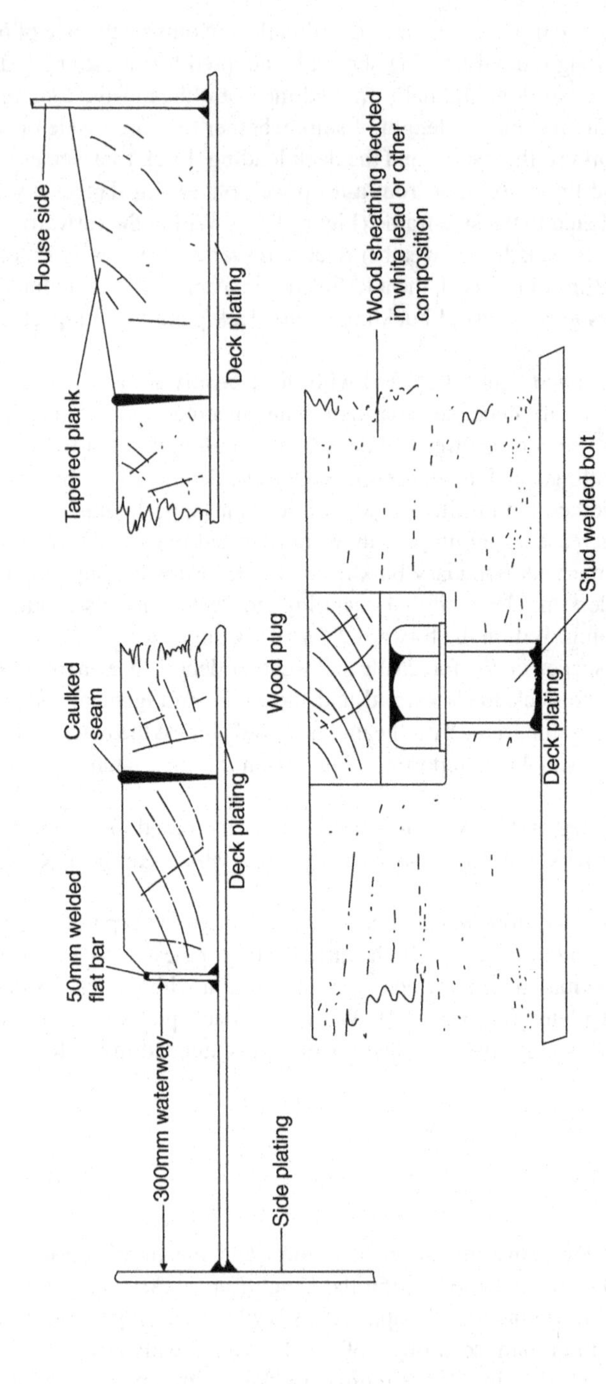

Figure 19.3 Deck construction.

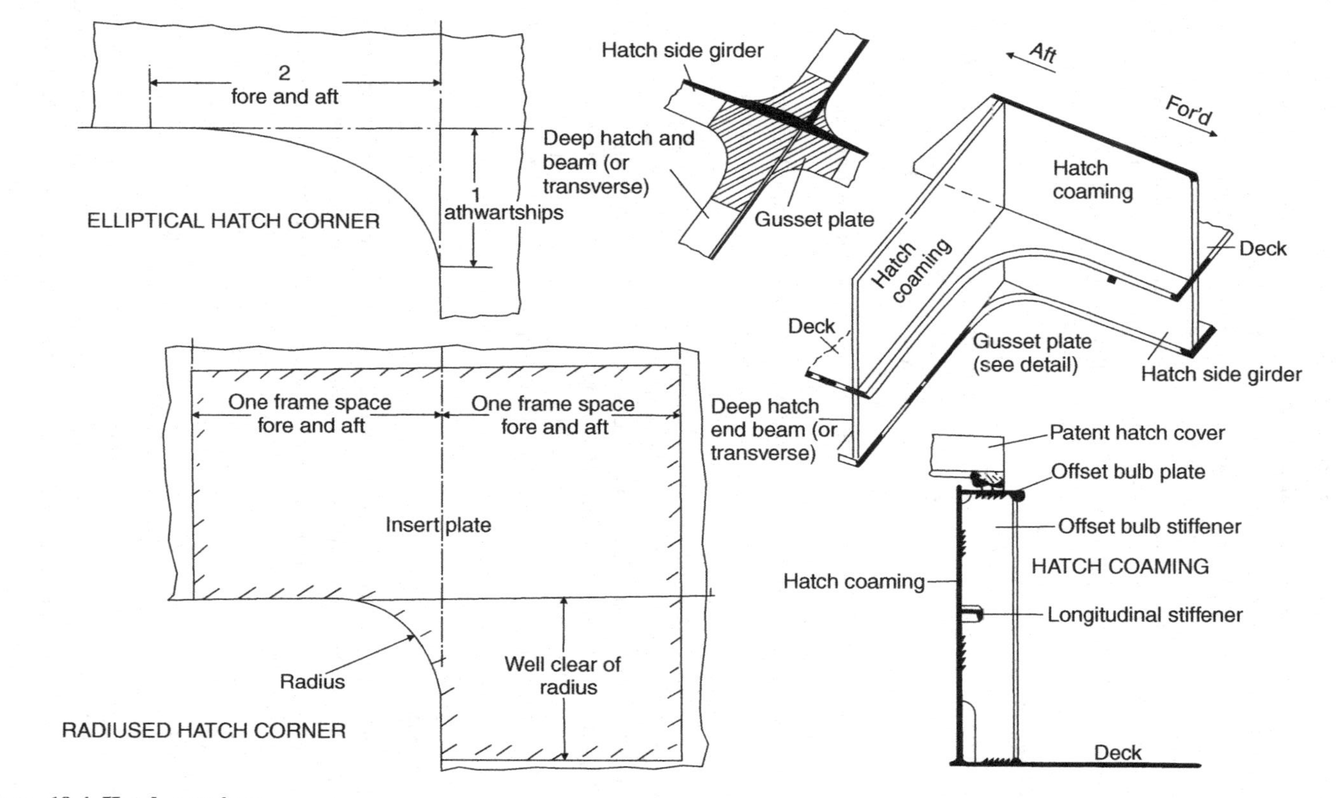

Figure 19.4 Hatch openings.

deflections of covers under specified loadings are given by the Load Line Convention. Under the Convention, ships fitted with approved steel covers having direct securing arrangements may have reduced B-100 or B-60 freeboards if they meet the subdivision requirements, but in general they are assigned standard cargo ship Type B freeboards. If steel pontoon-type covers that are self-supporting and have no direct securing arrangements are fitted, then the standard Type B freeboard only is assigned. The use of portable beams fitted with wood or light steel covers and tarpaulins as a traditional hatch covering has almost disappeared. If used, then the ship has an increased Type B freeboard, i.e. there is a draft penalty. This means that most ships are fitted exclusively with the stronger stiffened self-supporting steel covers.

Hatch coamings

Heights of coamings and cover closing arrangements in some instances depend on the hatch position. The positions differentiate between regions that are more exposed than others. Position 1 indicates that the hatch is on the exposed freeboard deck, raised quarter deck, or superstructure decks within 25% of the ship's length from forward. Position 2 indicates that the hatch is located on exposed superstructure decks abaft the forward 25% of the ship's length.

Hatches that are at Position 1 have coamings at least 600 mm high and those at Position 2 have coamings at least 450 mm high, the height being measured above any deck covering. Provision is made for lowering these heights or omitting the coaming altogether if directly secured steel covers are fitted and it can be shown that the safety of the ship would not be impaired in any sea condition. Where the coaming height is 600 mm or more the upper edge is stiffened by a horizontal bulb flat and supporting stays to the deck are fitted. Coamings less than 600 mm high are stiffened by a cope or similar bar at their upper edge. The steel coamings extend down to the lower edge of the deck beams, which are then effectively attached to the coamings (Figure 19.4).

Hatch covers

A number of patent steel covers, such as those manufactured by MacGregor Cargotec Group AB and TTS Marine ASA, are available that will comply with the requirements outlined by the International Conference on Load Lines 1966 and are in accordance with the requirements of the classification societies. The means of securing the hatches and maintaining their watertightness is tested initially and at periodic surveys. These patent covers vary in type, the principal ones being fore and aft single pull, folding, piggy-back, pontoon, and side rolling. These are illustrated in Figures 19.5 and 19.6. Single pull covers may be opened or closed by built-in electric motors in the leading cover panel (first out of stowage) that drive chain wheels, one on each outboard side of the panel. Each panel wheel is permanently engaged on a fixed chain located along each hatch side coaming. In operation the leading panel pushes the others into stowage and pulls them into the closed position. Alternatively, single pull covers are opened or closed by hydraulic or electric motors situated on the hatch

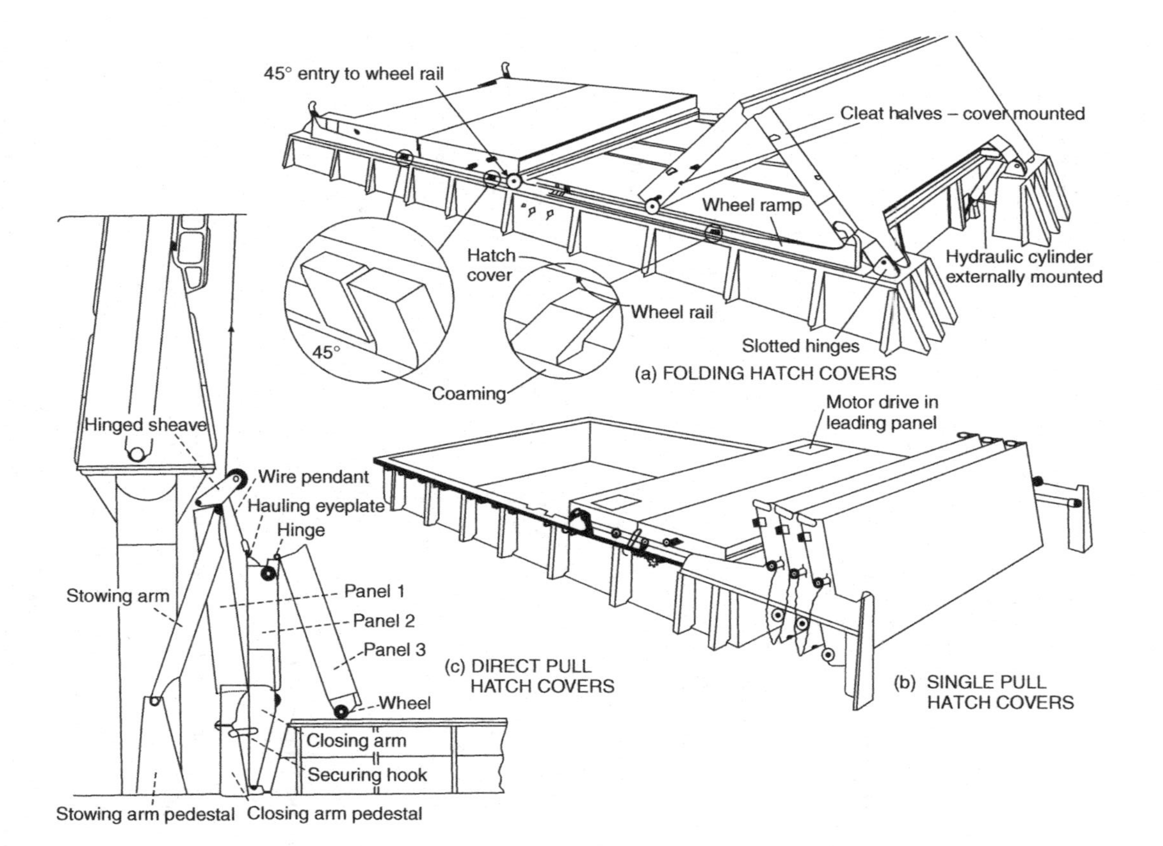

Figure 19.5 Hatch covers—1.

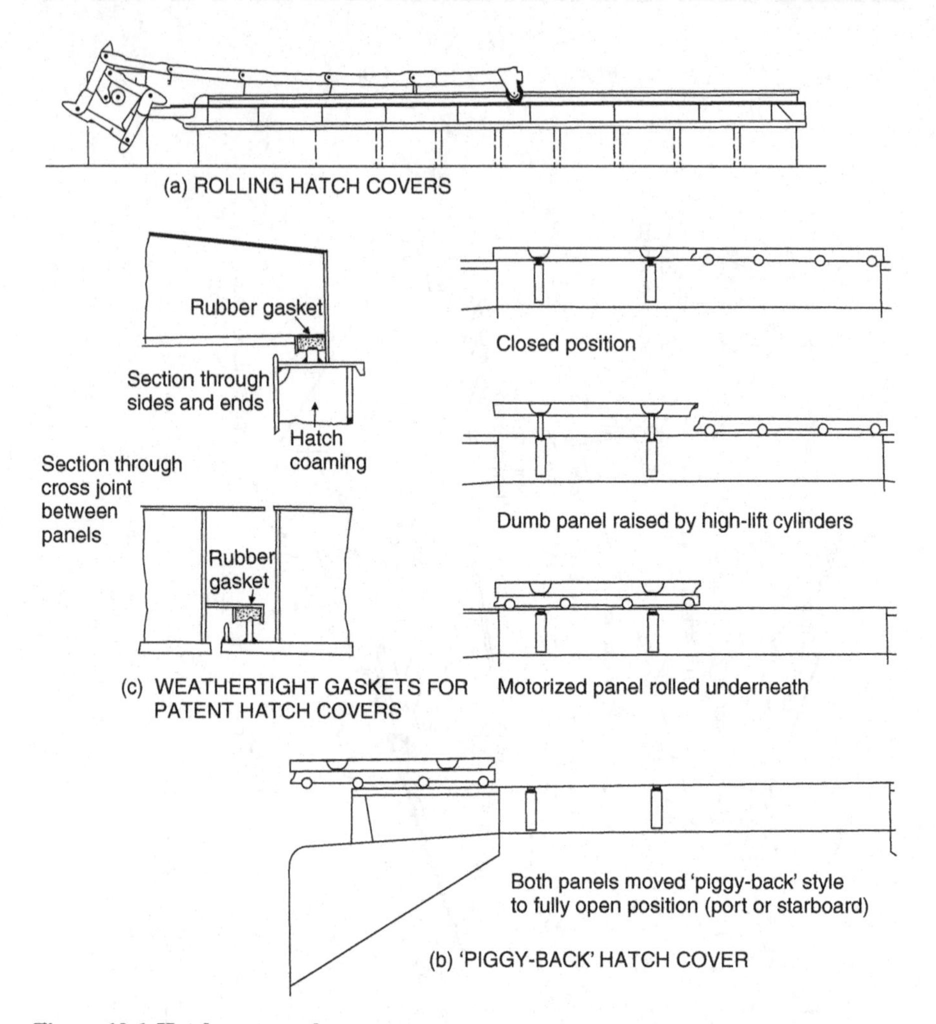

Figure 19.6 Hatch covers—2.

end coaming at the ship's center line, driving endless chains running along the full length of the hatch side coaming port and starboard and connected to the leading panel. Vertical stowage of panels is at one end of the hatch and covers may have a nesting characteristic if space is at a premium; also, on large hatches opening may be to both ends with vertical stowage at each end. Folding covers may be of direct pull type where suitable lifting gear is carried onboard or can be opened or closed by externally mounted hydraulic cylinders actuating the leading panels. Piggyback covers permit horizontal stowage of panels, avoiding fouling of lifting devices particularly in way of very large openings such as on bulk carriers and container ships, where the hatch need only be partially open for working. The covers consist of a dumb panel that is raised by high lift cylinders and a motorized panel that is rolled underneath the dumb panel. Both panels can then be moved 'piggy-back' style to the fully opened hatch position port or starboard or partially opened position fore and aft. Pontoon covers are commonly used on container ships, being lifted by the ships' or shore cranes with the container spreader. They are closed weathertight in a similar manner to the other patent covers. Side rollings are very common on large bulk carriers and can operate on similar principles to the single pull cover except that they remain in the horizontal stowed position when the hatch is open. Various other forms of cover are marketed, and tween deck steel covers are available to be fitted flush with the deck, which is essential nowadays when stowing cargoes in the tweens. To obtain weathertightness the patent covers have mating boundaries fitted with rubber gaskets; likewise at the hatch coamings, gaskets are fitted and hand or automatically operated cleats are provided to close the covers (see Figure 19.6). The gasket and cleat arrangements will vary with the type of cover.

Pontoon covers of steel with internal stiffening may be fitted, these being constructed to provide their own support without the use of portable beams. Each pontoon section may span the full hatch width, and cover perhaps one-quarter of the hatch length. They are strong enough to permit Type B freeboards to be assigned to the ship, but to satisfy the weathertightness requirements they are covered with tarpaulins and battening devices.

Where portable beams are fitted wood or stiffened steel plate covers may be used. These and the stiffened beams have the required statutory scantlings but an increase in the freeboard is the penalty for fitting such covers. The beams sit in sockets at the hatch coamings and the covers are protected by at least two tarpaulins. At the coaming the tarpaulins are made fast by battens and wedges fitted in cleats, and the sections of the cover are held down by locked bars or other securing arrangements (see Figure 19.7).

Bulwarks

Bulwarks fitted on weather decks are provided as protection for personnel and are not intended as a major structural feature. They are therefore of light scantlings, and their connections to the adjacent structures are of some importance if high stresses in the bulkwarks are to be avoided. Freeing ports are cut in bulwarks, forming wells on decks in order that water may quickly drain away. The required area of freeing ports is in accordance with the Load Line Convention requirement (see Chapter 31).

Construction of bulwarks

Bulwarks should be at least 1 m high on the exposed freeboard and superstructure decks, but a reduced height may be permitted if this interferes with the working of the ship. The bulwark consists of a vertical plate stiffened at its top by a strong rail section (often a bulb angle or plate) and is supported by stays from the deck (Figure 19.7). On the forecastle of ships assigned B-100 and B-60 freeboards the stays are more closely

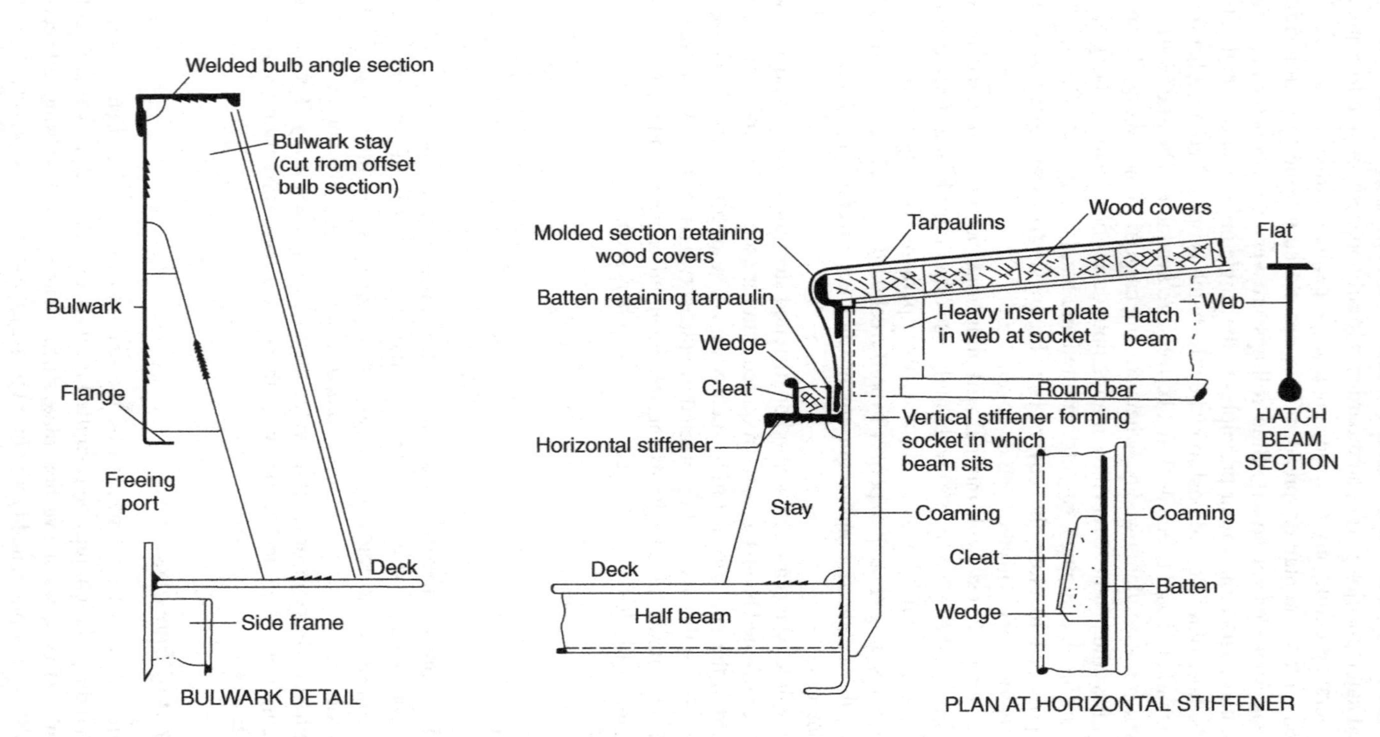

Figure 19.7 Bulwark detail and hatch with portable beams and wood covers.

spaced. Where the bulwark is cut for any reason, the corners are well rounded and compensation is also provided. No openings are permitted in bulwarks near the end of superstructures.

Superstructures and deckhouses

Superstructures may be defined as those erections above the freeboard deck that extend to the ship's side or almost to the side. Deckhouses are those erections on deck that are well within the line of the ship's side. Both structures are of importance in the assignment of the load line as they provide protection for the openings through the freeboard deck. Of particular importance in this respect are the end bulkheads of the superstructures, particularly the bridge front, which is to withstand the force of any seas shipped. The bridge structure amidships or the poop aft are, in accordance with statutory regulations, provided as protection for the machinery openings. It is possible, however, to dispense with these houses or superstructures and increase considerably the scantlings of the exposed machinery casing, but in other than very small vessels it is unlikely that such an arrangement would be adopted. Unless an excessive sheer is provided on the uppermost deck it is necessary to fit a forecastle forward to give added protection in a seaway. Each structure is utilized to the full, the after structure carrying virtually all the accommodation in modern ships. The crew may be located all aft in the poop structure or partly housed in any bridge structure with the navigating spaces. Passenger liners have considerable areas of superstructures covering tiers of decks and these will house the majority of passengers and some of the crew.

Of great structural importance is the strength of the vessel where superstructures and deckhouses terminate and are noncontinuous. At these discontinuities, discussed in Chapter 8, large stresses may arise and additional strengthening will be required locally as indicated in the following notes on the construction. Long superstructures exceeding 15% of the ship's length and extending within 50% of the vessel's length amidships receive special consideration as they contribute to the longitudinal strength of the ship, and as such must have scantlings consistent with the main hull strength members.

Forecastle

Sea-going ships must be fitted with a forecastle that extends at least 7% of the ship's length aft of the stem, and a minimum height of the bow at the forecastle deck above the summer load line is stipulated. By increasing the upper deck sheer at the forward end to obtain the same height of bow, the forecastle might be dispensed with, but in practice this construction is seldom found. The side and end plating of the forecastle has a thickness that is dependent on the ship's length and the frame and stiffener spacing adopted, the side plating being somewhat heavier than the aft end plating. If a long forecastle is fitted such that its end bulkhead comes within 50% of the ship's length amidships, additional stiffening is required.

Bridge structures

The side of bridge superstructures whose length exceeds 15% of the ship's length will have a greater thickness than the sides of other houses, the scantling being similar to that required for the ship's side shell. All bridge superstructures and midship deck-houses will have a heavily plated bridge front, and the aft end plating will be lighter than the front and sides. Likewise the stiffening members fitted at the forward end will have greater scantlings than those at the sides and aft end. Additional stiffening in the form of web frames or partial bulkheads will be found where there are large erections above the bridge deck. These are intended to support the sides and ends of the houses above and are preferably arranged over the watertight bulkheads below. Under concentrated loads on the superstructure decks, for example under lifeboat davits, web frames are also provided.

The longer bridge superstructure that is transmitting the main hull girder stresses requires considerable strengthening at the ends. At this major discontinuity, the upper deck sheerstrake thickness is increased by 20%, the upper deck stringer plate by 20%, and the bridge side plating by 25%. The latter plating is tapered into the upper deck sheerstrake with a generous radius, as shown in Figure 19.8a, stiffened at its upper edge, and supported by webs not more than 1.5 m apart. At the ends of short bridge superstructures less strengthening is required, but local stresses may still be high and therefore the upper deck sheerstrake thickness is still increased by 20% and the upper deck stringer by 20%.

Poop structure

Where there is no midship deckhouse or bridge superstructure the poop front will be heavily constructed, its scantlings being similar to those required for a bridge front. In other ships it is relatively exposed and therefore needs adequate strengthening in all cases. If the poop front comes within 50% of the ship's length amidships, the discontinuity formed in the main hull girder is to be considerably strengthened, as for a long bridge exceeding 15% of the ship's length. Where deckhouses are built above the poop deck these are supported by webs or short transverse bulkheads in the same manner as those houses fitted amidships. The after end of any poop house will have increased scantlings since it is more exposed than other aft end house bulkheads.

Passenger ship superstructures

It is shown in Chapter 8 that with conventional beam theory the bending stress distribution is linear, increasing from zero at the neutral axis to a maximum at the upper deck and bottom. If a long superstructure is fitted the stress distribution remains linear and the strength deck is above the upper deck in way of the superstructure deck. If a short superstructure is fitted the stress distribution will be broken at the upper deck, which is the strength deck, the stresses in the superstructure deck being less than those in the upper deck. The long superstructure is referred to as an 'effective superstructure', the erections contributing to the overall strength of the hull girder, and therefore they are substantially built.

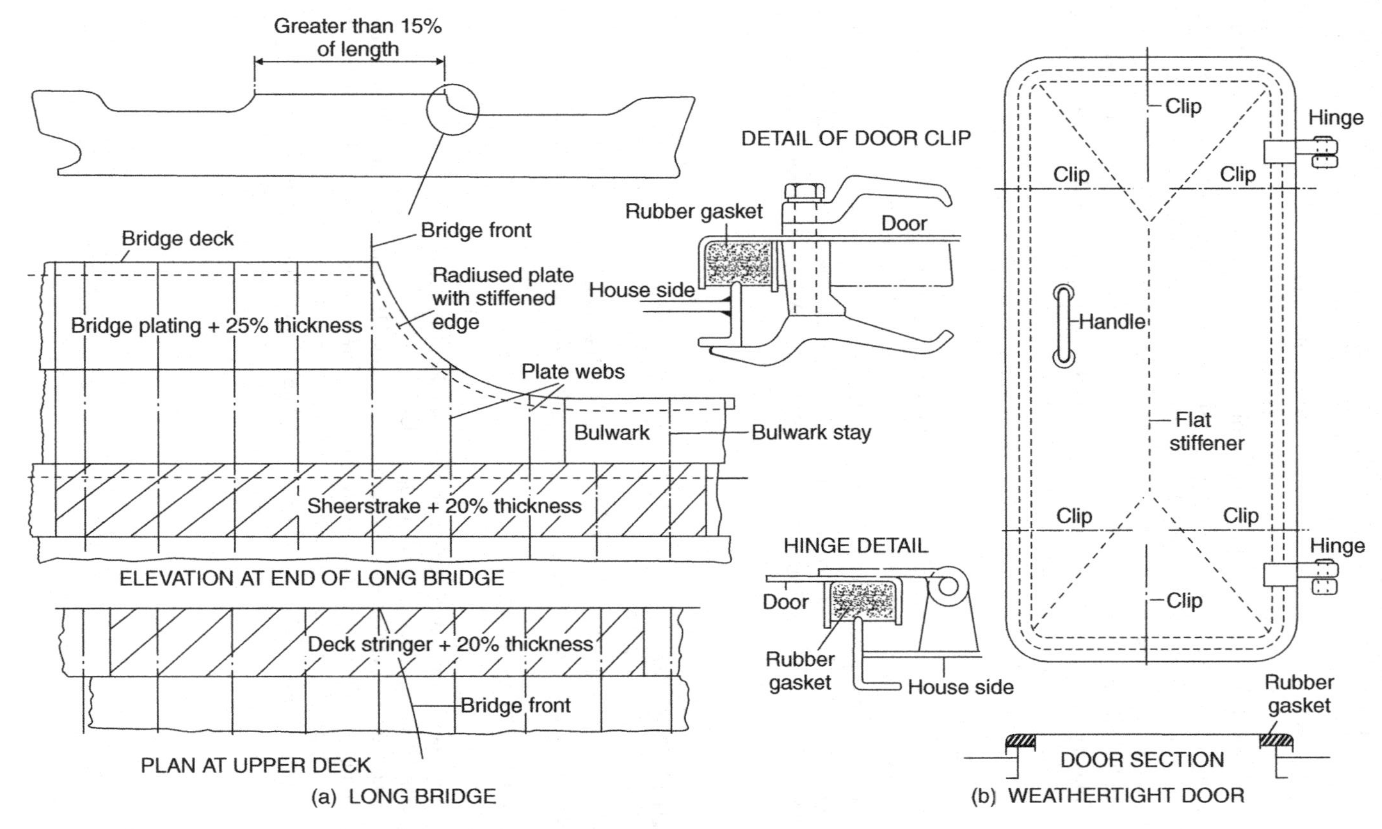

Figure 19.8 Effective superstructure connection and weathertight door.

In passenger ships with large superstructures the present practice is to make the structure effective with adequate scantlings. Some older ships have been fitted with expansion joints that are in effect transverse cuts introduced to relieve the hull bending stresses in the houses. It has been shown that at the end of deck erections the stresses do not conform to beam theory and the ends are ineffective in contributing to the longitudinal strength. The expansion joints were therefore so arranged that this 'end effect' extended from joint to joint, and lighter scantlings were then permitted for the superstructure. Unfortunately the expansion joint often provided an ideal 'notch' in the structure from which cracks initiated. Aluminum alloy superstructures offer an alternative to the use of expansion joints, since the low modulus of elasticity of the material results in lower stresses in the houses than would be the case with a steel superstructure, all other considerations being equal.

Weathertight doors

The integrity of houses on the freeboard and other decks that protect the openings in these decks must be maintained. Access openings must be provided to the houses and weathertight doors are fitted to these openings. These must comply with the requirements of the Load Line Convention (see Chapter 31) and are steel doors that may be secured and made watertight from either side. Weathertightness is maintained by a rubber gasket at the frame of the door (see Figure 19.8b).

Further reading

Common Structural Rules for Bulk Carriers. IACS publication.
International Conference on Load Lines, 1966. IMO publication (IMO 701E), 2005 edition.

Some useful websites

www.macgregor-group.com Hatch cover types and details.
www.tts-marine.com Hatch cover types and details.
www.iacs.org.uk See 'Guidelines and Recommendations': Recommendation 14 'Hatch cover securing and tightness'; Recommendation 15 'Care and survey of hatch covers of dry cargo ships—Guidance to owners'; 'Common structural rules for bulk carriers'.
http://exchange.dnv.com/publishing Includes requirements for large superstructures.

20 Fore end structure

Chapter Outline

Stem 241
Bulbous bows 243
Chain locker 244
Construction of chain locker 244
Hawse pipes 248
Bow steering arrangements 248
Bow thrust units 248
Some useful websites 248

Consideration is given in this chapter to the structure forward of the collision bulkhead. The chain locker is included as it is usually fitted forward of the collision bulkhead below the second deck or upper deck, or in the forecastle itself. An overall view of the fore end structure is shown in Figures 20.1 and 20.2, and it can be seen that the panting stiffening arrangements are of particular importance. These have already been dealt with in detail in Chapter 17 as they are closely associated with the shell plating.

On the forecastle deck the heavy windlass seating is securely fastened and given considerable support. The deck plating thickness is increased locally, and smaller pillars with heavier beams and local fore and aft intercostals, or a center-line pillar bulkhead, may be fitted below the windlass.

Stem

On many conventional ships a stem bar, which is a solid round bar, is fitted from the keel to the waterline region, and a radiused plate is fitted above the waterline to form the upper part of the stem. This forms what is referred to as a 'soft nose' stem, which in the event of a collision will buckle under load, keeping the impact damage to a minimum. Older ships had solid bar stems that were riveted and of square section, and as the stem had no rake it could cause considerable damage on impact because of its rigidity. Small ships such as tugs and trawlers may still have a solid stem bar extending to the top of the bow, and some existing large passenger ships may have steel castings or forgings forming the lower part of the stem. A specially designed bow is required for ships assigned 'Icebreaker' notation and additional scantlings are required for the stems of ships assigned other ice classes (see Chapter 17).

Ship Construction. DOI: 10.1016/B978-0-08-097239-8.00020-9

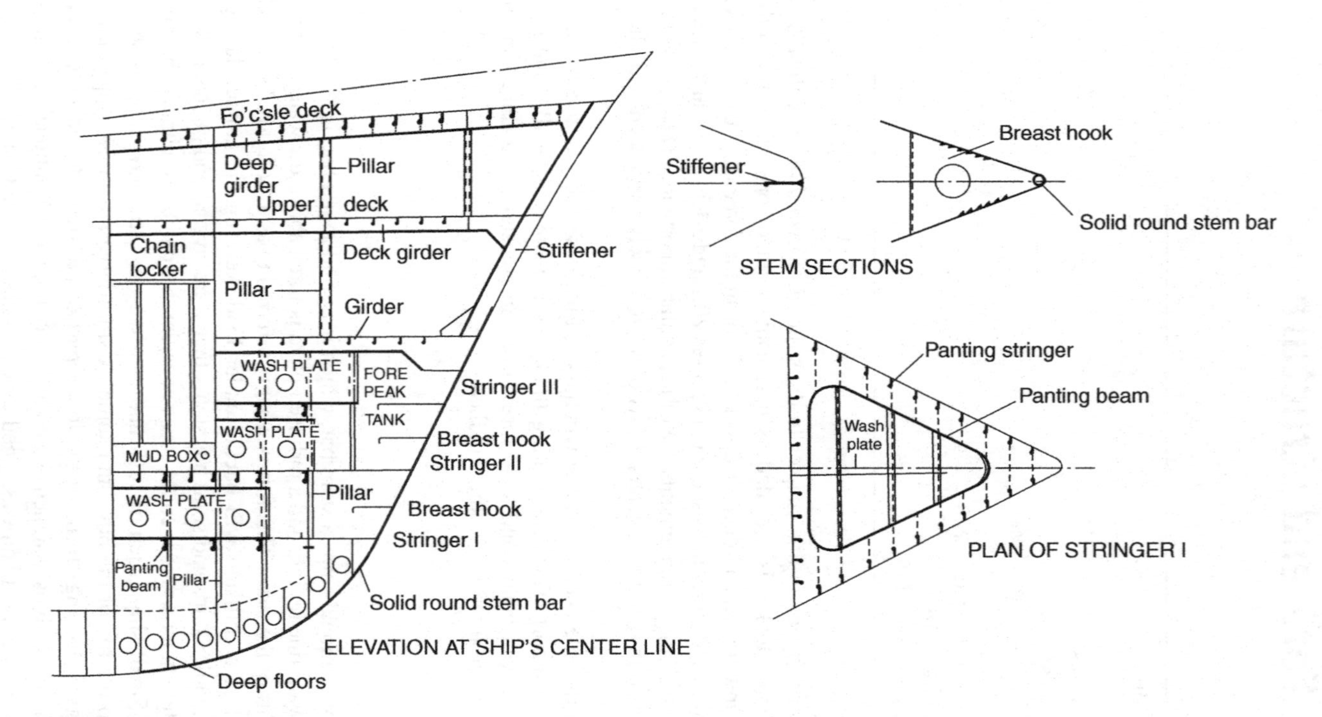
Fo'c'sle deck
Deep girder
Upper deck
Pillar
Chain locker
Pillar
Deck girder
Girder
Stiffener
WASH PLATE
FORE PEAK TANK
Stringer III
WASH PLATE
Breast hook
Stringer II
MUD BOX
WASH PLATE
Pillar
Breast hook
Stringer I
Panting beam
Pillar
Solid round stem bar
ELEVATION AT SHIP'S CENTER LINE
Deep floors
Stiffener
Breast hook
Solid round stem bar
STEM SECTIONS
Panting stringer
Panting beam
Wash plate
PLAN OF STRINGER I

Figure 20.1 Fore end construction.

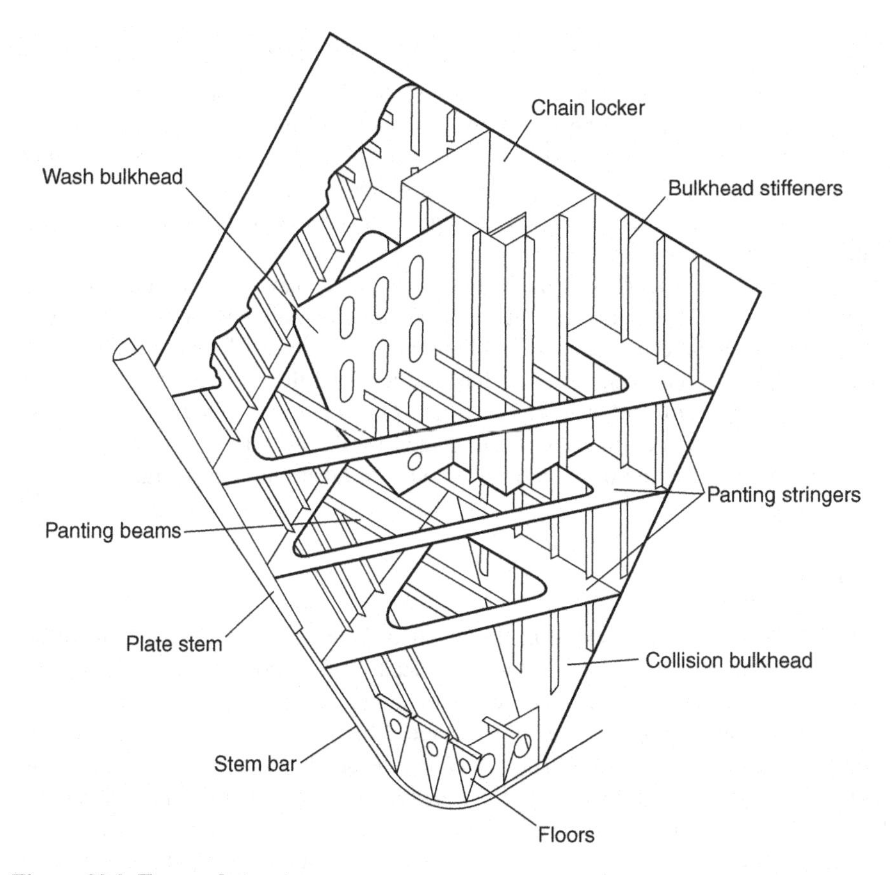

Figure 20.2 Fore end structure.

The solid round bar is welded inside the keel plate at its lower end, and inside the radiused stem plate at its upper end, the shell being welded each side (Figure 20.1). It is necessary to support that part of the stem that is formed by radiused plates with 'breast hooks', i.e. horizontal plate webs, between the decks and below the lowest deck, in order to reduce the unsupported span of the stem. Where the plate radius is large, further stiffening is provided by a vertical stiffener on the center line. The thickness of these plates will be in excess of that required for the side shell forward, but the thickness may taper to that of the side shell at the stem head.

Bulbous bows

Vessels operating at higher speeds, and those with high block coefficients, are often found to have a bulbous or protruding bow below the waterline. The arguments for and against fitting some form of bulbous bow are the province of textbooks on naval architecture, but it may be indicated that like most peculiarities of the immersed hull form this feature is usually intended to reduce the vessel's resistance to motion under certain conditions.

From the construction point of view the bulbous bow does not present any great difficulty if this aspect has been considered when the bulb form is designed. In general, however, a greater degree of plate curvature is involved, unless a rather convenient cylindrical form is adopted and fitted into the bow as a single unit. This has in fact been done successfully; but in general the protrusion forms a continuation of the side shell. Floors are fitted at every frame space in the bulb, and a center line wash bulkhead is introduced when the bulb is large. Transverses are fitted at about every fifth frame in long bulbs (see Figure 20.3). Smaller bulbs have a center-line web but not a wash bulkhead; and in all bulbous bows horizontal diaphragm plates are fitted. Shell plating covering the bulb has an increased thickness similar to that of a radiused plate stem below the waterline. This increased thickness should in particular cover any area likely to be damaged by the anchors and chains, and in designing the bow fouling of the anchors should be taken into consideration.

Chain locker

A chain locker is often arranged in the position forward of the collision bulkhead shown in Figure 20.1, below either the main deck or the second deck. It can also be fitted in the forecastle or aft of the collision bulkhead, in which case it must be watertight and have proper means of drainage. Chain locker dimensions are determined in relation to the length and size of cable, the depth being such that the cable is easily stowed, and a direct lead at all times is provided to the mouth of the chain pipe. Port and starboard cables are stowed separately in the locker, and the inboard ends of each are secured to the bottom of the center-line bulkhead or underside of deck (see Figure 20.4). It is desirable to have an arrangement for slipping the cable from outside the chain locker.

Construction of chain locker

The locker does not as a rule have the same breadth as the ship, but has conventionally stiffened forward and side bulkheads, the stiffeners being conveniently arranged outside the locker if possible to prevent their being damaged. A false bottom may be formed by perforated plates on bearers arranged at a height above the floor of the locker. Where fitted this provides a mudbox that can be cleaned and is drained by a center-line suction, the bottom plating sloping inboard. To separate the locker into port and starboard compartments a center-line bulkhead is fitted. This bulkhead does not extend to the crown of the locker, but allows working space above the two compartments. Access to the bottom of the locker is provided by means of foot holes cut in the bulkhead, and the stiffeners fitted to this bulkhead are of the vertical flush cope bar type. Any projections that would be damaged by the chains are thus avoided. The upper edge of the bulkhead is similarly stiffened and may provide a standing platform, with a short ladder leading from the hatch in the deck forming the crown of

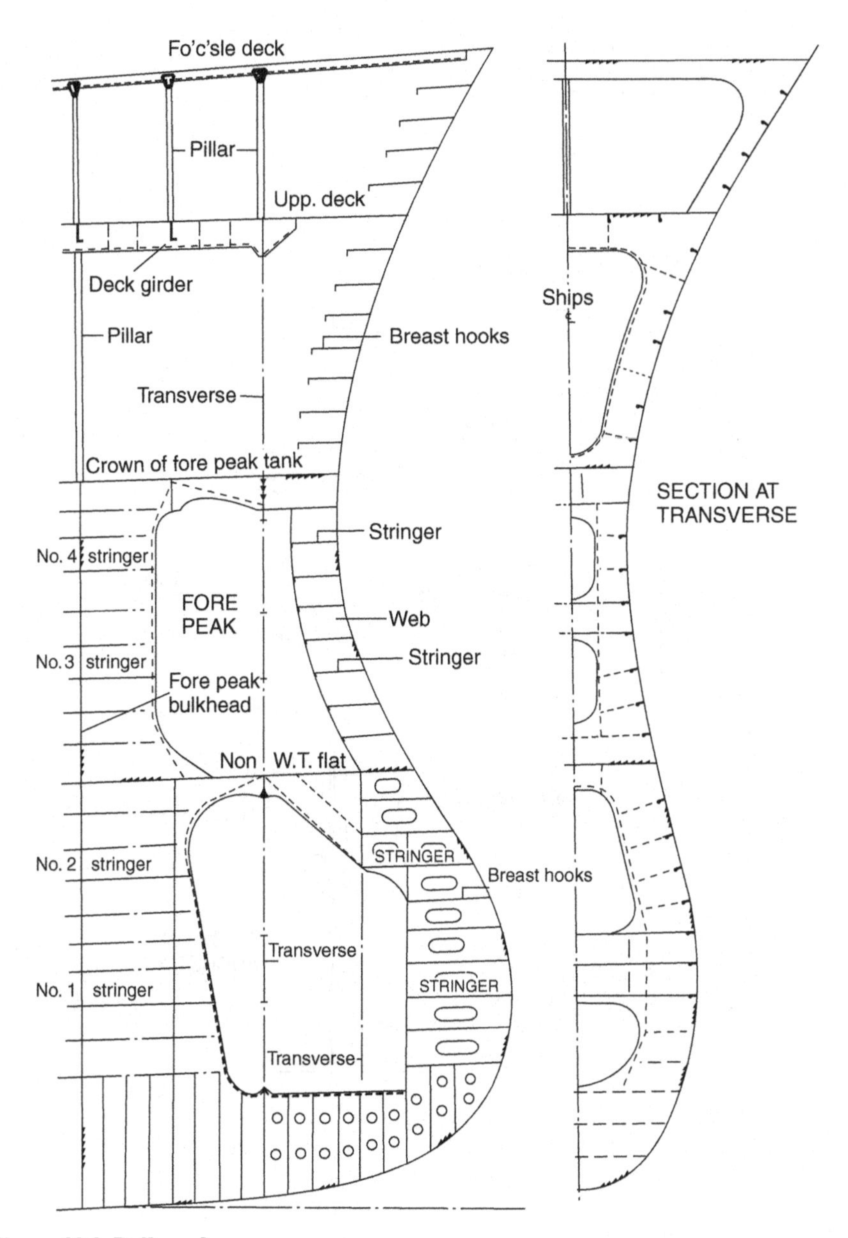

Figure 20.3 Bulbous bow.

the locker. Each cable is fed to the appropriate locker compartment through port and starboard chain pipes from the forecastle deck. These chain pipes or spurling pipes are of tubular construction with castings or other rounded end moldings to prevent chafing.

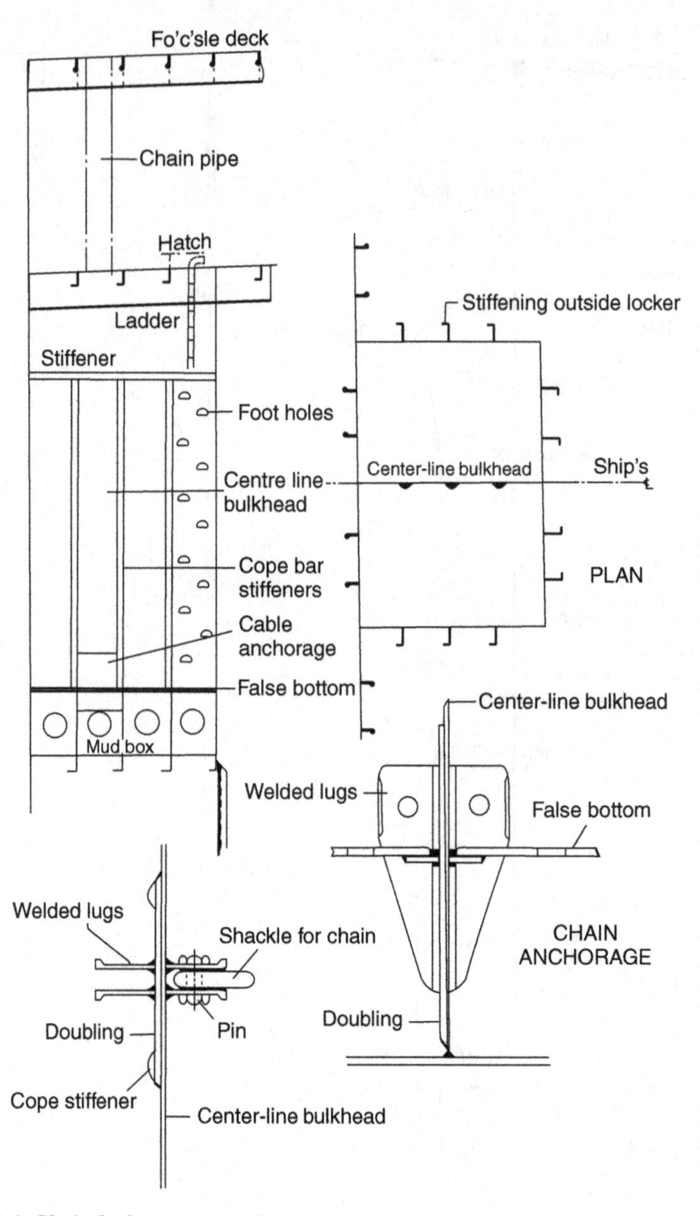

Figure 20.4 Chain locker construction.

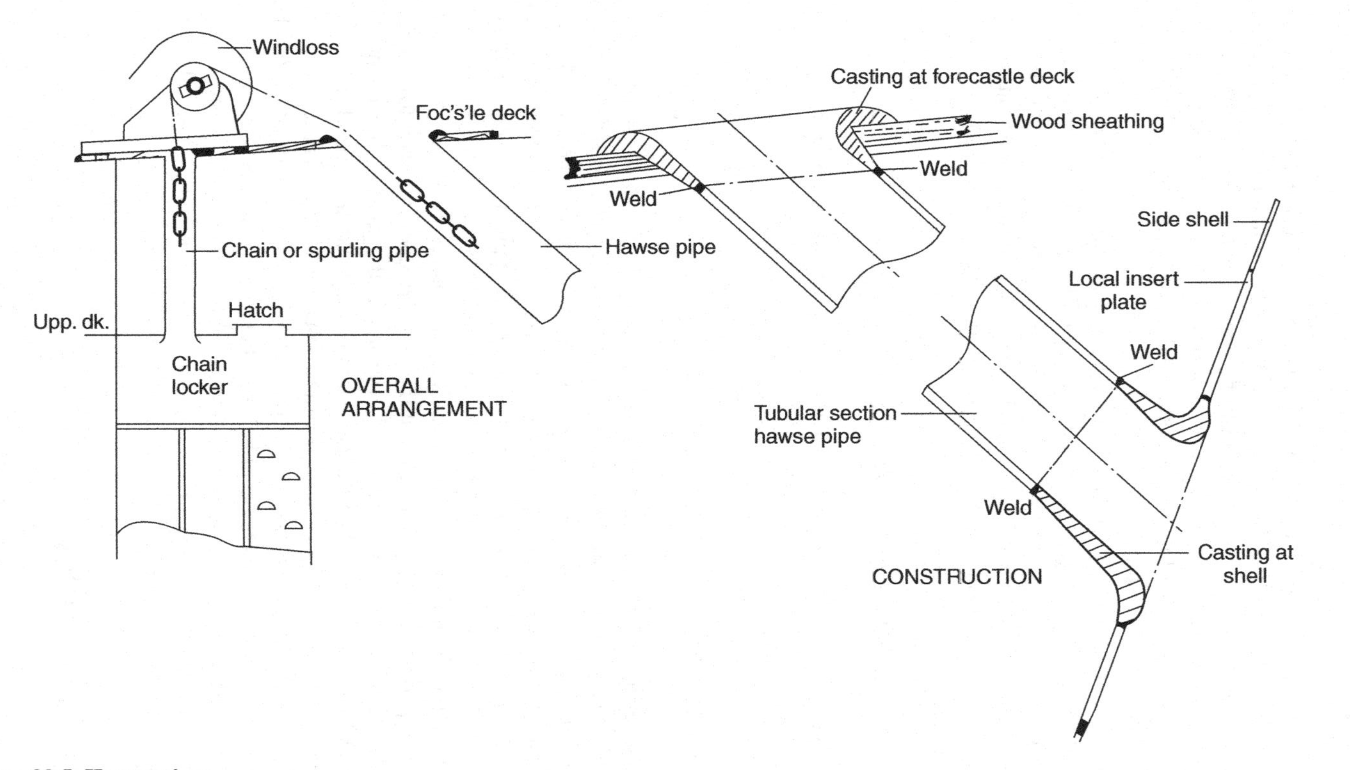

Figure 20.5 Hawse pipes.

Hawse pipes

To provide an easy lead for the cable from the windlass to the anchors, the hawse pipes must be carefully fitted. In the past it was not uncommon for a temporary scale model of the relevant fore end structure to be constructed, and the positions of the hawse pipes experimented with in order to obtain the best chain lead to ensure the anchor could be raised and lowered smoothly and housed properly. Today this can be achieved with the CAD hull model.

Tubular hawse pipes are generally fabricated, and castings are welded at the shell and deck to prevent chafing (see Figure 20.5). Additional stiffening in way of the hawse pipes is required at the side shell. On higher speed vessels a recess is often provided in the shell for anchor stowage; this helps to reduce any drag caused by the stowed anchor and prevents serious damage in the event of a collision.

Bow steering arrangements

Double-ended ferries are provided with a rudder at either end that is locked in position when it is at the fore end of the vessel under way.

Bow thrust units

For maneuvering in confined waters at low speeds, lateral bow thrust units are particularly useful. These are often found in research vessels, or drilling platform vessels where very accurate positioning must be maintained. They are also to be found in large ships and cross-channel vessels where they are provided as an aid to docking. The thrust unit consists as a rule of controllable pitch or reversible impeller fitted in an athwartships watertight tunnel. Control of the unit is from the bridge, but the driving motor is in way of the impeller. Thrust provided by the impellers is low; 16 tonnes is perhaps the largest fitted, but the unit size does not need to be large as small thrusts are very effective. It is true, however, that the greatest thrust is provided at zero speed and as the vessel gets under way the unit becomes much less effective.

From a construction point of view the most important feature is the provision of fairing sections at the ends of the athwartship tunnel in way of the shell. It has been shown that an appreciable increase in hull resistance and hence power may result if this detail is neglected. The best way of avoiding this is to close the tunnels at either end when they are not in use. This is possible, flush-mounted, butterfly action, hydraulically operated doors being available for this purpose.

Some useful websites

www.brunvoll.no For details of tunnel and azimuth bow thrust units.

21 Aft end structure

Chapter Outline

Stern construction 252
Stern frame 252
Rudders 254
Rudder construction 254
Rudder pintles 254
Rudder stock 256
Rudder bearing 256
Rudder trunk 256
Steering gear 256
Sterntube 258
Shaft bossing and 'A' brackets 258
Construction of bossing and 'A' brackets 260
Propellers 260
Controllable pitch propellers 260
Shrouded propellers 262
Electric podded propulsors 262
Further reading 264
Some useful websites 264

Considerable attention is paid to the stern in order to improve flow into and away from the propeller. The cruiser stern (see Figure 21.1) was for many years the favored stern type for ocean-going ships, but today most of these vessels have a transom stern (see Figure 21.2). A cruiser stern presents a more pleasant profile and is hydrodynamically efficient, but the transom stern offers a greater deck area aft, is a simpler construction, and can also provide improved flow around the stern. It is interesting to note that the *Queen Mary 2* has a combined stern, the upper part being of cruiser stern form for good seakeeping and the lower part being of transom form for efficient hydrodynamics.

Many forms of rudder are available and the type and form fitted is intended to give the best maneuvering characteristics. Both the shape of the stern and the rudder type will dictate the form of the stern frame, and this will be further influenced by the required propeller size. Of particular importance at the after end are the arrangements that permit both the propeller shaft and the rudder stock to pierce the intact watertight hull. The safety of the ship may depend on these arrangements. Where more than one screw propeller is to provide the thrust required to propel the ship, bossings or 'A'

Ship Construction. DOI: 10.1016/B978-0-08-097239-8.00021-0

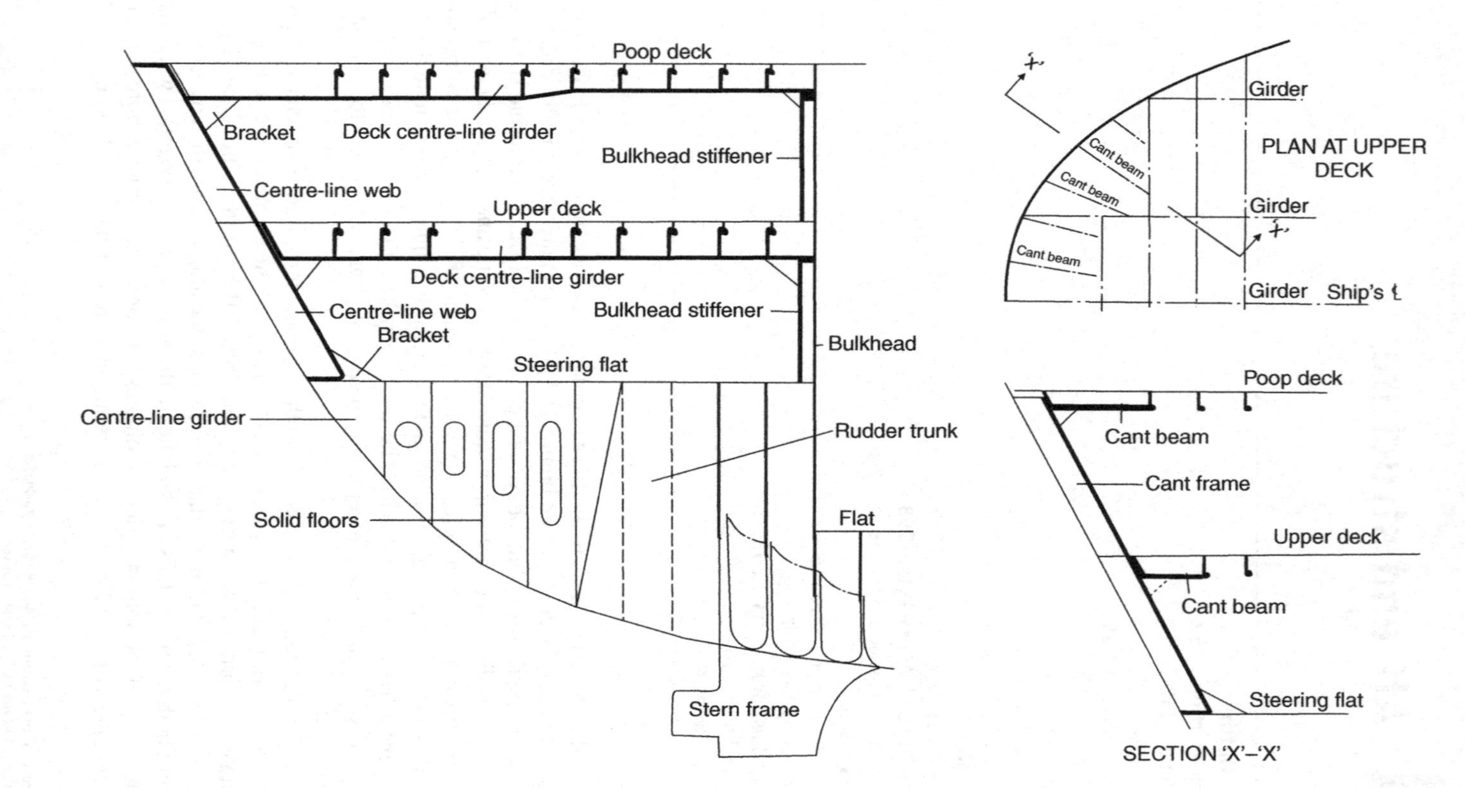

Figure 21.1 Cruiser stern.

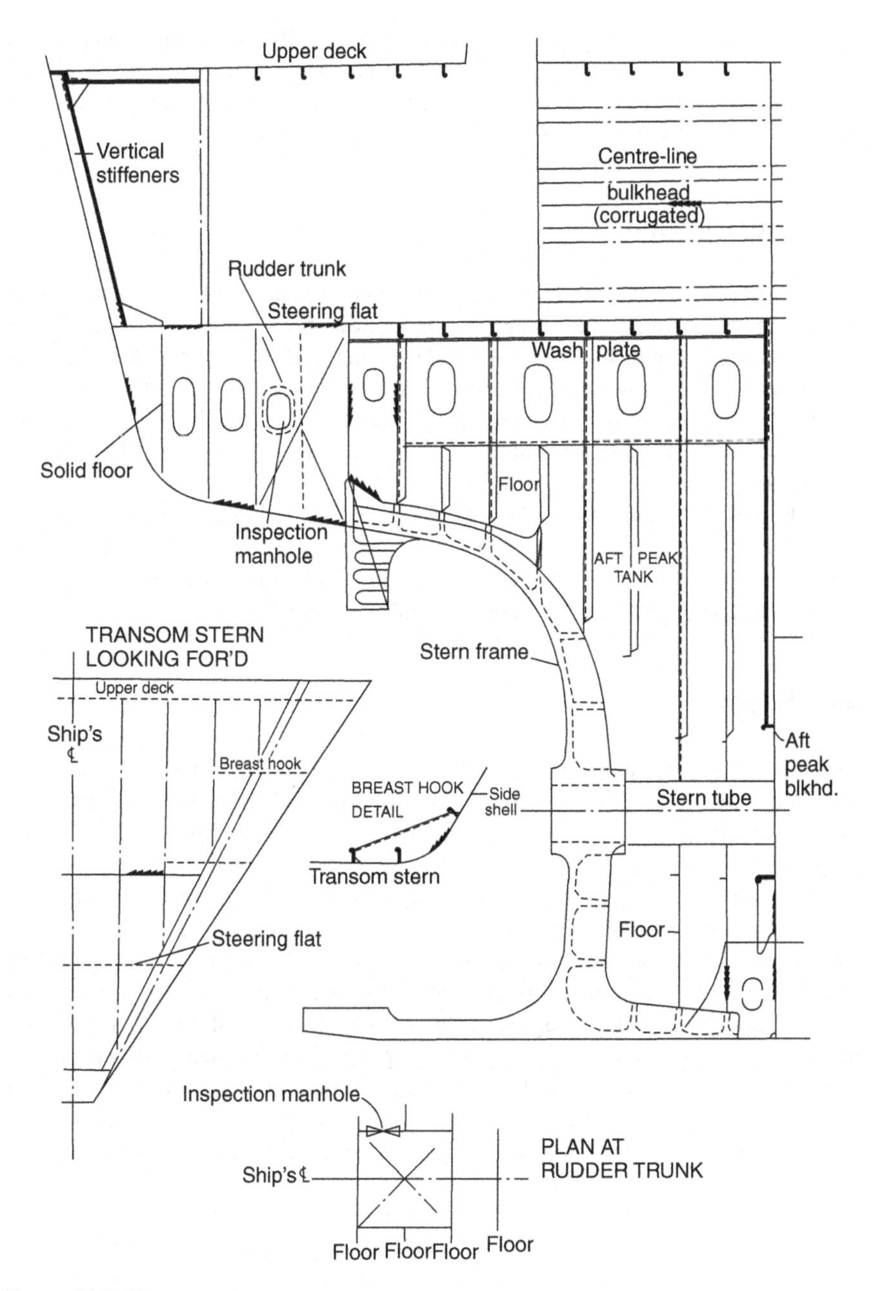

Figure 21.2 Transom stern.

brackets will be fitted to support the outboard shafts. Whilst most ships are propeller driven, other means of transmitting the power developed by the propulsion machinery into a thrust to propel the ship are not uncommon. Water jets, for example, are now a common feature on high-speed passenger and service craft.

Stern construction

As the cruiser stern overhang may be subjected to large slamming forces, a substantial construction with adequate stiffening is required. Solid floors are fitted at every frame space, and a heavy center-line girder is fitted right aft at the shell and decks. The stern plating is stiffened by cant frames or webs with short cant beams supporting the decks and led to the adjacent heavy transverse deck beam. Further stiffening of the plating is provided, or adopted in lieu of cant frames, by horizontal stringers extending to the first transverse frame.

Cant frames are not required where the transom stern is adopted, as the flat stern plating may be stiffened with vertical stiffeners (Figure 21.2). Deep floors and a center-line girder are provided at the lower region of the transom stern construction.

Panting arrangements at the aft end are dealt with in Chapter 17.

Stern frame

It has already been indicated that the form of the stern frame is influenced by the stern profile and rudder type. To prevent serious vibration at the after end there must be adequate clearances between the propeller and stern frame, and this will to a large extent dictate its overall size.

The stern frame of a ship may be cast, forged, or fabricated from steel plate and sections. On larger ships it is generally either cast or fabricated, the casting being undertaken by a specialist works outside the shipyard. To ease the casting problem with larger stern frames and also the transport problem, it may be cast in more than one piece and then welded together when erected in the shipyard. Fabricated stern frames are often produced by the shipyard itself, plates and bars being welded together to produce a form similar to that obtained by casting (see Figure 21.3). Forged stern frames are also produced by a specialist manufacturer and may also be made in more than one piece where the size is excessive or shape complicated.

Sternpost sections are of a streamline form, in order to prevent eddies being formed behind the posts, which can lead to an increase in the hull resistance. Welded joints in cast steel sections will need careful preparation and preheat. Both the cast and fabricated sections are supported by horizontal webs.

Two forms of stern frame are shown in Figure 21.3, one being a casting and the other fabricated, so that the similarity of the finished sections is indicated. Of particular interest is the connection of the stern frame to the hull structure for, if this is not substantial, the revolving propeller supported by the stern frame may set up

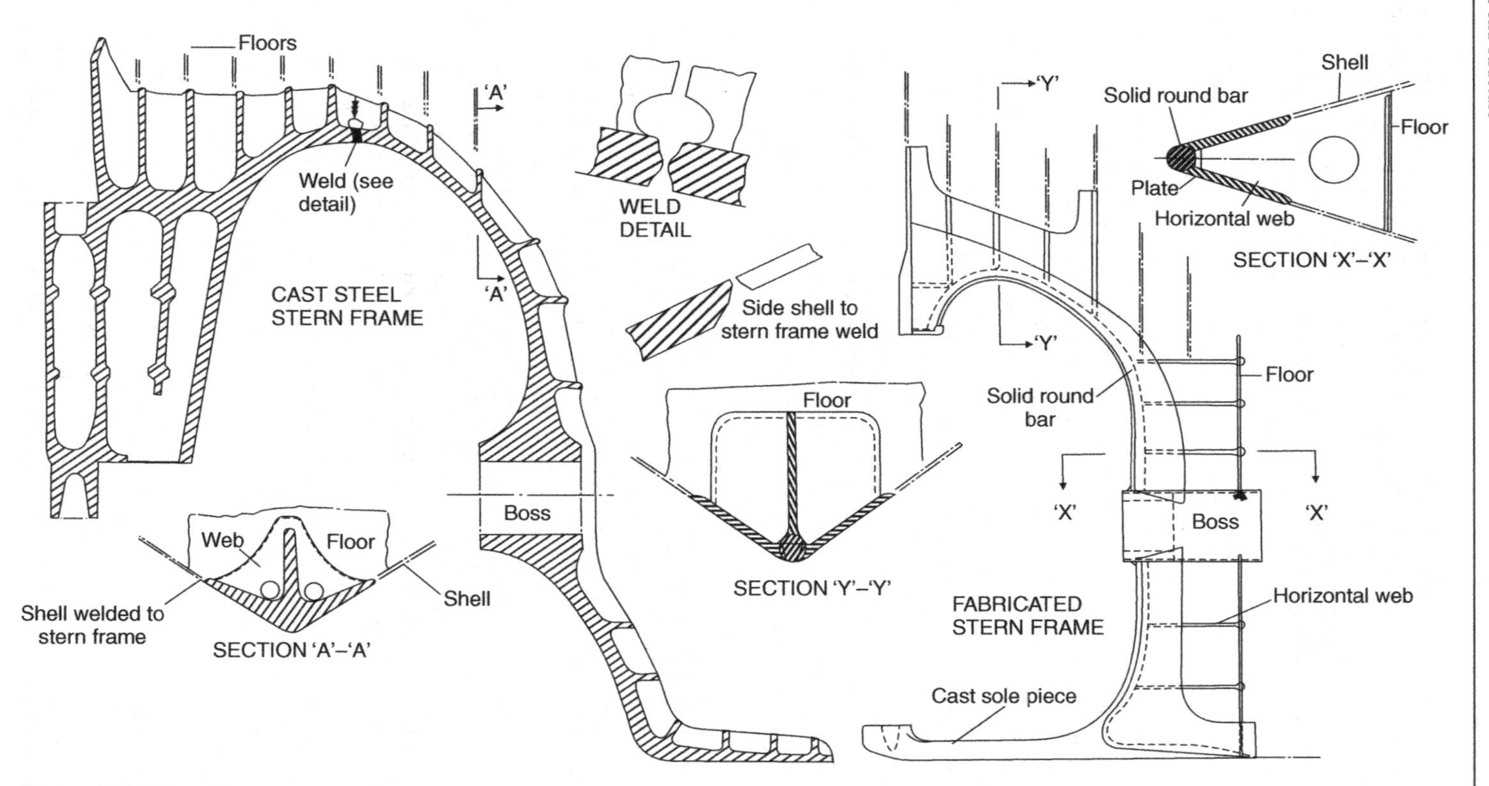

Figure 21.3 Stern frames.

serious vibrations. The rudder post is carried up into the main hull and connected to the transom floor, which has an increased plate thickness. Also, the propeller post may be extended into the hull and connected to a deep floor, the lower sole piece being carried forward and connected to the keel plate. Side shell plates are directly welded to the stern frame (Figure 21.3), a 'rabbet', i.e. a recess, sometimes being provided to allow the shell plate to fit flush with the sternpost section.

Rudders

Many of the rudders that are found on present-day ships are semi-balanced, i.e. they have a small proportion of their lateral area forward of the turning axis (less than 20%). Balanced rudders with a larger area forward of the axis (25–30%) and unbalanced rudders with the full area aft of the axis are also fitted. The object of balance is to achieve a reduction in torque, since the center of lateral pressure is brought nearer the turning axis. However, the fully balanced rudder will at low angles tend to drive the gear, which does not matter a great deal with power steering gears but is less satisfactory with any form of direct hand gear.

Designs of rudders are various, and patent types are available, all of which claim increased efficiencies of one form or another. Two common forms of rudder are shown in Figure 21.4, each being associated with one of the stern frames shown in Figure 21.3.

Rudder construction

Modern rudders are of streamlined form except those on small vessels, and are fabricated from steel plate, the plate sides being stiffened by internal webs. Where the rudder is fully fabricated, one side plate is prepared and the vertical and horizontal stiffening webs are welded to this plate. The other plate, often called the 'closing plate', is then welded to the internal framing from the exterior only. This may be achieved by welding flat bars to the webs prior to fitting the closing plate, and then slot welding the plate as shown in Figure 21.4. Other rudders may have a cast frame and webs with welded side and closing plates, which are also shown in Figure 21.4.

Minor features of the rudders are the provision of a drain hole at the bottom with a plug, and a lifting hole that can take the form of a short piece of tube welded through the rudder with doubling at the side and closing plates. To prevent internal corrosion the interior surfaces are suitably coated, and in some cases the rudder may be filled with an inert plastic foam. The rudder is tested when complete under a head of water 2.45 m above the top of the rudder.

Rudder pintles

Pintles on which the rudder turns in the gudgeons have a taper on the radius, and a bearing length that exceeds the diameter. Older ships may have a brass or bronze liner shrunk on the pintles, which turn in lignum vitae (hardwood) bearings fitted in

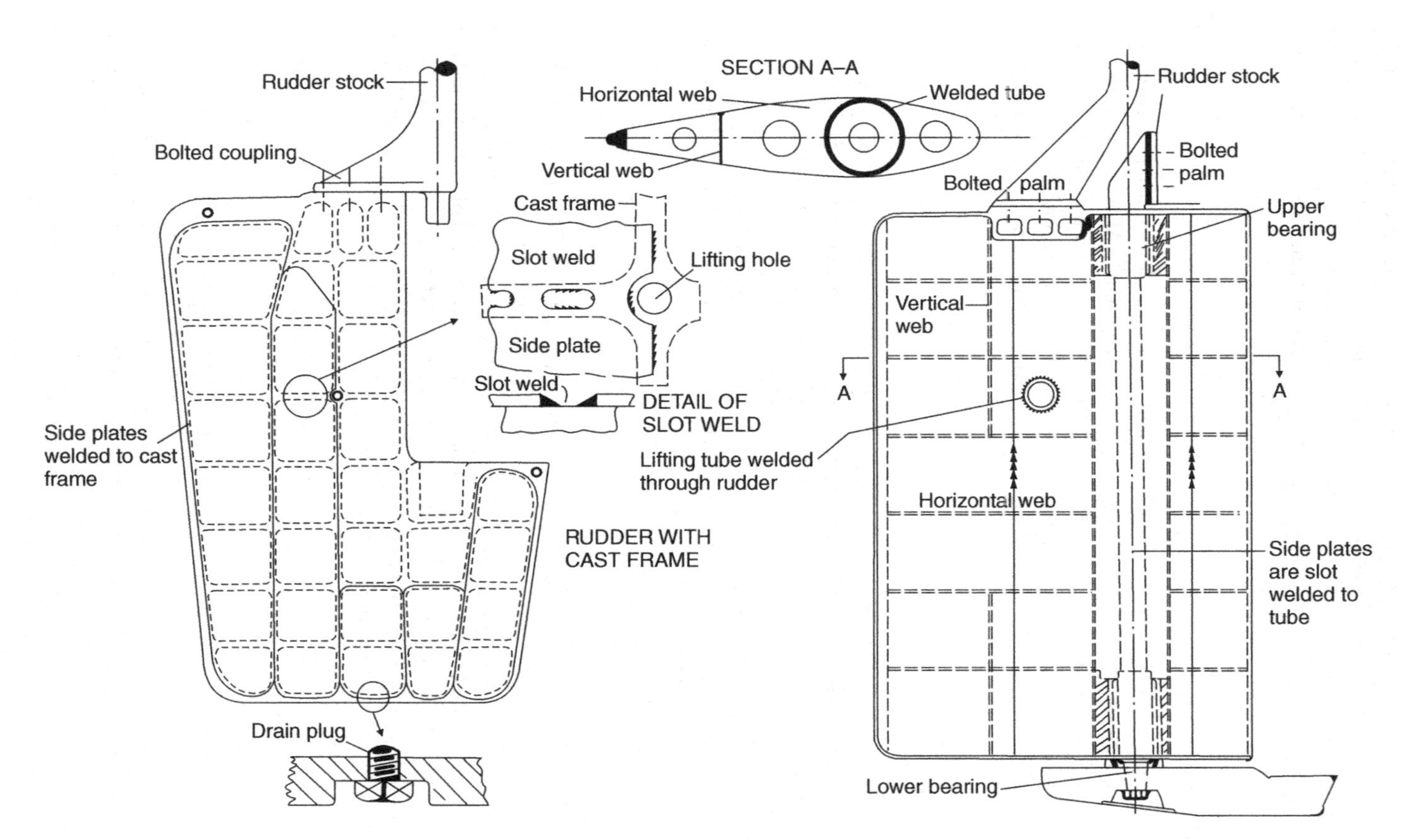

Rudder stock
Bolted coupling
Side plates welded to cast frame
Drain plug
SECTION A–A
Horizontal web
Welded tube
Vertical web
Cast frame
Slot weld
Lifting hole
Side plate
Slot weld
DETAIL OF SLOT WELD
Lifting tube welded through rudder
RUDDER WITH CAST FRAME
Rudder stock
Bolted palm
Bolted palm
Upper bearing
Vertical web
A
A
Horizontal web
Side plates are slot welded to tube
Lower bearing

Figure 21.4 Rudders.

the gudgeons. Modern practice is to use synthetic materials or composites for the bearings, and in some cases stainless steels for the liners. In either case lubrication of the bearing is provided by the water in which it is immersed. For many years it has not been found practicable to provide oil-lubricated metal bearings for the pintles, but a number of larger ships now have this innovation.

Rudder stock

Rudder stock may be of cast or forged steel, and its meter is determined in accordance with the torque and any bending moment it is to withstand. At its lower end it is connected to the rudder by a horizontal or vertical bolted coupling, the bolts having a cross-sectional area that is adequate to withstand the torque applied to the stock. This coupling enables the rudder to be lifted from the pintles for inspection and service.

Rudder bearing

The weight of the rudder may be carried partly by the lower pintle and partly by a rudder bearer within the hull. In some rudder types, for example the spade type that is only supported within the hull, the full weight is borne by the bearer. A rudder bearer may incorporate the watertight gland fitted at the upper end of the rudder trunk, as shown in Figure 21.5. Most of the rudder's weight may come onto the bearer if excessive wear-down of the lower pintle occurs, and the bearers illustrated have cast iron cones that limit their wear-down.

Rudder trunk

Rudder stocks are carried in the rudder trunk, which as a rule is not made watertight at its lower end, but a watertight gland is fitted at the top of the trunk where the stock enters the intact hull (Figure 21.5). This trunk is kept reasonably short so that the stock has a minimum unsupported length, and may be constructed of plates welded in a box form with the transom floor forming its forward end. A small opening with watertight cover may be provided on one side of the trunk, which allows inspection of the stock from inside the hull in an emergency.

Steering gear

Unless the main steering gear comprises two or more identical power units, every ship is to be provided with a main steering gear and an auxiliary steering gear. The main steering gear should be capable of putting the rudder over from 35° on one side to 35° on the other side with the ship at its deepest draft and running ahead at maximum service speed, and under the same conditions from 35° on either side to 30° on the other side in not more than 28 seconds. It is to be power operated where necessary to meet the above conditions and where the stock diameter exceeds

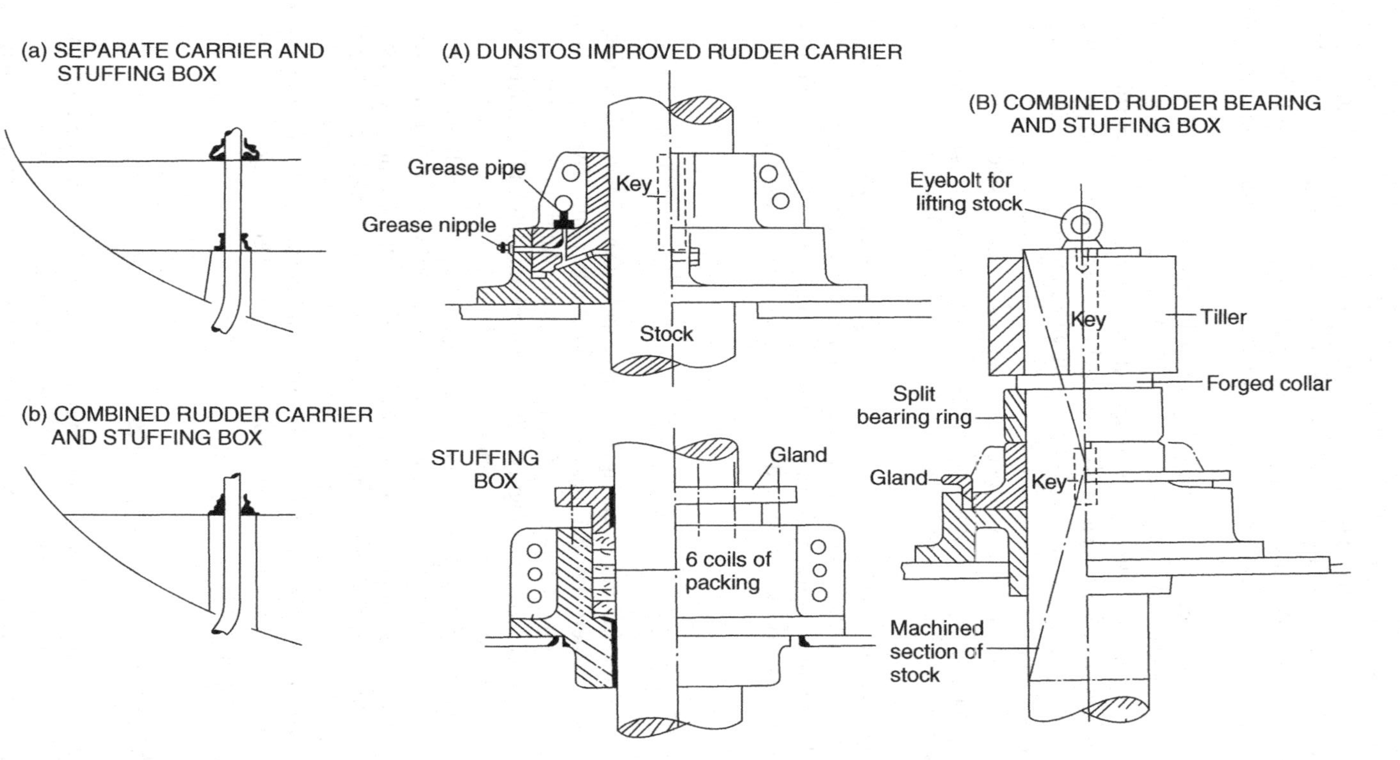

Figure 21.5 Rudder bearings.

120 mm. The auxiliary steering gear should be capable of putting the rudder over 15° on one side to 15° on the other side in not more than 60 seconds with the ship at its deepest draft and running ahead at half the maximum service speed or 7 knots, whichever is greater. Power-operated auxiliary steering gear is required if necessary to meet the forgoing requirement or where the rudder stock diameter exceeds 230 mm.

The main steering gear for oil tankers, chemical tankers or gas carriers of 10,000 gross tonnage or more and every other ship of 70,000 gross tonnage or more is to consist of two or more identical power units that are capable of operating the rudder as indicated for the main steering gear above and whilst operating with all power units. If a passenger ship, this requirement is to be met when any one of the power units is inoperable.

Steering gear control for power-operated main and auxiliary steering gears is from the bridge and steering gear compartment, the auxiliary steering gear control being independent of the main steering gear control (but not duplication of the wheel or steering lever).

Steering gear on ocean-going ships is generally of the electro-hydraulic type.

Where the rudder stock is greater than 230 mm, an alternative power supply should be provided automatically from the ship's emergency power supply or from an independent source of power located in the steering gear compartment.

Sterntube

A sterntube forms the after bearing for the propeller shaft, and incorporates the watertight gland where the shaft passes through the intact hull. Two forms of sterntube are in use, that most commonly fitted having water-lubricated bearings with the after end open to the sea. The other type is closed at both ends and has metal bearing surfaces lubricated by oil. In the former type the bearings were traditionally lignum vitae strips and the tail shaft (aft section of propeller shaft) was fitted with a brass liner. Today, composites are commonly used for water-lubricated sterntube bearings. The latter form of sterntube is preferred in many ships with machinery aft, where the short shaft is to be relatively stiff and only small deflections are tolerated. Where this patent oil-lubricated sterntube is fitted, glands are provided at both ends to retain the oil and prevent the ingress of water, white metal (high lead content) bearing surfaces being provided and the oil supplied from a reservoir. Both types of sterntube are illustrated in Figure 21.6.

Shaft bossing and 'A' brackets

Twin-screw or multi-screw vessels have propeller shafts that leave the line of shell at some distance forward of the stern. To support the shaft overhang, bossings or 'A' brackets may be fitted. Bossings are a common feature on the larger multiple-screw passenger ships and are in effect a molding of the shell that takes in the line of shaft

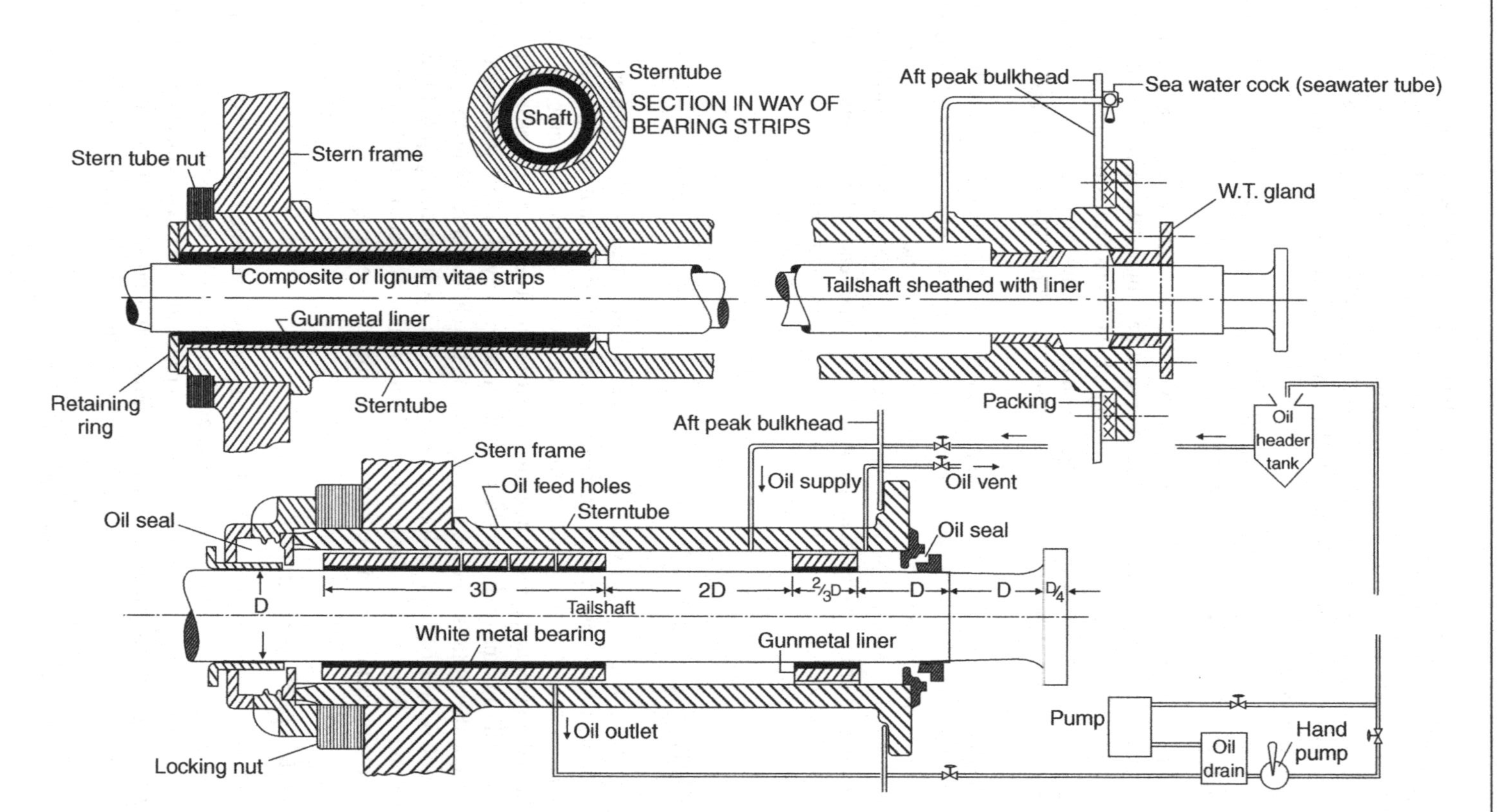

Figure 21.6 Sterntubes.

for some distance. Access from inside the hull is thus provided to the shaft over a great proportion of its length, and it is afforded greater protection. Many large liners having high speeds are shown to have benefitted by a decrease in resistance when bossings have been fitted rather than 'A' brackets. However, large liners of more recent design have in some instances had extended shafts solely supported by 'A' brackets of improved design.

Construction of bossing and 'A' brackets

The shaped frames and plating forming the bossing terminate in a casting known as the 'spectacle frame', which provides the aftermost bearing for the shaft. This may be cast or fabricated and forms a box-like section athwartships that is rigidly connected to heavy plate floors. The arms carrying the shafts extend from this section, which may be split in two or more parts in some instances to aid alignment when it is erected (see Figure 21.7).

'A' brackets may be cast or fabricated, particular attention being paid to the strut section to avoid increases in resistance and cavitation. The connections to the main hull are of particular importance since considerable rigidity of the structure is required. Although on smaller vessels the upper palms may simply be welded to a reinforcing pad at the shell, on larger vessels the upper ends of the struts enter the main hull and are connected to a heavy floor with additional local stiffening (Figure 21.7).

Propellers

Ship propellers may have from three to six similar blades, the number being consistent with the design requirements. It is important that the propeller is adequately immersed at the service drafts and that there are good clearances between its working diameter and the surrounding hull structure. The bore of the propeller boss is tapered to fit the tail shaft and the propeller may be keyed onto this shaft; a large locking nut is then fitted to secure the propeller on the shaft. For securing the propeller a patent nut with a built-in hydraulic jack providing a frictional grip between the propeller and tail shaft is available. This 'pilgrim nut' may also be used with keyless bore propellers. A fairing cone is provided to cover the securing nut.

Controllable pitch propellers

These are propellers in which the blades are separately mounted on the boss, and in which the pitch of the blades can be changed, and even reversed, by means of a mechanism in the boss, whilst the propeller is running. The pitch is mechanically or electromechanically adjusted to allow the engines' full power to be absorbed under different conditions of operation. It is incorrect to refer to such a propeller as a variable pitch propeller since virtually all merchant ship propellers have a fixed pitch variation from blade root to blade tip.

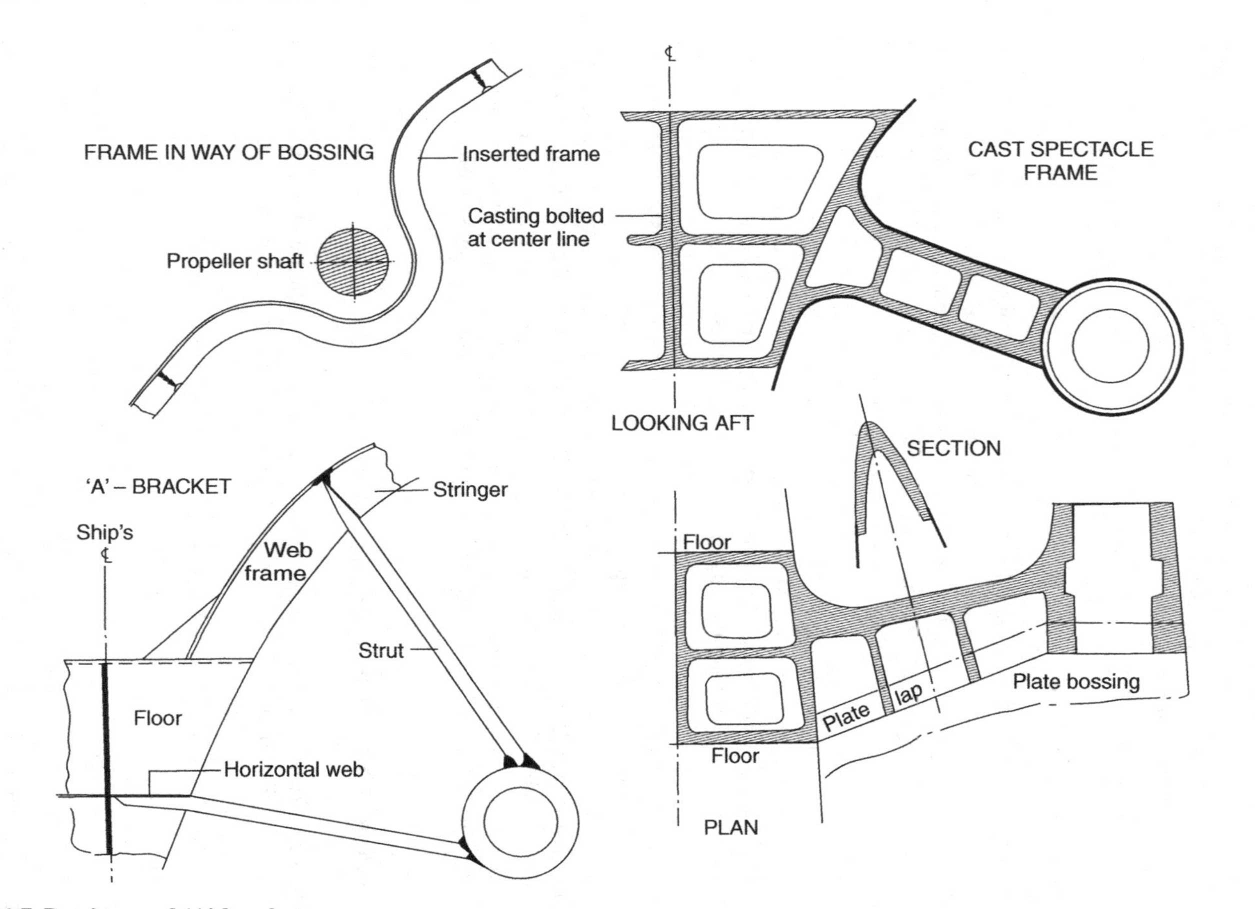

Figure 21.7 Bossings and 'A' brackets.

Propellers of this type are often found on diesel-engined tugs and trawlers where the propeller pitch may be changed to allow the full torque to be absorbed under towing or trawling conditions, and also when the vessel is running freely at full revolutions and a higher speed. It is possible to reverse the pitch in order to stop the vessel rapidly and go astern, with the propeller shaft and propeller still rotating in the one direction.

Large controllable pitch propellers have been fitted to large diesel-driven bulk carriers in recent years.

Shrouded propellers

To increase the thrust provided by a propeller of given diameter at low speeds and high slips it may be enclosed in a fixed nozzle. Single-screw tugs and trawlers are often fitted with the fixed patent 'Kort nozzle' where under a heavy tow the propeller is working at a high slip. This nozzle has a reducing diameter aft and is relatively short in relation to the diameter, to avoid increasing the directional stability, thus making steering difficult.

When running freely the slip is much lower, and it might appear that there was little advantage in fitting a shrouded propeller. However, serious consideration has been given to fitting shrouded propellers on very large single-screw ships where there is a problem of absorbing the large powers on the limited diameter.

Some ships are fitted with a steering nozzle that not only fulfills the purposes of a fixed nozzle but has a turning stock and fin, and pivots about the ship's longitudinal axis can be used to steer the ship.

The fabricated construction is similar for both the fixed and steering nozzles, and is shown in Figure 21.8 for a nozzle of 2.5–3 meters in diameter. The inner plating of the nozzle has a heavier 'shroud' plate in way of the propeller tips and this is carried some distance fore and aft of this position. At this diameter the nozzle section has two full ring webs and a half-depth ring web supporting the inner and outer plating.

Electric podded propulsors

In recent years a number of large cruise ships have been fitted with electric podded propulsors. Pod propulsion has also been considered for large container ships and fast ro-ro passenger ships.

The advantages of electric podded propulsors are:

1. Lengths of propeller shafting within the hull are eliminated, thus providing more revenue earning space.
2. The hull form can be designed for minimum resistance.
3. The thrusters provide good propeller clearances and can be aligned with the local waterflow to provide a clean flow of water to the pulling propeller.
4. Lower noise and vibrations are claimed but it is reported that with some installations this has not always been the case.
5. There is improved maneuverability, which is a major advantage for cruise ships visiting a number of ports.

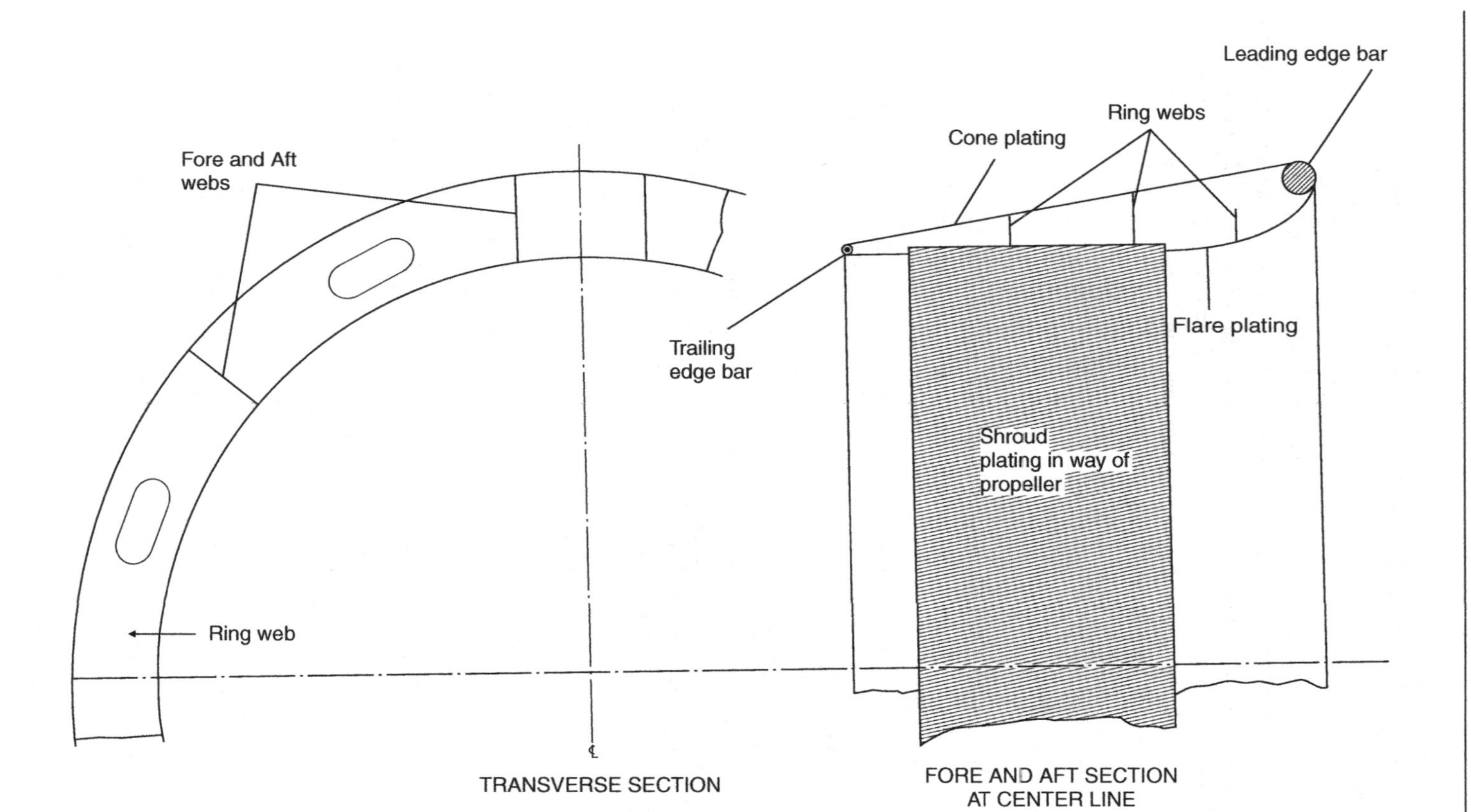

Figure 21.8 Propeller nozzle.

Various podded propulsor arrangements are installed; a smaller vessel may have a single unit providing both propulsion and steering, whereas a large ship like the *Queen Mary 2* has two fixed podded propulsors, ahead of and further outboard, than the two rotating podded propulsors that provide the steering force. Each of these podded propulsors has a 6-meter-diameter four-bladed propeller.

For each podded propulsor a seating is welded into the ship's bottom structure to distribute the loads from the pod over a wide area of the hull. The fixed propulsors are bolted directly to their seating whilst the steering units are carried by a slewing ring in their seatings. Power from the ship's main machinery space is transferred to the motors in the steering pods via a slipring unit.

Further reading

Developments in water-lubricated bearing technology, *The Naval Architect* (February 2004).

Marine Power and Propulsion—Solutions for Naval Architects, The Royal Institution of Naval Architects, 2005.

Queen Mary 2: Genesis of a Queen, The Royal Institution of Naval Architects, 2004.

Some useful websites

www.becker-marine-systems.com See 'rudders' and 'Kort nozzles'.

www.hundestedpropeller.dk See 'multi-pitch propellers' and 'bow/stern thrusters'.

www.schottel.de See 'rudder propeller', 'CP propeller', 'transverse thruster' and 'pod drives'.

www.stonemanganese.co.uk See 'propeller manufacture' and 'notable propellers'. Also details 'nickel–aluminium bronze alloy' and most widely used propeller material.

22 Tanker construction

Chapter Outline

Oil tankers 265
Materials for tanker construction 267
Mild steel 267
Higher tensile steel 267
Construction in tank spaces 268
Longitudinal framing 269
Floors and transverses 269
Bottom and deck girders 269
Bulkheads 271
Hatchways 271
Testing tanks 272
Fore end structure 272
Deep tank 272
Forepeak 273
Ice strengthening 273
After end structure 273
Superstructures 273
Floating production, storage, and offloading vessels 274
Chemical tankers 276
Further reading 277
Some useful websites 277

Ships designed specifically to carry bulk liquid cargoes are generally referred to as tankers. Tankers are commonly associated with the carriage of oil, but a wide variety of liquids are carried in smaller tank vessels and there are a considerable number of larger tank vessels dedicated to carrying chemicals in bulk.

Oil tankers

Small tankers not exceeding 75 meters in length, involved principally in the coastal trade, have a single longitudinal bulkhead on the center line providing two athwartships tanks. The machinery is aft, and an expansion trunk, if fitted, is on the center line in way of the tank spaces (see Figure 22.1).

This chapter is concerned with the construction of the larger ocean-going type, which may be considered in two classes. There are those ships that carry refined oil

Ship Construction. DOI: 10.1016/B978-0-08-097239-8.00022-2

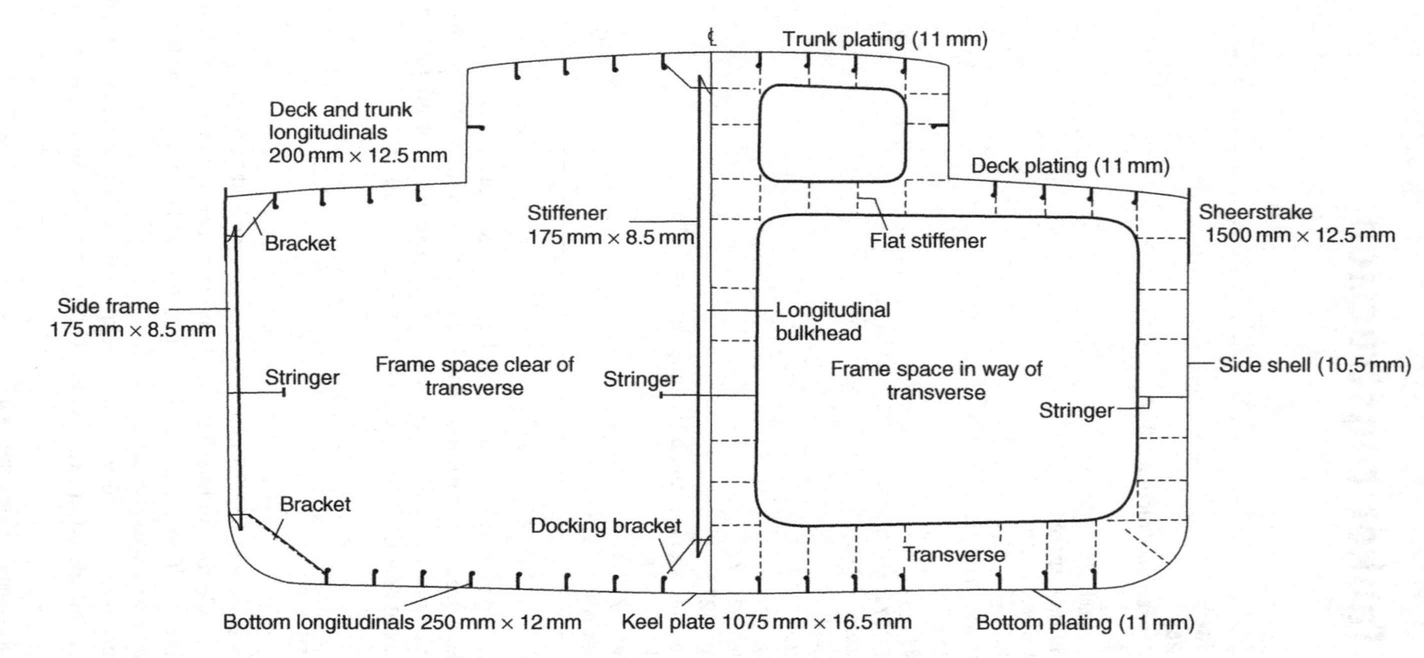

Figure 22.1 Midship section of coastal tanker of less than 5000 tonnes deadweight with trunk.

products, and perhaps some other cargoes like molasses, which tend to be in the smaller 12,000–50,000 tonnes deadweight range. Then there are the crude oil carriers that extend to the 500,000 tonnes deadweight range. The former vessels, the smaller of which may have a single center-line longitudinal bulkhead (see Figure 22.4), have a greater number of tanks, and more complicated pumping arrangements that permit the carriage of a number of different products on a single voyage.

Both types of ship have traditionally been single flush deck ships with longitudinal bulkheads and a structure within the tank spaces consisting of a grillage of longitudinal and transverse members. The structural arrangement over the cargo tank length is dictated by the requirements of the MARPOL convention (see Chapter 29). Since 1980 new crude tankers of 20,000 tonnes deadweight or more and new products carriers of 30,000 tonnes deadweight or more were required to be provided with segregated ballast tanks (SBTs), the capacity of the SBTs being so determined that the ship could operate safely on ballast voyages without recourse to the use of cargo tanks for ballast water. These SBTs were to be located within the cargo tank length and arranged to provide a measure of protection against oil outflow in the event of grounding and or collision. The protective location of these tanks had the effect of providing full or partial double bottoms and/or side tanks in way of the cargo tank space. Subsequent amendments to MARPOL required every tanker of 5000 tonnes deadweight or more that commenced construction after 1993 to have the entire cargo tank length protected by ballast tanks or spaces other than cargo and fuel oil tanks, or other provisions offering equivalent protection against oil pollution. As we have seen in Chapter 3 double hulls have been required for Category 1 oil tankers from 2005 and for Category 2 and 3 oil tankers from 2010. (There is, however, provision for the flag state to permit some oil tankers of Category 2 and 3 to operate beyond these dates to 2015 or until their 25th anniversary, but port states may deny such ships entry to their ports).

Materials for tanker construction

Mild steel is used throughout the structure, but higher tensile steels may also be introduced in the more highly stressed regions of the larger vessels.

Mild steel

As with dry cargo ships it is a requirement that Grade B, D, and E steels be used for the heavier plating of the main hull strength members where the greatest stresses arise in tankers. These requirements are the same as those indicated in Table 17.1. Grade E plates, which we have equated with the 'crack arrester strake' concept (see Chapter 8), will be seen to be required over the midship region in ships exceeding 250 meters in length and are also required as shown in Table 22.1.

Higher tensile steel

Higher tensile steels are often used for the deck and bottom regions of the larger tankers, Grades DH and EH being used for the heavier plating. As indicated in

Table 22.1 Use of Grade E steel in tankers

Location	Thickness
Stringer plate, sheerstrake, rounded gunwale	Greater than 15 mm
Bilge strake, deck strake in way of longitudinal bulkhead	Greater than 25 mm
Main deck plating, bottom plating, keel, upper strake of longitudinal bulkhead	Greater than 40 mm

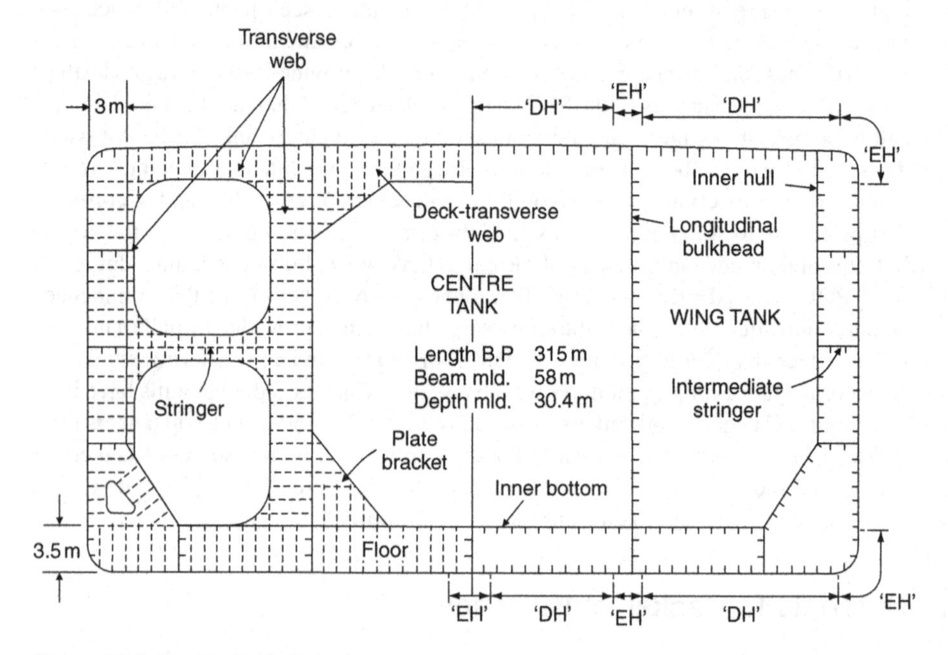

Figure 22.2 Double-hull oil tanker.

Chapter 5, the use of higher tensile steel leads to a reduction in the scantlings of these structural items with advantages both for the shipbuilder and owner. There is some concern that the thinner material may lead to problems if corrosion occurs, and that the thinner plate may also lead to additional fatigue problems due to excess movement of the structure under the forces encountered. The extent of this plating and section material is indicated in Figure 22.2.

Construction in tank spaces

Today's ocean-going crude oil tankers are generally large and in way of the tank spaces have a longitudinally framed double bottom supported by transverse floors and a longitudinally framed deck supported by deck transverses. The longitudinally framed side shell, inner hull, and longitudinal bulkheads are all supported by transverse webs aligned with the transverse floors and deck transverses (see Figure 22.3).

Longitudinal framing

Deck and bottom longitudinals have the greatest scantlings since they are stiffening the more highly stressed flanges of the hull girder. At the side shell the upper longitudinals have the least scantlings, and a uniform increase in size occurs down the side shell until the bilge is reached. The bilge longitudinal size then approaches that of the bottom shell. For products carriers the deck longitudinals may be fitted above the deck to provide a flush internal tank surface for cleaning (see Figure 22.4). An important feature of the longitudinal framing is that continuity of strength is maintained, particularly at the bulkheads forming the ends of the tanks. This feature is increasingly important as the ship length is extended, the bottom and deck longitudinals being continuous through the bulkhead where the ship length is excessive, unless an alternative arrangement is permitted by the classification society (see Figure 22.3). Higher tensile steel longitudinals are to be continuous irrespective of ship length.

The longitudinals can be offset bulb plates that may be built up to give the required scantling on large ships. It is not, however, uncommon to find on many tankers specially fabricated tee longitudinals having a web and symmetrical flat plate flange.

Floors and transverses

To support the bottom and inner bottom longitudinals solid plate floors are fitted. Stiffened transverse plate webs support the side shell and inner shell longitudinals, deck longitudinals, and longitudinal bulkhead stiffening. In way of the center tank a bracketed deck transverse supports the deck longitudinals and large plate brackets are fitted between the longitudinal bulkheads and inner bottom (see Figure 22.2).

Transverses are as a rule built of a plate web and heavier flat face bar, the depth being adequate to allow sufficient material abreast the slots through which the longitudinals pass. Within the double-hull side space they form a solid web and are supported by horizontal stringers (see Figure 22.2). In the wing tank space they support the deck longitudinals and fore and aft bulkhead longitudinals, and in turn are supported by a mid-height longitudinal plate stringer (see Figure 22.2).

Bottom and deck girders

A center-line girder is required in the double-bottom space, which with the heavy keel plate constitutes the immediate structure through which docking loads are transmitted when the vessel is placed on the keel blocks.

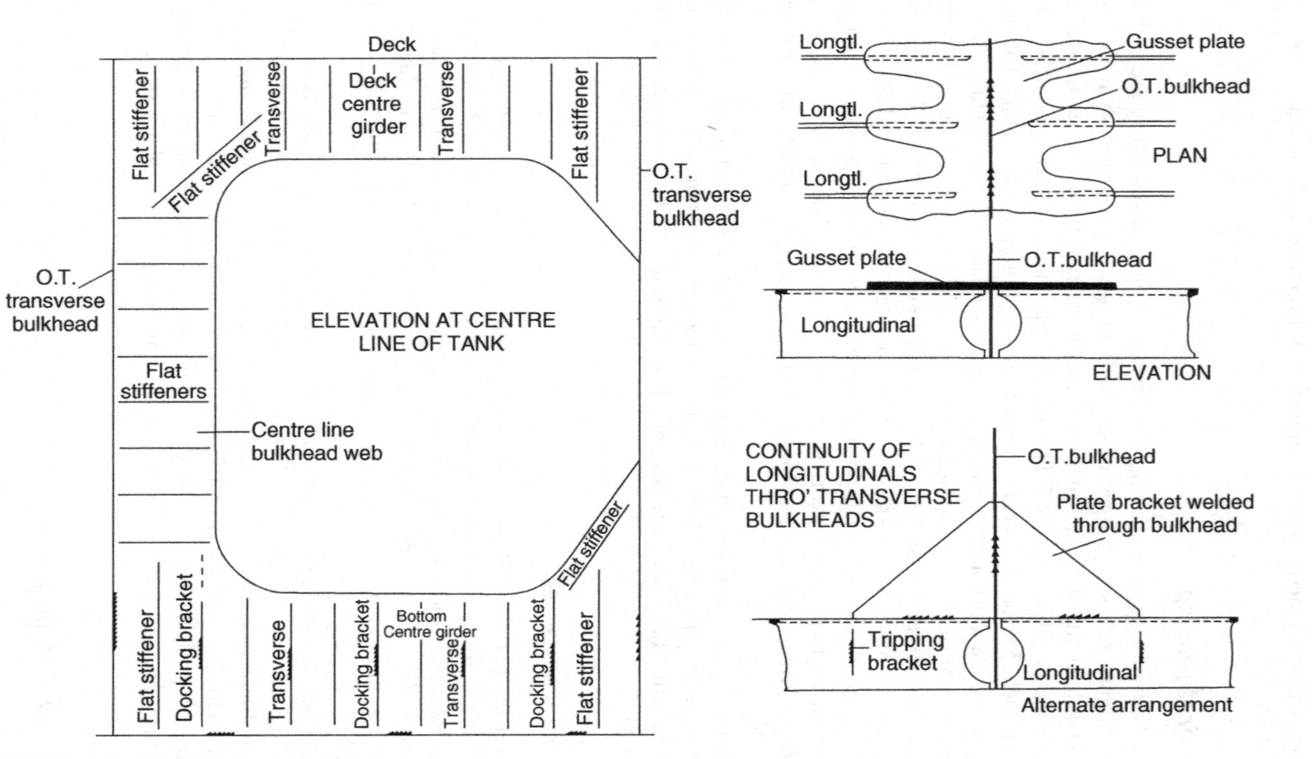

Figure 22.3 Longitudinal framing of oil tankers.

Bulkheads

Bulkhead spacing throughout the cargo tank space is determined by the permissible length of cargo tanks. MARPOL requires that the length of each cargo tank shall not exceed the greater of 10 meters or a length expressed as a percentage of the ship's length that is dependent on the number of longitudinal bulkheads fitted and the minimum distance from the ship's side of the outer longitudinal bulkhead. Tankers with two or more longitudinal bulkheads may have wing and center tank lengths up to 20% of the ship's length. Lloyd's Register require the disposition of transverse bulkheads to comply with the stipulations given in Table 18.1, as applicable to ships with machinery aft. Cofferdams, which may be formed with two adjacent oiltight transverse bulkheads at least 760 mm apart, are required at the ends of the cargo space. However, in many cases a pump room is fitted at the after end of the cargo space (also forward on some products carriers) and a ballast tank is fitted at the forward end, each of these compartments being accepted in lieu of a cofferdam. A cofferdam is also provided between any accommodation and oil cargo tanks.

Construction of the transverse bulkheads is similar to that in other ships, the bulkhead being oiltight. Vertical stiffeners are fitted, or corrugated plating is provided with the corrugations running either vertically or horizontally. Horizontal stringers support the vertical stiffeners and corrugations, and vertical webs support any horizontal corrugations. Further support is provided by the vertical center-line web, which is as a rule deeper on one side of the bulkhead than on the other, unless the tank is very long and the web may then be symmetrical either side of the bulkhead.

Longitudinal bulkheads, which are oiltight, may be conventionally stiffened or may be corrugated with the corrugations running horizontally. Vertical corrugated center-line bulkheads may be fitted. Conventional stiffening is arranged vertically where the side framing is vertical and arranged longitudinally when the side is longitudinally framed. Vertical webs are fitted to the longitudinal bulkhead when this is corrugated or longitudinally framed. Corrugated longitudinal bulkheads are only permitted in ships of less than 200 m in length.

Hatchways

Oiltight hatchways provide access to the tank spaces at the exposed deck. The openings for these are kept as small as possible, and the corners are well rounded, circular openings being not uncommon. Coamings provided for the openings should be of steel and at least 600 mm high, and suitably fastened steel or other approved material covers are fitted. Patent oiltight hatches are available and approved with both steel- and fiber-reinforced plastic covers.

Access to any cofferdams and water ballast tanks may be by similar hatches in the deck, or alternatively a watertight manhole may be fitted with a cover of suitable thickness. Other openings are provided in the deck for ullage plugs and tank cleaning, these being on the open deck, and not within enclosed deck spaces.

Testing tanks

Each cargo tank and cofferdam may be tested separately when complete by filling the tank with water to a head 2.45 m above the highest point of the tank excluding the hatchways, and by filling the cofferdam to the top of the hatch. Water testing on the building berth or dry dock may be undesirable owing to the size of flooded tanks, which gives rise to large stresses on the supporting material and structure. Testing afloat is therefore permitted, each tank being filled separately until about half the tanks are full, when the bottom and lower side shell in the empty tanks are examined. Water is then transferred to the empty tanks, and the remainder of the bottom and side shell is inspected. This testing may take place after the application of protective coatings, provided that welds have been carefully examined beforehand.

In practice, a combination of a structural water test and air leak test is often used. The air leak test is carried out on the building berth, the tanks being subject to an air pressure similar to that required for testing double-bottom tanks. A water pressure test is carried out on one center tank and two wing tanks selected by the surveyors.

Clean water ballast tanks are tested in the same manner as the cargo tanks, but bunkers and deep tank test requirements are similar to those in dry cargo ships, i.e. a head of water 2.45 m above the crown of the tank is applied. Any bulkhead not forming a tank boundary is hose tested.

Fore end structure

Forward of the tank space deep tanks may be fitted. Framing throughout this space and the forepeak may be transverse, longitudinal, or a combination of both.

Deep tank

If transverse framing is adopted, forward floors are fitted at every frame space in conjunction with a center-line girder or center-line bulkhead and intercostal side girders not more than three times the transverse frame spacing apart. If longitudinal bulkheads are fitted port and starboard in the cargo tanks, these may be extended to the fore side of the deep tank in lieu of a center-line bulkhead. Above the floors the transverse frames are supported by stringers spaced not more than 5 m apart, which are either supported by web frames connected to deep beams to form a vertical ring frame, or connected to longitudinal stringers on the transverse bulkheads to form horizontal ring frames. Alternatively, in narrow tanks perforated flats may be fitted at 5 m spacings.

Longitudinally framed deep tanks are supported by side transverses five frame spaces apart. Where the depth of the tank exceeds 16 m the side transverses, and the web frames in a transversely framed tank, either have one or more deep stringers fitted, cross ties, or perforated flats with deep beams in way of the transverses or webs.

A longitudinal bulkhead generally must be provided if the tank width exceeds 50% of the ship's beam and may solely be a wash bulkhead on the center line. Where the breadth of tank exceeds 70% of the ship's beam, at least one solid bulkhead on the center line is recommended.

Forepeak

A transversely framed forepeak has a similar construction to that of a conventional cargo ship and includes the usual panting arrangements (see Chapter 20). If the forepeak is longitudinally framed the side transverses have a maximum spacing of between 2.5 and 3.5 m depending on the length of ship, and are connected to transverses arranged under the decks in line with the side transverses. Forecastles have web frames to support any longitudinal side framing, and the forecastle deck may be supported by pillars at its center line.

Ice strengthening

With the growing export of oil from Russia an increasing number of new tankers are being provided with ice strengthening forward (see Chapter 17) so that they can trade to parts of the Baltic Sea that can be frozen in winter. These include several 20,000 tonne deadweight ice-breaking oil tankers.

After end structure

The machinery is arranged aft in ocean-going tankers, and a transversely framed double-bottom structure is adopted in way of the machinery space. Constructional details of this double bottom are similar to those of the conventional dry cargo ship, with floors at each frame space, additional side girders, and the engine seating integral with the bottom stiffening members.

Transverse or longitudinal side and deck framing may be adopted in way of the engine room and aft of this space. If transversely framed, web frames are fitted not more than five frame spaces apart below the lowest deck and may be supported by side stringers. The web frames may be extended into the poop, and where an all-houses aft arrangement is adopted they may also extend into the superstructure. Similar transverse webs are introduced to support any longitudinal framing adopted in a machinery space. Transverse webs have the same spacing as web frames except in tween decks above the aft peak, where the maximum spacing is four frame spaces.

The aft peak and stern construction follows that of other merchant ship types, a center-line bulkhead being provided in the aft peak. Any ice strengthening is similar to that required in other ships (see Chapter 17).

Superstructures

To permit the assignment of deeper freeboards for oil tankers, the conditions of assignment of load line (see Chapter 31) stipulate the requirements for protective housings enclosing openings in the freeboard and other decks. They also require provision of a forecastle covering 7% of the ship's length forward.

Structurally, the houses are similar to those of other vessels, special attention being paid to the discontinuities in way of breaks at the ends of the houses. Particular attention must be paid to any endings in the midship length of the hull. As the machinery is aft the poop front merits special attention, and is in fact structurally similar to the bridge front, since its integrity is essential.

A particular feature of the tanker with its lower freeboard is the requirement for an access gangway at the level of the first tier of superstructure between accommodation spaces. This is still often found on vessels with all accommodation aft, although the regulations would permit a rail or similar safety arrangement at the deck level. The provision of a gangway in the latter case is at the owner's wish, and is an added safety factor at greater initial cost, requiring additional maintenance. Another feature is the absence of bulwarks on the main decks, the regulations requiring the provision of open rails over at least half the length of the wells, which are often awash in heavy weather.

Floating production, storage, and offloading vessels

Floating production, storage, and offloading vessels (FPSOs) and floating storage units (FSUs) are now a common feature of offshore oil installations.

A typical FSU is of 100,000 tons deadweight with a storage capacity of 112,000 cubic meters representing 10 days production. Crude oil from a floating production unit (FPU) is transferred to the FSU, where it is loaded via a submerged turret loading (STL) system. From the FSU the crude oil is transferred to shore by shuttle tankers, which are loaded by flexible hoses from the stern of the FSU. An FPSO combines the functions of the FPU and FSU in a single floating unit.

Many oil tankers, more recently those with a single hull that are barred from some trades, have been converted to FPSOs and FSUs, but there are also numbers of new design and build contracts for such vessels. The basic hull construction is similar to that for conventional tankers but there are arrangements and structural features peculiar to these vessels. The vessels are moored using a turret system forward, which allows the hull to weathervane into the wind and waves to reduce dynamic loading on the hull girder. However, because the waves will be coming mainly towards the bow for the whole of its life, the hull girder will experience greater fatigue problems than would be the case for a conventional tanker. The conventional tanker carries little in the way of loading on its main deck, whereas an FPSO is fitted with a substantial oil production facility, which can weigh in excess of 20,000 tons. This loading needs to be transferred into the hull structure, which means that vertical structural components are more complex and substantial than those of a tanker. Further, because both FSUs and FPSOs are expected to remain on station for their lifetime and withstand arduous conditions without dry-docking, the exposed hull construction is generally upgraded and strengthened with increased corrosion protection.

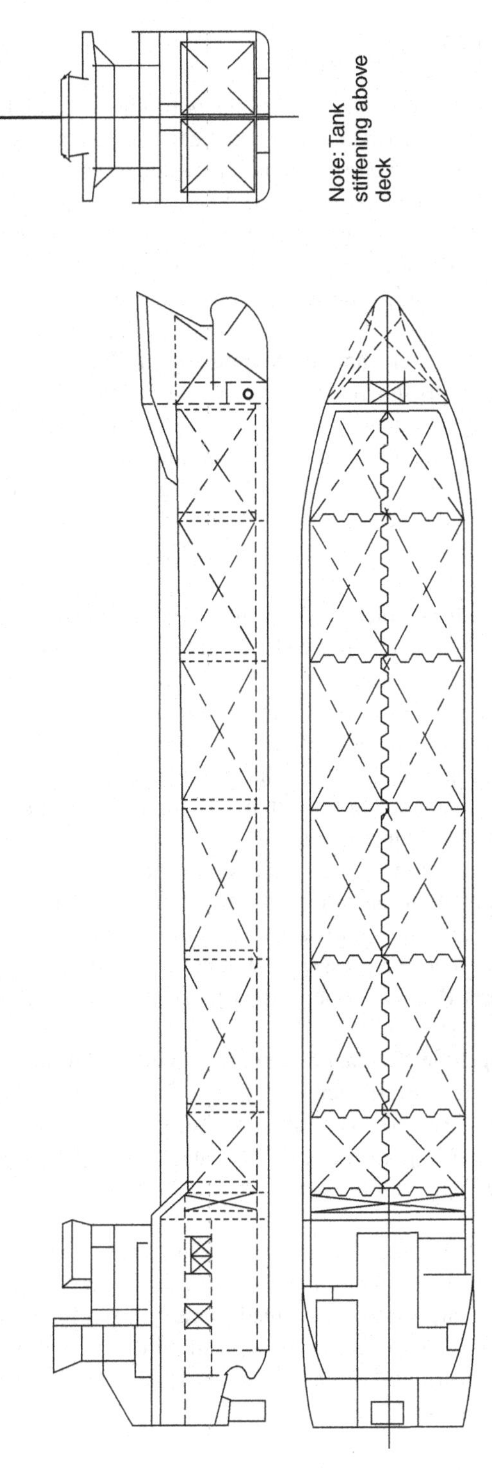

Figure 22.4 Oil products/chemical tanker of 12,700 tonnes deadweight.

Chemical tankers

The structural configurations and arrangements of chemical tankers often are basically similar to those described for oil tankers, particularly where the chemical product is not required to be carried in an independent tank. For some cargoes tanks constructed of stainless steel are preferred. Where the chemical product is required to be carried in an independent tank the structure and arrangements may be similar to ships carrying liquefied gases described in Chapter 23.

Regardless of size, ships built or converted on or after 1 July 1986 and engaged in the carriage of bulk cargoes of dangerous or noxious liquid chemical substances, other than petroleum or similar flammable products, are required to comply with the International Code for the Construction and Equipment of Ships Carrying Dangerous Chemicals in Bulk (IBC Code). Such ships built or converted before that date are to comply with the earlier Code for the Construction and Equipment of Ships Carrying Dangerous Chemicals in Bulk (BCH Code). Ships that comply with these requirements are issued with an 'International Certificate of Fitness for the Carriage of Dangerous Chemicals in Bulk'.

Under the IBC Code chemical tankers are designed and constructed to one of three specified standards. A type 1 ship is a chemical tanker intended for the transportation of products considered to present the greatest overall hazard. It should be capable of surviving the most severe standard of damage and its cargo tanks should be located at the maximum prescribed distance inboard from the shell plating. Type 2 and 3 ships carry products of progressively lesser hazards. Where a ship is intended to carry a range of products it is assigned the standard applicable to the product having the most stringent ship type requirement.

The IBC Code also defines cargo tank types. Tank type 1 is an independent tank that is not contiguous with, or part of, the hull structure. Tank type 2 is an integral tank, i.e. it is part of the ship's hull structure. A gravity tank (G) is an independent or integral tank that has a design pressure of not more than 0.7 bar gage at the top of the tank. A pressure tank (P) is an independent tank that has a design pressure of more than 0.7 bar gage.

The IBC Code specifies for each individual product to be carried the ship type and required tank type, e.g. sulfuric acid—ship type 3—tank type 2G.

Also, requirements for the materials of construction, freeboard, stability, general arrangement, piping arrangements, electrical and environmental arrangements, tank venting and gas freeing, fire protection, etc. of chemical tankers are covered by the IBC Code.

Figure 22.4 shows a 130 m LOA type II oil products/chemical tanker of 12,700 tonnes deadweight. The 130 meter length overall by 20 meter breadth by 9.75 meter depth vessel has 12 cargo tanks, equipped to carry six different types or grades of cargo simultaneously. Each tank has a deepwell pump rated at 300 cubic meters per hour and the ship's maximum discharge rate is 1800 cubic meters per hour with six lines working at the same time.

Further reading

Code for the Construction and Equipment of Ships Carrying Dangerous Chemicals in Bulk (BCH Code). IMO publication (IB772E), 2005 edition.

Common Structural Rules For Double Hull Oil Tankers. IACS publication.

Design and operation of double hull tankers. Conference Proceedings and Papers, Royal Institution of Naval Architects Publications, 2004.

International Code for the Construction and Equipment of Ships Carrying Dangerous Chemicals in Bulk (IBC Code). IMO publication (IMO-100E), 1998 edition.

MARPOL 73/78 Consolidated Edition, IMO publication (IMO 1B520E), 2002.

MARPOL 2005, Amendments 2005. IMO publication (IMO 1525E)

Some useful websites

www.intertanko.com Current regulatory and technical issues relating to oil and chemical tankers.

www.imo.org ‘Safety’; see ‘tanker safety’.

www.iacs.org.uk See ‘Common structural rules for double hull oil tankers’.

23 Liquefied gas carriers

Chapter Outline

Liquefied petroleum gas (LPG) 279
Liquefied natural gas (LNG) 280
The IMO international gas carrier code 280
Integral tanks 280
Membrane tanks 281
Semi-membrane tanks 281
Independent tanks 281
Internal insulation tanks 281
Secondary barrier protection 282
Liquefied petroleum gas ships 282
Fully pressurized tanks 282
Semi-pressurized (or semi-refrigerated) tanks 282
Fully refrigerated tanks 283
Liquefied natural gas ships 283
Independent Type A tanks 283
Independent Type B tanks 286
Membrane tanks 286
Semi-membrane Type B tanks 286
General arrangement of gas carriers 287
Lloyd's classification 288
Further reading 289
Some useful websites 289

A large number of ships are in service that are designed to carry gases in liquid form in bulk. Many of the smaller ships are designed to carry liquefied petroleum gas (LPG), whilst a smaller number of ships are designed to carry liquefied natural gas (LNG). However, there has been a significant increase in LNG ships of large size as gas has become a preferred source of electrical power generation and new resources have been exploited.

Liquefied petroleum gas (LPG)

LPG is the name originally given by the oil industry to a mixture of petroleum hydrocarbons, principally propane and butane and mixtures of the two. LPG is used as

Ship Construction. DOI: 10.1016/B978-0-08-097239-8.00023-4

a clean fuel for domestic and industrial purposes. These gases may be converted to the liquid form and transported in one of three conditions:

1. Solely under pressure at ambient temperature
2. Fully refrigerated at their boiling point (−30 to −48 °C)
3. Semi-refrigerated at reduced temperature and elevated pressure.

A number of other gases with similar physical properties, such as ammonia, propylene, and ethylene, are commonly shipped on LPG carriers. These gases are liquefied and transported in the same conditions as LPG except ethylene, which boils at a much lower temperature (−104 °C) and which is therefore carried in the fully refrigerated or semi-refrigerated condition.

Liquefied natural gas (LNG)

LNG is natural gas from which most of the impurities such as sulfur and carbon dioxide have been removed. It is cooled to or near its boiling point of −165 °C at or near atmospheric pressure and is transported in this form as predominantly liquid methane. Methane has a critical pressure of 45.6 kg/cm^2 at a critical temperature of −82.5 °C, i.e. the pressure and temperature above which liquification cannot occur, so that methane can only be liquefied by pressure at very low temperatures.

The IMO international gas carrier code

In 1975 the Ninth Assembly of the IMO adopted the Code for the Construction and Equipment of Ships Carrying Liquefied Gases in Bulk, A.328 (IX), which provides international standards for ships that transport liquefied gases in bulk. It became mandatory in 1986 and is generally referred to as the IMO International Gas Carrier Code. The requirements of this code are incorporated in the rules for ships carrying liquefied gases published by Lloyd's Register and other classification societies.

The code covers damage limitations to cargo tanks and ship survival in the event of collision or grounding, ship arrangements for safety, cargo containment and handling, materials of construction, environmental controls, fire protection, use of cargo as fuel, etc. Of particular interest in the context of ship construction is the section on cargo containment, which defines the basic cargo container types and indicates if a secondary barrier is required, i.e. a lining outside the cargo containment that protects the ship's hull structure from the embrittling effect of the low temperature should cargo leak from the primary tank structure. The cargo containment types are described below.

Integral tanks

These tanks form a structural part of the ship's hull and are influenced in the same manner and by the same loads that stress the adjacent hull structure. These are used for the carriage of LPG at or near atmospheric conditions, butane for example, where no provision for thermal expansion and contraction of the tank is necessary.

Membrane tanks

These are non-self-supporting tanks consisting of a thin layer (membrane) supported through insulation by the adjacent hull structure. The membrane is designed in such a way that thermal and other expansion or contraction is compensated for without undue stressing of the membrane. Membrane tanks are primarily used for LNG cargoes (see Figure 23.4).

Semi-membrane tanks

These are non-self-supporting tanks in the load condition. The flat portions of the tank are supported, transferring the weight and dynamic forces through the hull, but the rounded corners and edges are not supported so that tank expansion and contraction is accommodated. Such tanks were developed for the carriage of LNG, but have been used for a few LPG ships.

Independent tanks

These are self-supporting and independent of the hull. There are three types:

- 'Type A', which are designed primarily using standard traditional methods of ship structural analysis. LPG at or near atmospheric pressure or LNG may be carried in such tanks (see Figure 23.2).
- 'Type B', which are designed using more sophisticated analytical tools and methods to determine stress levels, fatigue life, and crack propagation characteristics. The overall design concept of these tanks is based on the so-called 'crack detection before failure principle', which permits their use with a reduced secondary barrier (see Figure 23.3). LNG is normally carried in such tanks.
- 'Type C', which are designed as pressure vessels, the dominant design criteria being the vapor pressure. Normally used for LPG and occasionally ethylene.

Internal insulation tanks

These are non-self-supporting and consist of thermal insulation materials, the inner surface of which is exposed to the cargo supported by the adjacent inner hull or an independent tank. There are two types:

- 'Type 1', where the insulation or combination of insulation and one or more liners act only as the primary barrier. The inner hull or independent tank forms the secondary barrier.
- 'Type 2', where the insulation or combination of insulation and one or more liners act as both the primary and secondary barriers and are clearly distinguishable as such.

The liners on their own do not act as liquid barriers and therefore differ from membranes. These tanks are a later addition to the Code and Type 1 is known to have been used for the carriage of LPG.

Secondary barrier protection

The requirements for secondary barrier protection are given in Table 23.1.

Liquefied petroleum gas ships

Ships carrying LPG are categorized by their cargo containment system. Figure 23.1 shows the three types discussed below.

Fully pressurized tanks

The capacity of fully pressurized ships is usually less than 2000 m^3 of propane, butane, or anhydrous ammonia carried in two to six uninsulated horizontal cylindrical pressure vessels arranged below or partly below deck. These independent tanks of Type C are normally designed for working pressures up to 17.5 kg/cm^2, which corresponds to the vapor pressure of propane at 45 °C, the maximum ambient temperature the vessel is likely to operate in. The tanks can be constructed from ordinary grades of steel, are mounted in cradle-shaped foundations, and if below deck are fitted with domes protruding through the deck to which are fitted all connections. Wash bulkheads are fitted in very long tanks. The shape of the tanks generally prevents good utilization of the underdeck space.

Semi-pressurized (or semi-refrigerated) tanks

The capacity of semi-pressurized ships ranges up to about 5000 m^3, the cargoes carried being similar to fully pressurized ships. The independent Type C tanks are generally constructed of ordinary grades of steel suitable for a temperature of −5 °C

Table 23.1 Requirements for secondary barrier protection

	Cargo temperature at atmospheric pressure		
Basic tank type	***−10 °C and above*** **No secondary barrier required**	***Below −10 °C down to −55 °C*** **Hull may act as secondary barrier**	***Below −55 °C*** **Separate secondary barrier required**
Integral			Tank type not normally allowed
Membrane			Complete secondary barrier
Semi-membrane			Complete secondary barrier
Independent			
Type A			Complete secondary barrier
Type B			Partial secondary barrier
Type C			No secondary barrier
Internal insulation			
Type 1			Complete secondary barrier
Type 2			Complete secondary barrier incorporated

and are designed for a maximum pressure of about 8 kg/cm^2. The outer surface of the tank is insulated and refrigeration or reliquefication plant cools the cargo and maintains the working pressure. Cargo tanks are often horizontal cylinders mounted on two saddle supports and many designs (see Figure 23.1) incorporate bio-lobe tanks to better utilize the underdeck space and improve payload.

Fully refrigerated tanks

The capacity of fully refrigerated ships ranges from 10,000 to 100,000 m^3, the smaller ships in the range being multi-product carriers whilst the larger vessels tend to be single-product carriers on a permanent route. Tanks fall almost exclusively into the prismatic, independent Type A category with tops sloped to reduce free surface and bottom corners sloped to suit the bilge structure. In most cases they are subdivided along the center line by a liquid-tight bulkhead that extends to the underside of the dome projecting through the deck that is used for access and piping connections, etc. The tanks sit on insulated bearing blocks so that surfaces are accessible for inspection and are located by anti-roll and pitch keys in such a manner that expansion and contraction can take place relative to the ships structure. Antiflotation chocks are provided to prevent the tank floating off the bearings if the hold were flooded. Tanks are constructed of a notch ductile steel for the normal minimum operating temperature of −43 °C, the boiling point of propane.

The ship has a double hull extending over the bottom and bilge area, the secondary barrier being provided by low-temperature (notch ductile) steel at the inner bottom, sloping bilge tank, part side shell, and sloping bottom of topside tank. Transverse bulkheads may be single- or double-plate (cofferdam) type between cargo holds. Insulation can be either on the tank or the secondary barrier for this type of ship.

Liquefied natural gas ships

There are over 20 approved patent designs of containment vessel for LNG ships, the majority of which fall into the membrane or independent tank categories. Those types that have been or are more commonly found in service are described below. A feature of LNG ships is their double-hull construction, within which are fitted the cargo tanks and the secondary barrier system.

At the beginning of 2011, some 350 large LNG ships were trading; older ships have independent Type B tanks of the Kvaerner-Moss design with most being of the membrane type. The GAZ Transport membrane system is twice as prevalent as other membrane systems.

Independent Type A tanks

Early LNG ships such as the 'Methane Princess' and 'Methane Progress' were fitted with self-supporting tanks of aluminum alloy having center-line bulkheads (see Figure 23.2). The balsa wood insulation system was attached to the inner hull

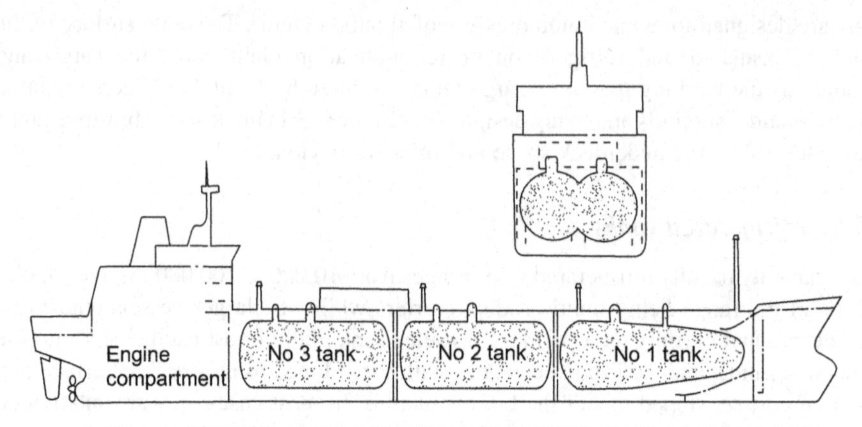

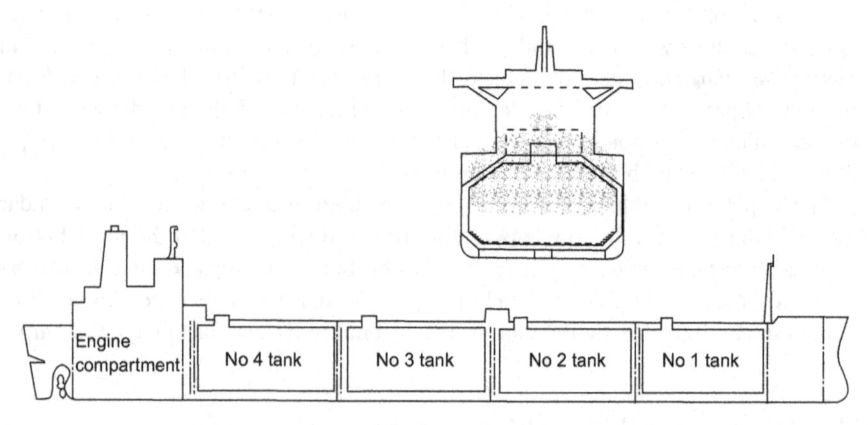

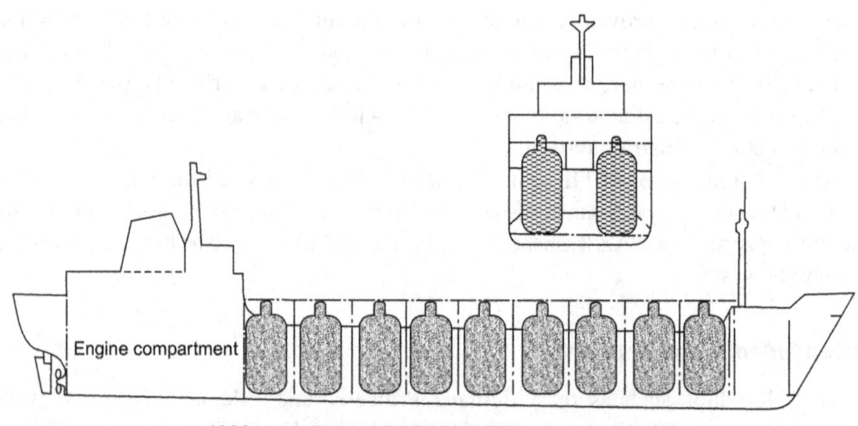

Figure 23.1 Liquefied gas carriers.

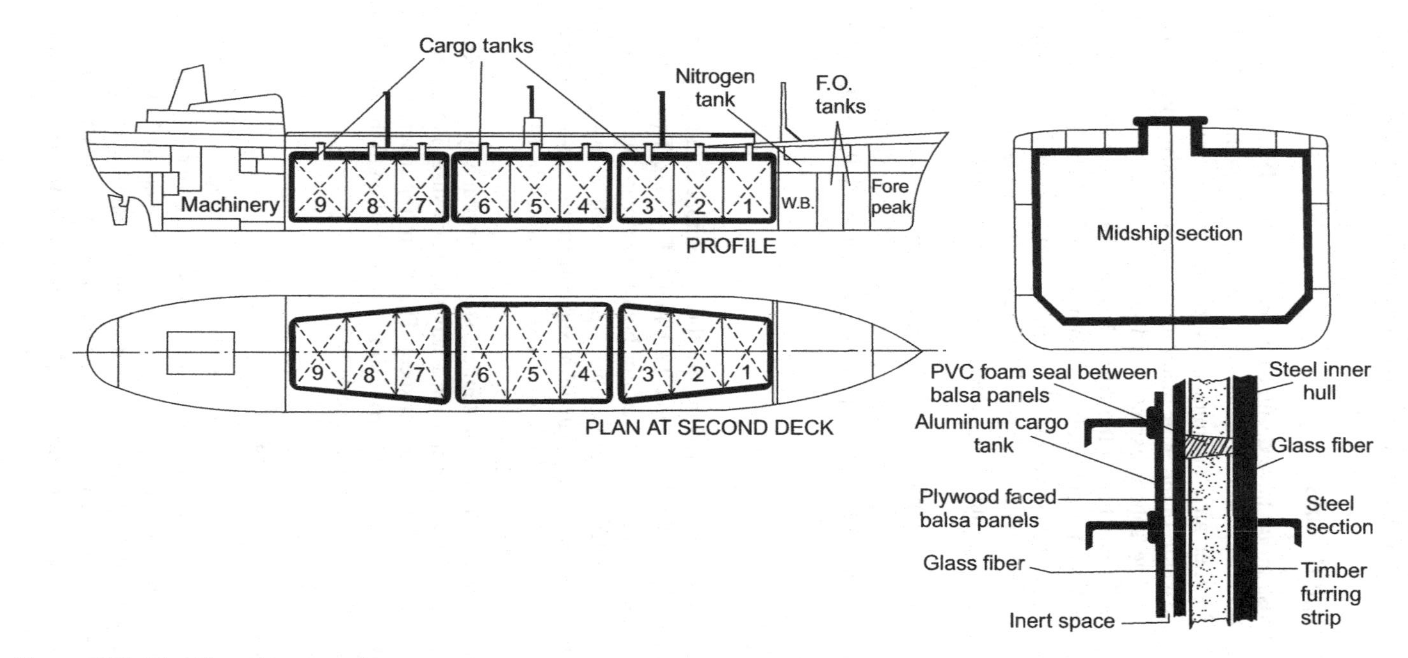

Figure 23.2 Liquid methane carrier.

(secondary barrier) and each insulated hold contained three tanks. Later vessels built with tanks of this category have adopted a prismatic tank design.

Independent Type B tanks

The Kvaerner-Moss group have designed an independent Type B tank containment system that has been well accepted and is installed in a good number of LNG ships. Tanks consist of either an aluminum alloy or 9% nickel steel sphere welded to a vertical cylindrical skirt of the same material, which is its only connection to the hull (see Figure 23.3). The sphere expands and contracts freely all movements being compensated for in the top half of the skirt. The outer surface of the sphere and part of the skirt is covered with a polyurethane foam insulation. The system is fitted with a partial secondary barrier consisting of a drip tray under the tank and splash shields at the sides. In accordance with its Type B notation, each tank is provided with sensors that will detect leakage and allow timely repairs before any crack reaches critical proportions.

Spherical tanks make poor use of available cargo space, a substantial hull being required to house, say, five large spheres providing a cargo-carrying capacity of 125,000 m^3. Above deck the spheres are protected by substantial weather covers.

Membrane tanks

Two common membrane tank designs are those developed and associated with the French companies Gaz Transport and Technigaz. The Gaz Transport system uses a 36% nickel–iron alloy called 'Invar' for both the primary and secondary barriers. Invar has a very low coefficient of thermal expansion, which makes any corrugations in the tank structure unnecessary. The Invar sheet membrane used is only 0.5–0.7 mm thick, which makes for a very light structure. Insulation consists of plywood boxes filled with perlite (see Figure 23.4).

The Technigaz system utilizes a stainless steel membrane system where tanks are constructed of corrugated sheet in such a way that each sheet is free to contract and expand independently of the adjacent sheet. This forms the inner primary barrier and a balsa insulation and secondary barrier similar to that fitted to the Independent Type A tanks described earlier is fitted (see Figure 23.4).

Semi-membrane Type B tanks

The Japanese ship builder IHI has designed a semi-membrane, Type B tank to carry LNG cargoes that has been used in a number of LPG carriers. The rectangular tank consists of plane unstiffened walls with moderately sloped roof and rounded edges and corners that are not supported so that expansion and contraction is accommodated. The tank is of 15- to 25-mm-thick aluminum alloy supported on a layer of PVC insulation and the partial secondary barrier is made of plywood 25 mm thick integral with a PVC foam insulation.

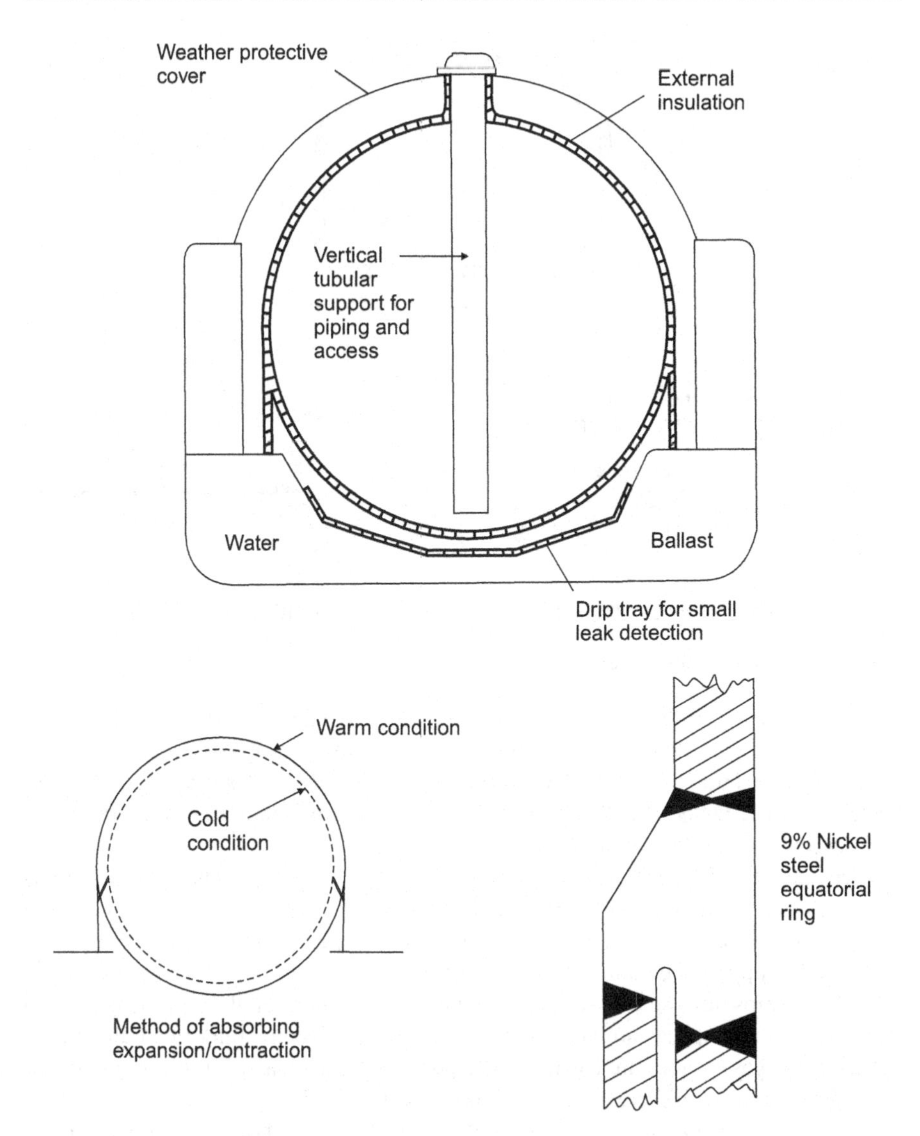

Figure 23.3 Kvaerner-Moss spherical tank.

General arrangement of gas carriers

Gas carriers have a similar overall arrangement to tankers in that their machinery and accommodation are aft and the cargo containment is spread over the rest of the ship to forward where the forecastle is fitted.

Specific gravity of LPG cargoes can vary from 0.58 to 0.97, whilst LNG ships are often designed for a cargo specific gravity of 0.5 so that a characteristic of LNG ships in particular and most LPG ships is their low draft and high freeboards. Water ballast

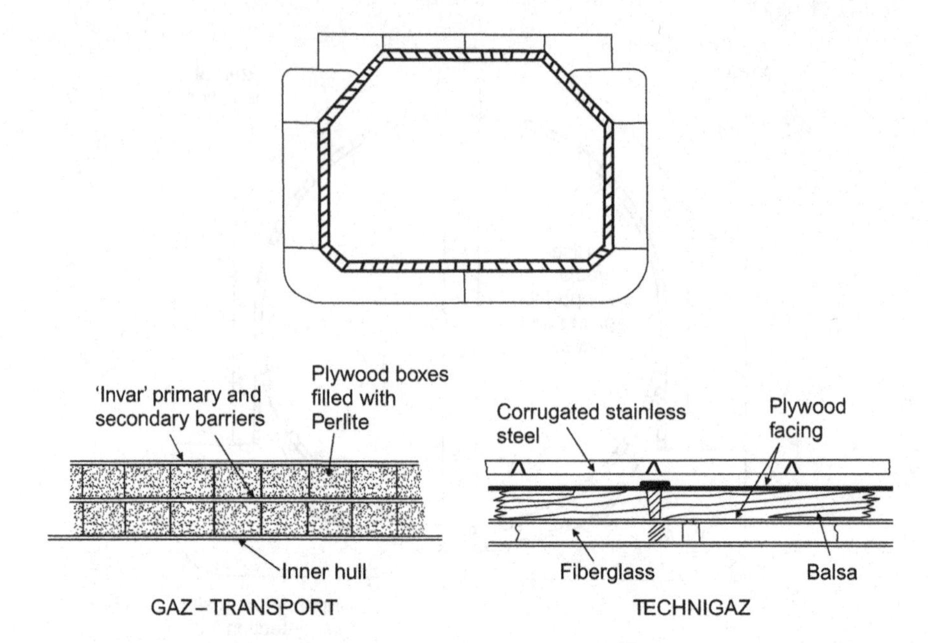

Figure 23.4 Membrane systems.

cannot be carried in the cargo tanks so adequate provision is made for it within the double-hull spaces, double-bottom, bilge tank, and upper wing tank spaces.

The double-hull feature of LNG carriers and many LPG ships is a required safety feature and the tanks of LPG ships that do not have this feature are required to be a minimum distance inboard of the shell.

Fore end and aft end structure is similar to that for other ships. The cargo section is transversely or longitudinally framed, depending primarily on size, in the same manner as other cargo ships, the inner hull receiving special consideration where it is required to support the containment system.

All gas ships have spaces around the tanks that are monitored for gas leaks and in many ships these spaces are also inerted, an inert gas system being fitted aboard the ship. Liquid gas cargoes are carried under positive pressure at all times so that no air can enter the tanks and create a flammable mixture.

Liquefaction equipment is provided aboard LPG ships; 'boil-off' vapor from the tanks due to any heat ingress is drawn into the liquefaction plant and returned to the tank. Boil-off vapor from LNG ship tanks can be utilized as a boiler fuel in steam ships, otherwise it is vented to atmosphere, although this is not permitted in many ports, and several other solutions have been developed to overcome this problem.

Lloyd's classification

For liquefied gas ships Lloyd's Register may assign either one of two classes, namely '100A liquefied gas tanker' where the vessel is designed to carry liquefied gases in bulk in integral or membrane tanks, or '100A1 liquefied gas carrier' where the vessel is designed to carry liquefied gases in bulk in independent tanks. Class notations in respect of the type of tanks, names of gases carried, maximum vapor pressure, minimum cargo temperature, etc. may be added.

Further reading

Code for the Construction and Equipment of Ships Carrying Liquefied Gases in Bulk. IMO publication (IMO-782E), 1993.

Design and operation of gas carriers, 2004, Royal Institution of Naval Architects Conference Proceedings.

Design and operation of gas carriers, Supplement to *The Naval Architect,* September 2004.

Ffooks R: *Natural Gas by Sea: The Development of a New Technology,* ed 2. 1993, Witherby & Co. Ltd.

Gas ships—Trends and technology, *The Naval Architect,* 2006.

Gas ships—Trends and technology, Supplement to *The Naval Architect,* October 2006.

Harris S: *Fully Refrigerated LPG Carriers.* Witherby & Co. Ltd.

Harper I: *Future Development Options for LNG Marine Transportation,* Maldon, UK. American Institute of Chemical Engineers, Spring National Meeting, March 2002, Wavespec Limited.

International Code for the Construction and Equipment of Ships Carrying Liquefied Gases in Bulk (IGC Code), 1993, IMO publication (IMO-104E).

SOLAS Consolidated Edition *see Chapter VII, Part C: Construction and equipment of ships carrying liquefied gases in bulk,* 2004, IMO publication (IMO-110E).

Some useful websites

www.wavespec.com A copy of *Future Development Options for LNG Marine Transportation* mentioned in the list above is available from this website.

http://www.thedigitalship.com/powerpoints/norship05/lng/Jim%20MacDonald%20LR.pdf

Part Six

Outfit

24 Cargo lifting arrangements

Chapter Outline

Shipboard cranes 293
Masts and Sampson posts 294
Mast construction and stiffening 297
Derrick rigs 297
Forces in derrick rigs 300
Initial tests and re-tests of derrick rigs 304
Further reading 305
Some useful websites 305

The past several decades have seen many changes in cargoes and thus in cargo handing. Significant changes have been in the use of containers for mixed cargoes, replacing traditional break-bulk dry cargo ships, bulk carriers for coal, ore, grain and similar cargoes, and roll-on roll-off ships. Faster loading and discharge of cargoes has demanded faster handling, with cargoes in greater units, or continuous loading for example for bulk carriers. There has been a very considerable reduction in the number of shipyard-built and rigged derricks as shipboard lifting devices. Various systems have replaced these, in particular cranes. Cranes have greater capacity, are less labor intensive, faster operating, more easily controlled, and require less deck area. They do, however, require a higher degree of onboard maintenance.

When ordering a new ship the shipowner normally specifies the number, safe working load, position, and any special features of the cargo lifting devices to be fitted. The shipbuilder or a specialized supplier will determine the equipment and arrangements required to satisfy the owner's requirements.

Shipboard cranes

Many ships rely on shore-based loading and unloading, especially the larger ships that require high discharge rates for a rapid turnaround. Larger ships also are more likely to run on fixed routes, between larger terminals that are fully equipped for rapid loading and unloading. Minimizing port time is an important economic requirement.

Panamax and Capesize bulk carriers usually have no cranes and are described as gearless. Larger ships would require larger capacity cranes with a long outreach, which would be more expensive. Such cranes are available, but shore-based cranes

Ship Construction. DOI: 10.1016/B978-0-08-097239-8.00024-6

have higher productivity. There are also shore-based continuous loading/unloading systems, for example grain elevators and ore conveyors.

Handy sized bulk carriers and those not trading on fixed routes often have center-line-mounted deck cranes that are usually geared, typically with three or four cranes. These may be of 30 tonnes capacity and outreach up to 30 meters. Heavy-duty cranes can have lifting capacities from 25 to 50 tonnes when using a cargo grab and have an outreach up to 36 meters. The bulk carrier cranes may also be used for lifting hatch covers if these are not of the sliding or side rolling types.

Figure 24.1 shows in section the arrangement of one of three 30-tonne safe working load (SWL) wire luffing cranes fitted to one side of a 70,000-tonne deadweight bulk carrier. This crane installation allows the ship to transsport cargoes outside draft restricted ports, the three cranes in this instance being capable of handling somewhat more than 10,000 tonnes in a 24-hour period.

Feeder container ships may be fitted with cranes, with an SWL of up to 45 tonnes, to facilitate loading/discharging containers in ports where container handling facilities are not available ashore (see Figure 24.2). Typically, two or three cranes per ship are provided with lifting capacities in the range of 36–45 tonnes, with outreaches between 26 and 40 m. With ship cranes, cargo spotting may easily account for as much as 40% of total crane cycle time, especially when slotting unit loads. Newer cranes have electronic control systems that are claimed to significantly reduce the time taken, increasing productivity. These cranes may also be used for handling pontoon hatch covers and are generally fitted to one side of the ship (see Figure 24.2) with the structure below substantially reinforced to carry the loadings.

A wide range of standard shipboard general-purpose cranes with SWLs from 5 to 60 tonnes are found on general cargo and reefer ships. The electrohydraulically powered cranes may have single or twin jibs and wire or ram luffing arrangements. These cranes may be positioned on the ship's center line, but this may require an extremely long jib when the ship's beam is large and a reasonable outreach is desired. Pairs of cranes at one end of the hatch or at opposite corners of the hatch and with an ability to work through 360° can provide a full range of load/discharge options for a ship.

There has been a recent move to replace hydraulic power with electric power. This is said to reduce the costs of outfitting the ship, and also to reduce the potential for environmental problems caused by loss of oil. The electric crane is supplied as a packaged unit, which effectively only has to be bolted into place on its support and the electric power connections made. The energy consumption of electric cranes is reportedly lower. The cranes are also easier to control, making for faster cargo handling.

Tailor-made heavy lift cranes of 400- to 800-tonne capacity have been supplied for offshore and heavy-lift ships. In some cases the latter have been installed on one side of the hull in order to perform a tandem lift with a maximum capacity of 1600 tonnes.

Masts and Sampson posts

Masts on a general cargo ship may fulfill a number of functions, but their prime use in modern ships is to carry and support the derricks used for cargo handling. Single

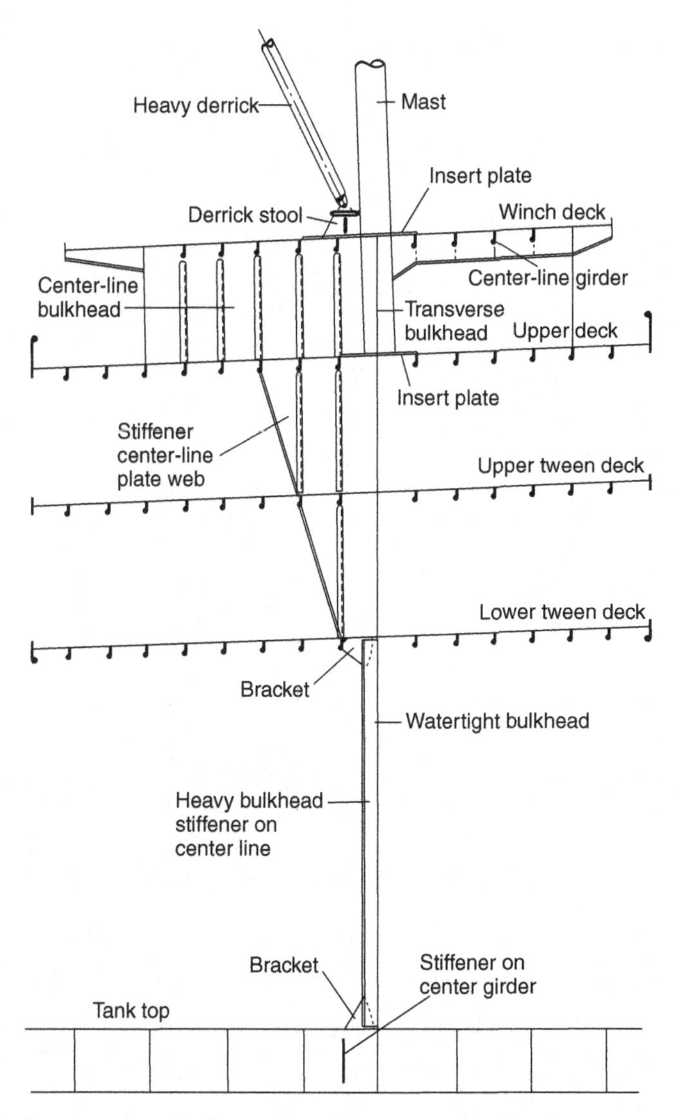

Figure 24.1 Cross-section of bulk carrier with side mounted 30-tonne wire luffing cranes.

masts are often fitted, but many ships now have various forms of bipod mast, which are often more suitable for supporting derricks, although some types can restrict the view from the bridge. Sampson posts are often fitted at the ends of houses and may also be found at the other hatches.

The strength of masts and Sampson posts is indicated by the classification societies. As a result of the span loads and derrick boom thrusts, a single mast or post may be considered similar to a built-in cantilever with axial and bending loads. Some torque may also be allowed for where the post has a cross-tree arrangement to an

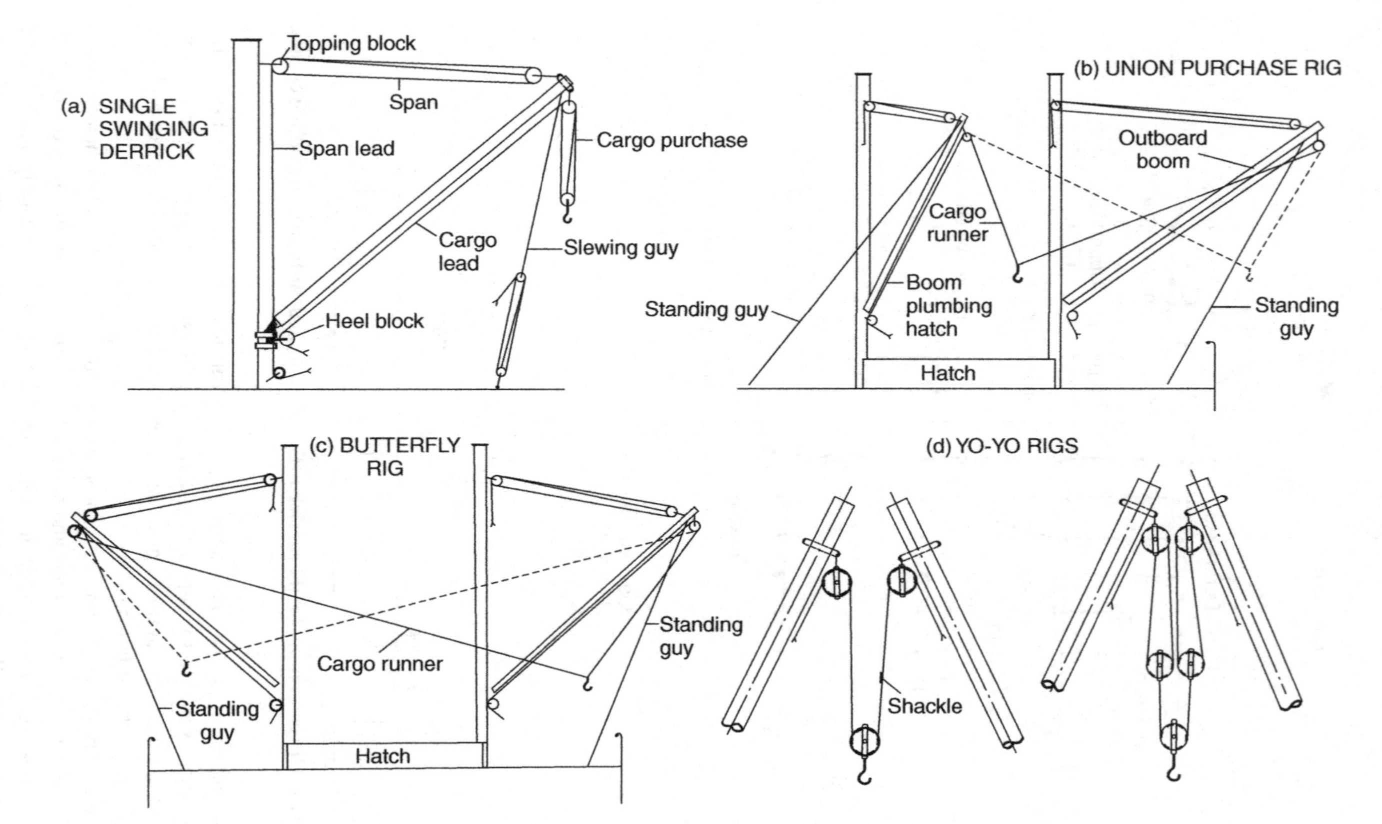

Figure 24.2 520 TEU container feeder ship fitted with two 40-tonne ram luffing cranes.

adjacent post. Where shrouds and preventers are fitted these must be allowed for, which makes the calculations somewhat more difficult. In modern ships there is a tendency to simplify the rigging, which can restrict cargo handling. Shrouds are often dispensed with and preventers may only be rigged when heavy derricks are used. Each mast or post has adequate scantlings so that they may remain unstayed.

Mast construction and stiffening

Tubular steel sections are commonly used in mast and post construction, the sections being rolled in short lengths and welded in the shipyard. The short lengths may be tapered and are of different plate thicknesses to allow for the greater stresses experienced at the base of the mast. Where connections are made for fittings such as the gooseneck and a masthead span swivel, doubling or welded reinforcing pads may be provided. To obtain the necessary mast scantlings, excessive doubling or internal stiffeners are rarely found in modern practice, except where a heavier derrick than that for which the mast was originally designed is carried. Higher tensile steels are often used to advantage in mast construction, giving less weight high up in the ship and dispensing with the need for any form of support, without excessive scantlings.

Cross-trees, mast tables, etc. may be fabricated from welded steel plates and sections.

Derrick booms are as a rule welded lengths of seamless tubular steel. The middle length may have a greater diameter to allow for the bending moment, to which the boom is subject in addition to the axial thrust.

At the base of the mast adequate rigidity must be provided, the amount of additional structural stiffening increasing with the size of derricks carried by the mast. Many cargo ships have mast houses into which the masts are built, the house being suitably strengthened. These houses need not be designed to support the mast, the structure being of light scantlings and the support provided by stiffening in the tweens. Where the house is strengthened the masts or posts generally land on the upper deck, but where heavy derricks are installed the mast may then land on the upper tween deck. Since the derricks and mast are as a rule midway between holds they land over the hold transverse bulkheads, which lend further support.

Heavy derrick masts will require extensive stiffening arrangements in the mast house, and also in the tweens, with support for the transverse bulkhead so that the loads are transmitted through the structure to the ship's bottom. Partial longitudinal and transverse bulkheads with deck girders may provide the mast house stiffening. Stiffened plate webs at the ship's center line in the tweens and heavier stiffeners on the transverse bulkhead in the hold then provide the additional strengthening below decks (see Figure 24.3). Heavy insert plates are fitted in way of the mast at the various decks.

Derrick rigs

Various forms of derrick rig may be used aboard the cargo ship, the commonest use of the single derrick being as a 'single swinging derrick' (see Figure 24.4a). Adjacent

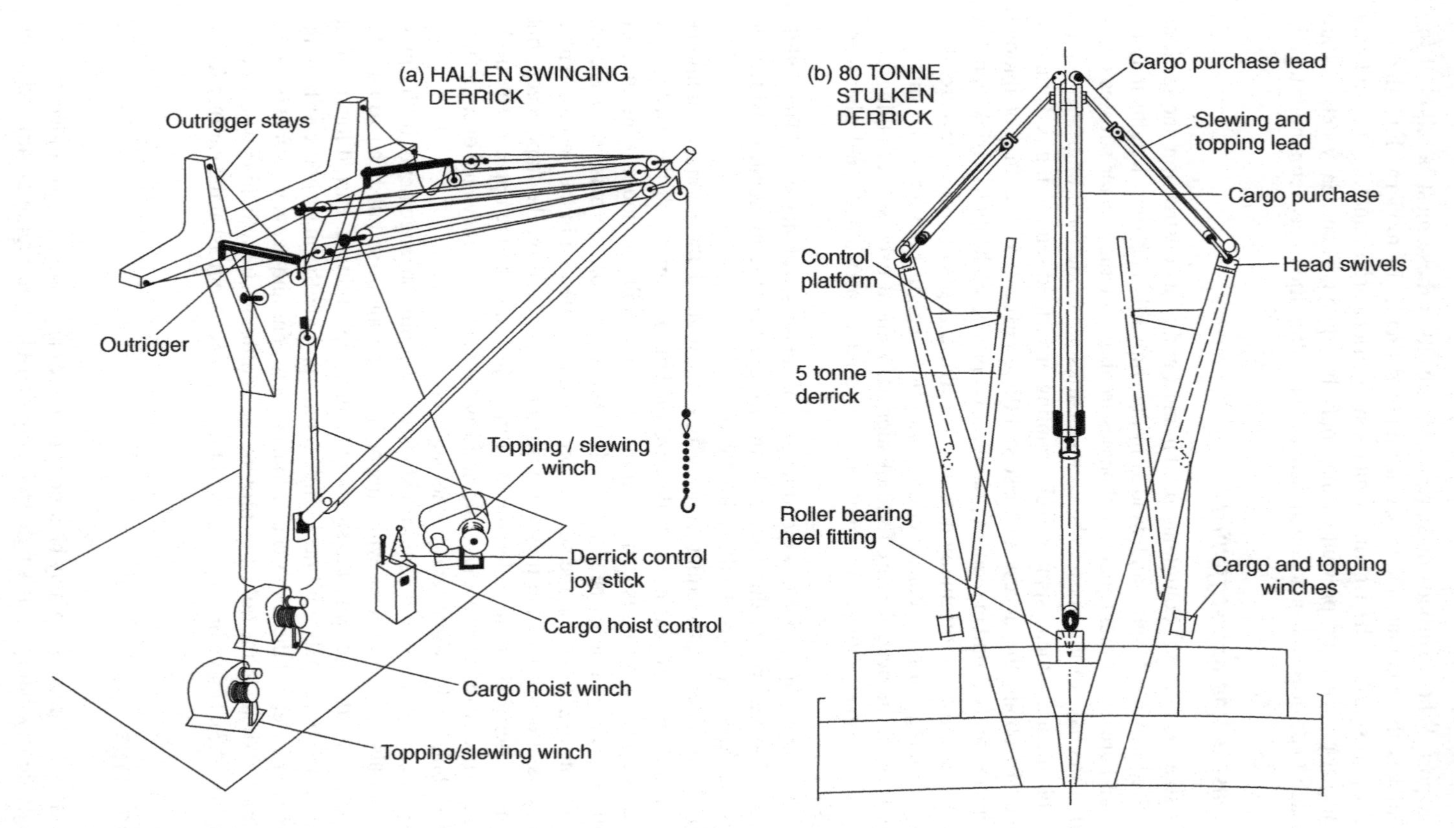

Figure 24.3 Stiffening in way of mast and heavy derrick.

derrick booms may be used in 'union purchase' (Figure 24.4b), the booms being fixed in the overboard and inboard positions. Cargo is lifted from the hatch and swung outboard by the operator controlling the winches for the cargo runner. Variations on this rig are often adopted, for example the 'butterfly' rig (Figure 24.4c), which is used where cargo is discharged from a hold to both sides of the ship.

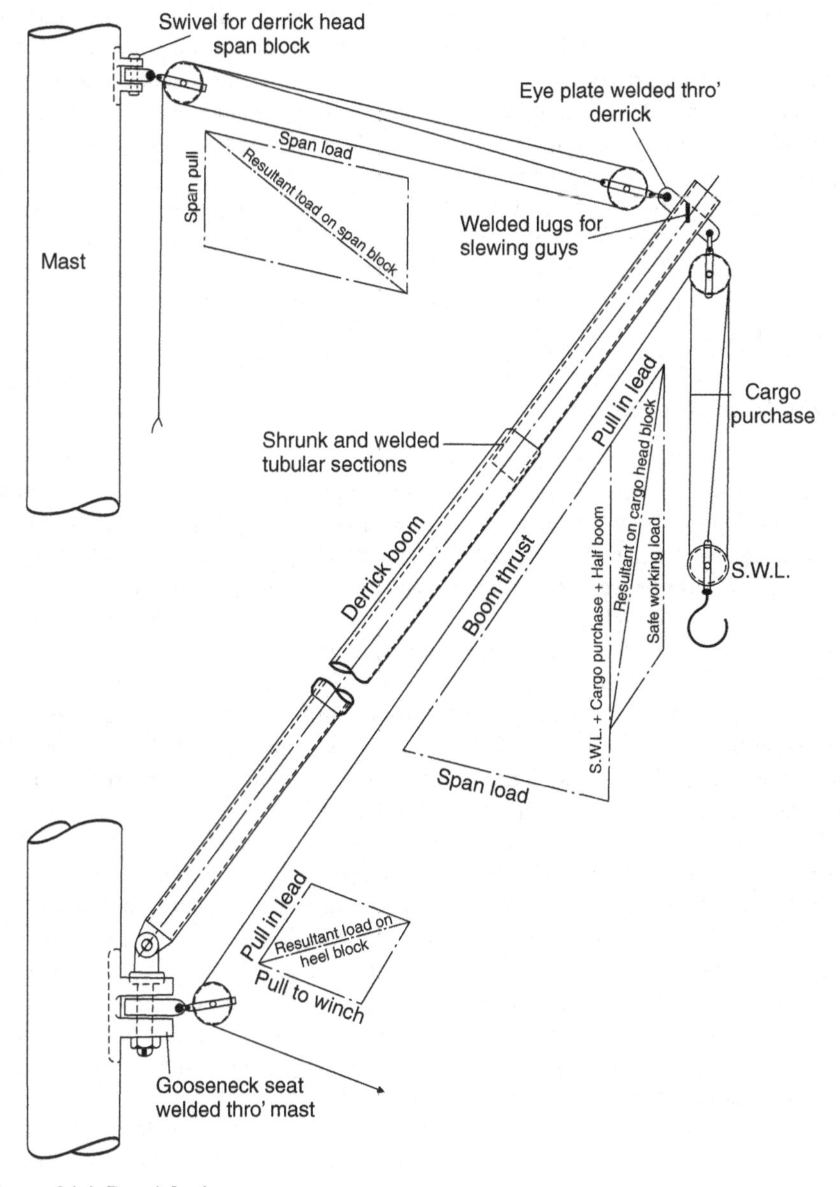

Figure 24.4 Derrick rigs.

Where a weight exceeding the safe working load of the derricks is to be lifted, the 'yo-yo' rig is sometimes adopted (Figure 24.4d). The heads of the derricks are brought together, and a traveling block may be arranged on the cargo runner or a wire connecting the two cargo purchases as illustrated.

Both the 'butterfly' and 'yo-yo' rigs give a load pattern similar to the 'union purchase' and 'single swinging derrick' rigs for which calculations are made, but the guy loads with each can be particularly severe.

Patent derricks are generally of the single swinging type with some form of powered slewing. The Hallen swinging derrick is shown diagrammatically in Figure 24.5a. This type of derrick may be installed at the ship's center line to reach outboard on both sides of the ship and is controlled by a single operator in a manner not unlike the operation of a mechanical crane. As a rule the safe working load of this type of derrick is between 10 and 80 tonnes.

Of particular note in the very heavy lifting range is the patent Stülken derrick (Figure 24.5b) marketed by Blohm and Voss AG, which may have a safe working load of between 80 and 300 tonnes. One advantage that this derrick has is its ability to serve two hatches, the boom swinging through an arc between the posts in the fore and aft directions.

Forces in derrick rigs

The geometry of the derrick rig will to a large extent influence the loads carried by the rig components. Those dimensions that have the greatest influence are the length of boom, the distance between the boom heel and the masthead span connection (height of suspension), and the angle at which the boom is topped.

When the ratio between boom length and height of suspension is increased, the boom thrust will be higher; therefore, should a long boom be required the height of suspension must be adequate. It is not unusual, however, for shipowners to object to having posts at the bridge front and if the height of suspension is then restricted there is some limitation on the boom length, which can make working cargo from that position difficult. The angle at which the derrick is topped has no effect on the axial thrust, but the lead from the cargo purchase often increases the thrust as it is led parallel to the boom on all except heavy lift derricks.

Loads carried by the span are dependent on both the ratio of boom length to height of suspension and the angle at which the derrick is topped. The span load is greater at a lower angle to the horizontal, and increases with longer booms for a given suspension height.

To determine these forces simple space and force diagrams may be drawn and the resultant forces determined to give the required wire sizes, block and connection safe working loads, and the thrust experienced by the boom. The horizontal and vertical components of the span load and boom thrust are also used to determine the mast scantlings. Force diagrams are shown for the rig components of the single swinging derrick illustrated in Figure 24.6.

For a safe working load of 15 tonnes or less the forces may be calculated with the derrick at angles of 30° and 70° to the horizontal unless the owner specifies that the derrick is to be used at a lower angle (not less than 15°). At safe working loads greater

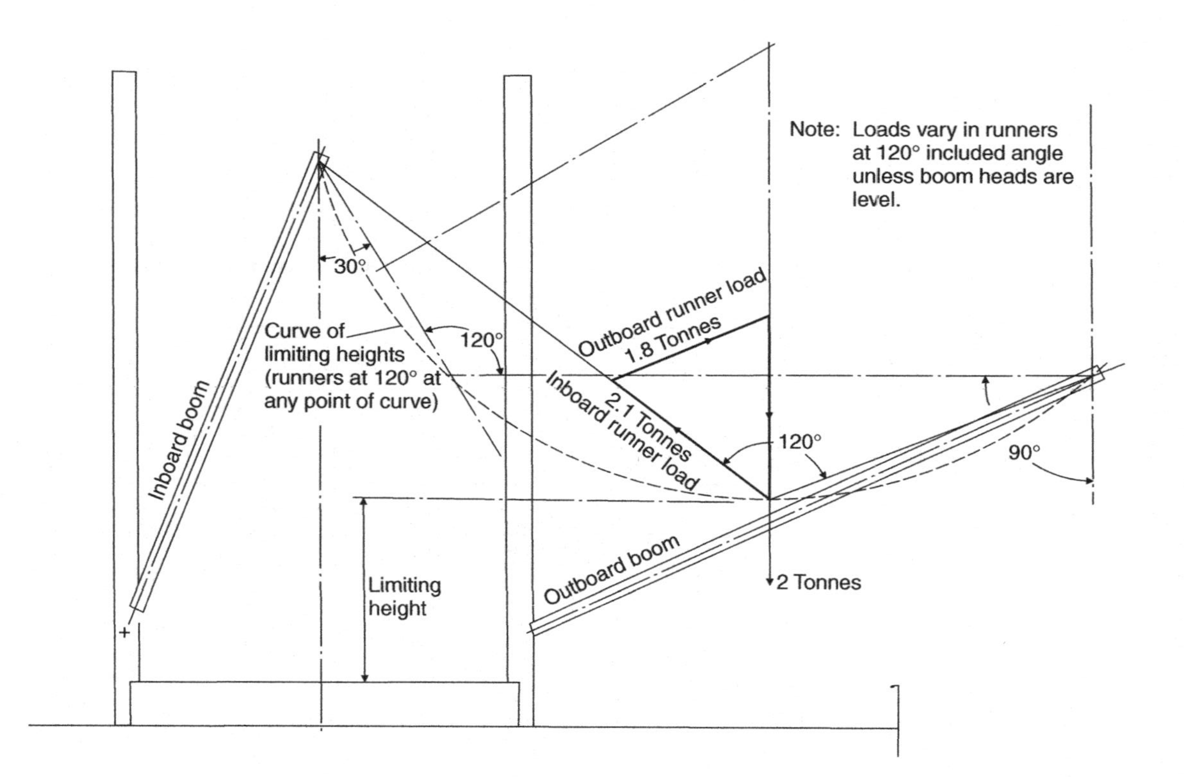

Figure 24.5 Patent derricks.

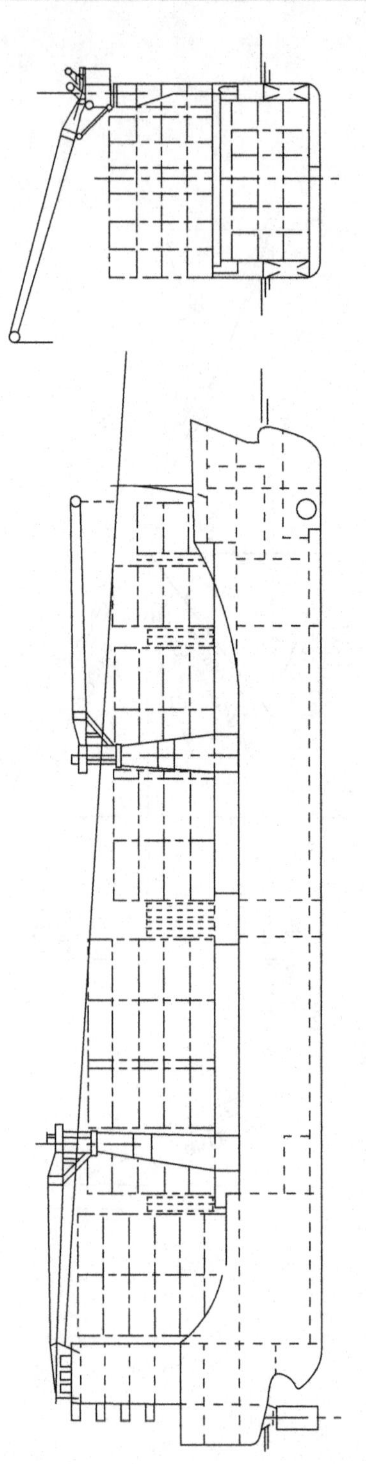

Figure 24.6 Forces in single swinging derrick rig.

than 15 tonnes the forces may be calculated at an angle of 45° to the horizontal. The loads on all the blocks except the lower block of a cargo purchase will be the resultant of the two forces to which the block is subjected. A single-sheave block has a safe working load that is half the resultant, and multi-sheave blocks have a safe working load that is the same as the resultant.

In determining the span loads and boom thrusts, not only is the derrick safe working load considered to be supported by the span, but also the weight of the cargo purchase and half the boom weight. The other half of the boom weight is supported by the gooseneck fitting.

Allowances must be made for the frictional resistance of the blocks when determining the forces. This includes an allowance for the rope friction, i.e. the effort required to bend and unbend the rope around the pulley, as well as an allowance for journal friction. Shipbuilders using British Standards adopt the following assumed cumulative friction values:

Small and medium sheaves	8% sheave with bushed plain bearings
	5% sheave with ball or roller bearings
Large diameter sheaves	6% sheave with bushed plain bearings
	4% sheave with ball or roller bearings
Derrick exceeding 80 tonnes SWL	5% sheave with bushed plain bearings
	3% sheave with ball or roller bearings

Force diagrams that are more involved than those for the single swinging derrick are prepared for the union purchase rig. These diagrams indicate the safe working load of the rig, the 'limiting height', the boom thrusts (which are greater with this rig), and the optimum guy leads. The 'limiting height' is that height below which all positions of the lifted weight will result in an included angle between the outboard and inboard runners of less than 120°. At 120°, if the boom heads are level the inboard and outboard runners will experience a force equivalent to the cargo weight (see Figure 24.7). Usually the runner size determines the safe working load in union purchase, but the thrust experienced by the derrick boom can determine this value where only light derricks are fitted. The positioning of the guys may be important to the loads experienced by the span and the guys themselves. If these are at too narrow an angle to the boom, excessive tension in the guys will result; a good lead is therefore essential. Unfortunately, in practice the magnitude of the guy loads is not always appreciated, but more attention has been paid to this problem of late and preventers are now often set up to reduce the load in the guy. There is available a suitable preventer for this purpose; the use of old runners etc. as preventers should not be tolerated.

In union purchase rigs it is possible to obtain a condition where the load comes off the outboard span, and the boom may then close to the mast under load. This condition is referred to as 'jack-knifing', and may be apparent from the force diagram prepared for the rig, since the triangle of forces does not close. At the design stage the guy positions can be adjusted to avoid this happening. In practice this condition appears to occur occasionally where derricks are used in union purchase at the bridge front. Here the positioning of the guys is made difficult by the presence of the bridge structure, but the correct placing of a suitable preventer should overcome this problem.

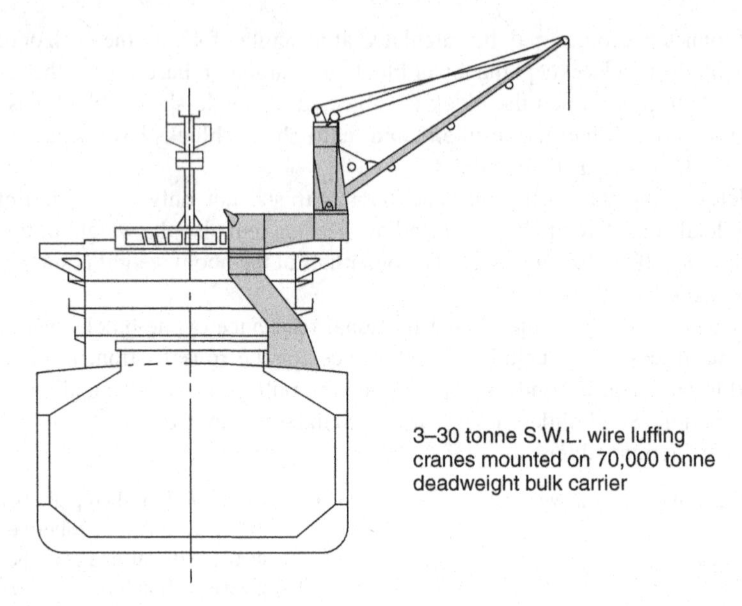

3–30 tonne S.W.L. wire luffing
cranes mounted on 70,000 tonne
deadweight bulk carrier

Figure 24.7 Forces in cargo runners of union purchase rig.

Initial tests and re-tests of derrick rigs

To comply with the national and class regulations, ships' derricks designed to operate as single swinging derricks are initially tested with a proof load that exceeds the specified safe working load of the derrick by the following amounts:

SWL less than 20 tonnes—25% in excess of SWL
SWL 20–50 tonnes—5 tonnes in excess of SWL
SWL over 50 tonnes—10% in excess of SWL.

Heavy-lift derricks are tested at an angle of not more than 45° to the horizontal and other derricks at an angle of not more than 30° to the horizontal. During the test the boom is swung as far as possible in both directions, and any derrick intended to be raised by power under load is raised to its maximum working angle at the outermost position.

Before the test for a heavy derrick it is usual to ensure that the vessel has adequate transverse stability. Before, during, and after all tests it is necessary to ensure that none of the components of the rig show signs of any failure, and it is good practice to have a preventer rigged during the test as a precaution against any of the span gear carrying away. On completion of the test the heel of the derrick boom is clearly marked with:

1. Its safe working load in single purchase.
2. Its safe working load in double purchase, if it is designed for that purpose.
3. Its safe working load in union purchase, if it is designed for that purpose, the letter 'U' preceding the safe working load.

For example,

SWL 3/5 tonnes SWL (U) 2 tonnes

and a certificate of test and examination is issued in an approved form.

Re-tests are required if the rig is substantially modified or a major part is damaged and repaired.

The International Labour Organization (ILO) Convention, 152 adopted, 25 June 1979, requires thorough examination by a competent person once every 12 months and re-testing at least once every 5 years.

Further reading

British Standards Institution, BS MA 48:1976, *Code of Practice for Design and Operation of Ships Derrick Rigs.*

British Standards Institution, BS MA 47:1977, *Code of Practice for Ships Cargo Blocks.*

British Standards Institution, BS MA 81:1980, *Specification for Ships Derrick Fittings.*

Code for Lifting Appliances in a Marine Environment, Lloyd's Register, 2003.

House DJ: *Cargo Work for Maritime Operations.* Elsevier.

ISO 8431:1988, Shipbuilding—Fixed jib cranes—Ship-mounted type for general cargo handling, 2007.

Some useful websites

www.macgregor-group.com See 'General cargo ships—Cranes'.

25 Cargo access, handling, and restraint

Chapter Outline

Stern and bow doors 307
Ramps 308
Side doors and loaders 309
Portable decks 311
Scissors lift 312
Cargo restraint 312
Further reading 314
Some useful websites 314

To speed cargo handling and storage in modern ships, apart from changes in ship design (Chapter 3), the introduction of mechanically handled hatch covers (Chapter 19) and improved lifting devices (Chapter 24), various patented or specially manufactured items may be brought into the shipyard and fitted to the ship by the shipbuilder. Some notable items that fall into this category are described in this chapter. These primarily relate to cargo access handling and restraint in ro-ro ships, container ships, car carriers, and vessels in which palletized cargo is carried.

Stern and bow doors

Roll-on roll-off (ro-ro) vessels are used primarily on short sea crossings and are designed to carry road vehicles. They provide a link in a road journey, often overnight, and the speed of loading and unloading is often critical. The vehicles are driven on and off using large doors and a series of ramps to access different deck levels.

Ro-ro vessels may be fitted with stern doors of the hinge-down or hinge-up type, which if large are articulated. Bow doors are either of the visor type or of the side-hinged type ('barn door' type). These are situated above the freeboard deck and where the bow doors lead to a complete or long forward enclosed superstructure, Lloyd's require an inner door to be fitted that is part of the collision bulkhead. This would also be in keeping with the SOLAS requirements for passenger ships, where the collision bulkhead is to be extended weathertight to the deck next above the bulkhead deck, but need not be fitted directly above that bulkhead. A sloping weathertight vehicle ramp may be fitted in some ships to form the collision bulkhead above the freeboard deck

Ship Construction. DOI: 10.1016/B978-0-08-097239-8.00025-8

and the inner door is omitted. This ramp may extend forward of the specified limit for the collision bulkhead above a height of more than 2.3 m above the bulkhead deck, i.e. above the height of a conventional tween deck space. Stern and bow door strengths are equivalent to the strength of the surrounding structure and where they give access to enclosed superstructures they are required to close weathertight.

Stern doors and bow visors can be mechanically raised and lowered with wire rope and purchase arrangements, but in general they and the side-hinged bow doors are hydraulically opened and closed (see Figure 25.1). These weathertight doors are casketed and cleated.

Ramps

Ro-ro ships fitted with ramps usually have a stern ramp (see Figure 25.1b), but some vessels fitted with bow doors may also have a bow ramp that doubles as the inner weathertight door and is lowered onto a linkspan when the bow visor or side-hinged doors have been opened. Ramps may also be fitted internally to give access from deck to deck. These can be hydraulically or mechanically tilted to serve more than one deck and can be fixed in the horizontal position to serve as decks themselves (see Figure 25.2). In some ships they can even be raised into the hatch space and serve as weathertight covers.

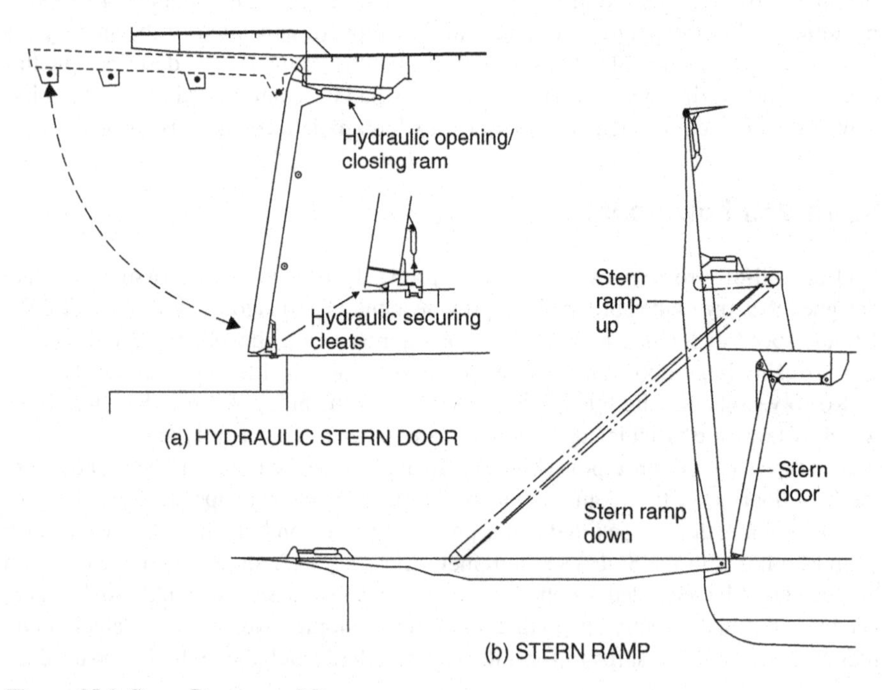

Figure 25.1 Stern Ramps and Doors.

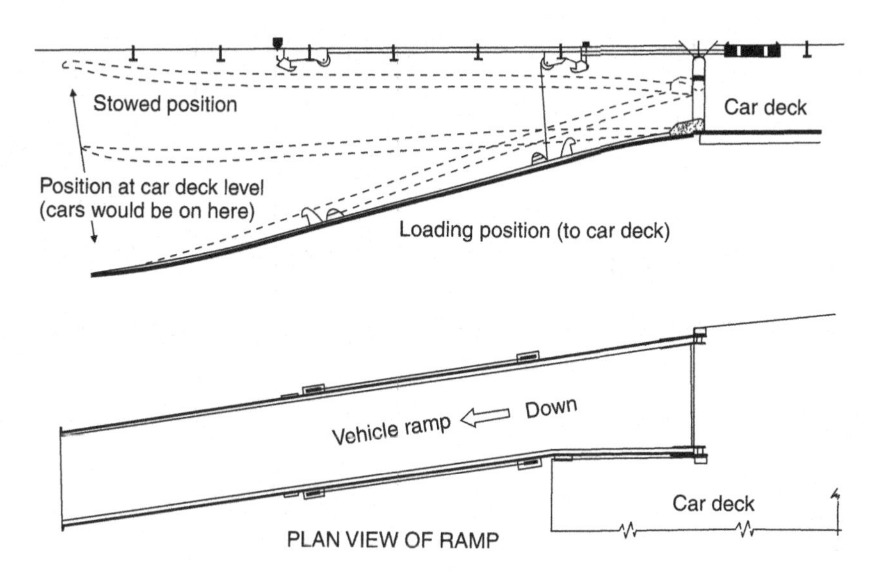

Figure 25.2 Internal Deck Access Ramps.

Stern ramps can be fixed axial ramps, fixed quarter ramps, slewing ramps, or semi-slewing quarter ramps (see Figure 25.3). The axial stern ramp may also serve as the stern door and can be lowered or raised hydraulically or by wire rope arrangements. The quarter ramp was designed for ro-ro ships using ports that are not provided with right-angled quays or link span connections. The large articulated quarter ramp is raised and lowered by wire rope purchase arrangements to hydraulic winches. Slewing ramps serve a similar purpose to the quarter ramp, but are more flexible. The slewing ramp moves around the stern on a curved guide rail, the movement being affected by the lifting and lowering wire purchases, which are led to hydraulic winches.

Side doors and loaders

Side door/ramps are available for ro-ro operations and are similar to stern door/ramp installations. Most side door installations, however, are intended for quayside fork-lift operations with palletized cargo being loaded onto a platform at the door by the quayside fork-lift and stowed in the ship by another fork-lift truck. Instead of a loading platform on ships trading to ports with high tidal ranges a ramp onto which the quayside fork-lift truck drives may be fitted. Elevator platforms may be fitted immediately inboard of the side door to service various tween decks and the hold. A particular type of elevator system is that developed for the transportation of paper products, especially newsprint. The quayside fork-lift places the newsprint rolls on the height-adjustable loading platform, which together with the elevator platform is fitted with roller conveyors. Movement of the roller conveyors is automatic, the

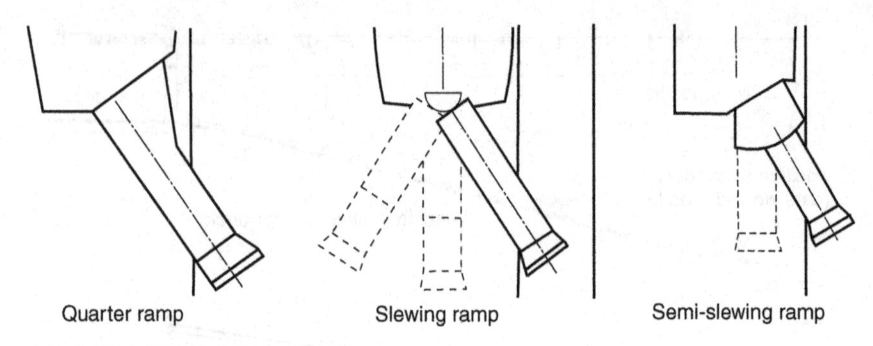

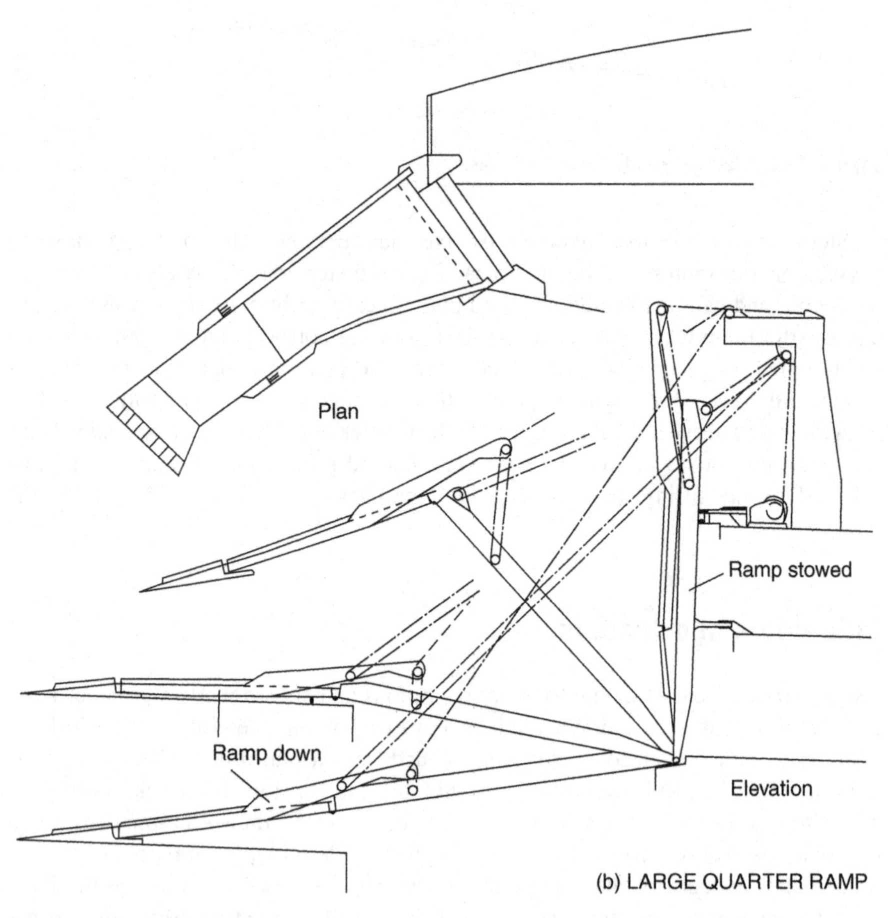

Figure 25.3 Stern and Quarter Access Ramps.

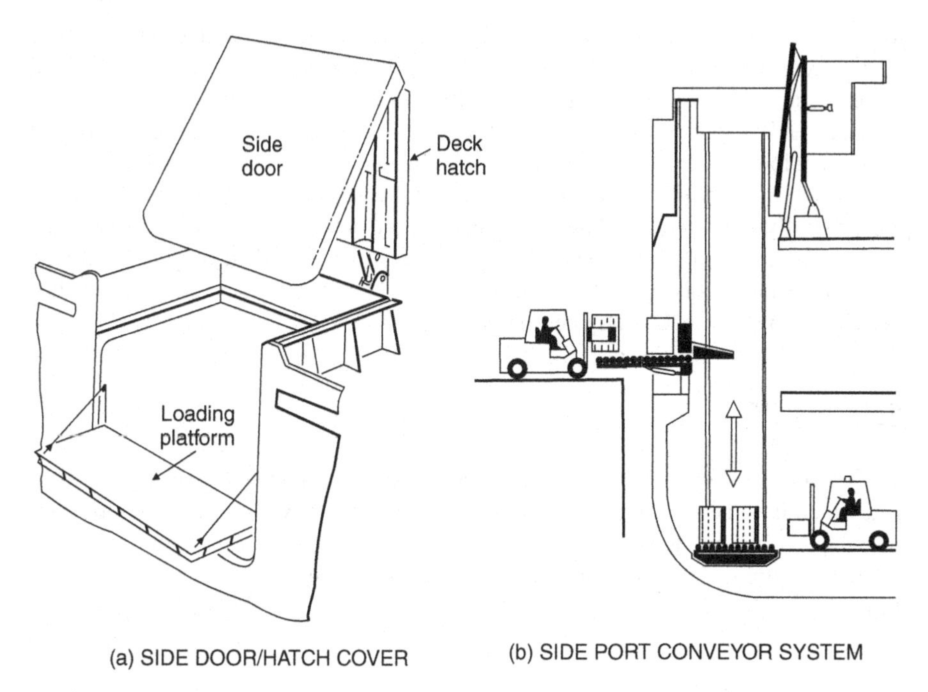

Figure 25.4 Side Doors and Elevators.

newsprint rolls being transferred from the loading platform to the elevator platform, which travels to the preselected deck or hold level for unloading (see Figure 25.4).

Upward folding doors with hydraulic cylinders actuating the hinge are usually fitted to the side opening, the load platform being fitted inside the door and hinged at the bottom of the opening, automatically being lowered when the door is opened. Combined side door/hatch covers are fitted in designs where the ship is low in the water relative to the height of quay in order to provide sufficient head room for fork-lift truck operation (see Figure 25.4). With the side port elevator system referred to above, a combined door/hatch is fitted to the hatch carrying part of the tower that houses the upper part of the cargo elevator.

A side loader that dispenses with the need for a side door is the MacGregor-Navire International AB 'Rotoloader'. This can be a fixed or portable installation. The unit load is raised from the quay to a point above the ship's side, swung inboard through 180° on a rotating frame unit and lowered through the hatch to the hold or tween.

Portable decks

Portable decks are fitted in a variety of ships, permitting flexibility of stowage arrangements and allowing totally different cargoes to be carried on different voyages. An extreme example is a 50,000-tonne deadweight bulk carrier fitted with hoistable car decks stowed under the hold wing tanks when taking ore from Australia to Japan, and lowered for the return voyage when 3000 cars are carried. The car deck is the most common form of portable deck and common in ro-ro ferries. Either commercial vehicles can be loaded using the full height between decks or, with the portable deck in place, two levels of cars can be carried in the same space. Hoistable decks are lowered from and stowed at the deckhead by hoist wires led through a hydraulic jigger winch. Folding decks stow at the sides and ends of ship spaces and are generally hydraulically lowered into the horizontal position. Lloyd's Register includes requirements for movable decks in their rules and if the ship is fitted with portable decks complying with these rules at the owner's or builder's request the class notation 'movable decks' may be assigned.

Scissors lift

Cargo can be lowered or raised between decks or to the hold by means of a scissors lift, which is sometimes fitted in ro-ro ships as an alternative to internal ramps, as it takes up less room. However, the scissor lift is slower in operation than ramps. The hydraulic cylinder-powered scissors lift is also often designed to transfer heavy unit loads.

Cargo restraint

In ro-ro and container ships the lashing of cargo is an important safety consideration and usually calls for fittings that will permit rapid and easy but effective securing of the cargo because of short ship turnaround times. The shipbuilder is responsible for the deck and perhaps hatch fittings for the securing devices, and will look to the ship operator for guidance on their type and positions. On the decks of ro-ro ships where the direction of lashing is unpredictable and vehicles must transverse the fitting, a cloverleaf deck socket in conjunction with an elephant's foot type of end lashing is popular (see Figure 25.5).

Containers have very little strength in any direction other than vertically through the corner posts, thus it is necessary to provide substantial support to the containers when they are on the ship. Stowage of containers is with their longer dimension fore and aft since the ship motion transmitted to cargo is greater in rolling than pitching and it is therefore prudent to limit any possible cargo movement within the container to the shorter transverse dimension. Also, of course, when offloading the fore and aft container is more easily received by road or rail transport. Below-decks containers are restrained in vertical cell guides, which are typically $150 \times 150 \times 12$ angles and they are structurally supported so that any dynamic forces other than purely vertical are transmitted as much as possible through the ship's structure and not into the containers. The cell guides are not to form an integral part of the ship's structure, they are to be so designed that they do not carry the main hull stresses. Where four container corners are adjacent the cell guides may be built into a composite pillar (see Figure 25.5). The clearance between container and cell guide is critical. If it is too small the container will

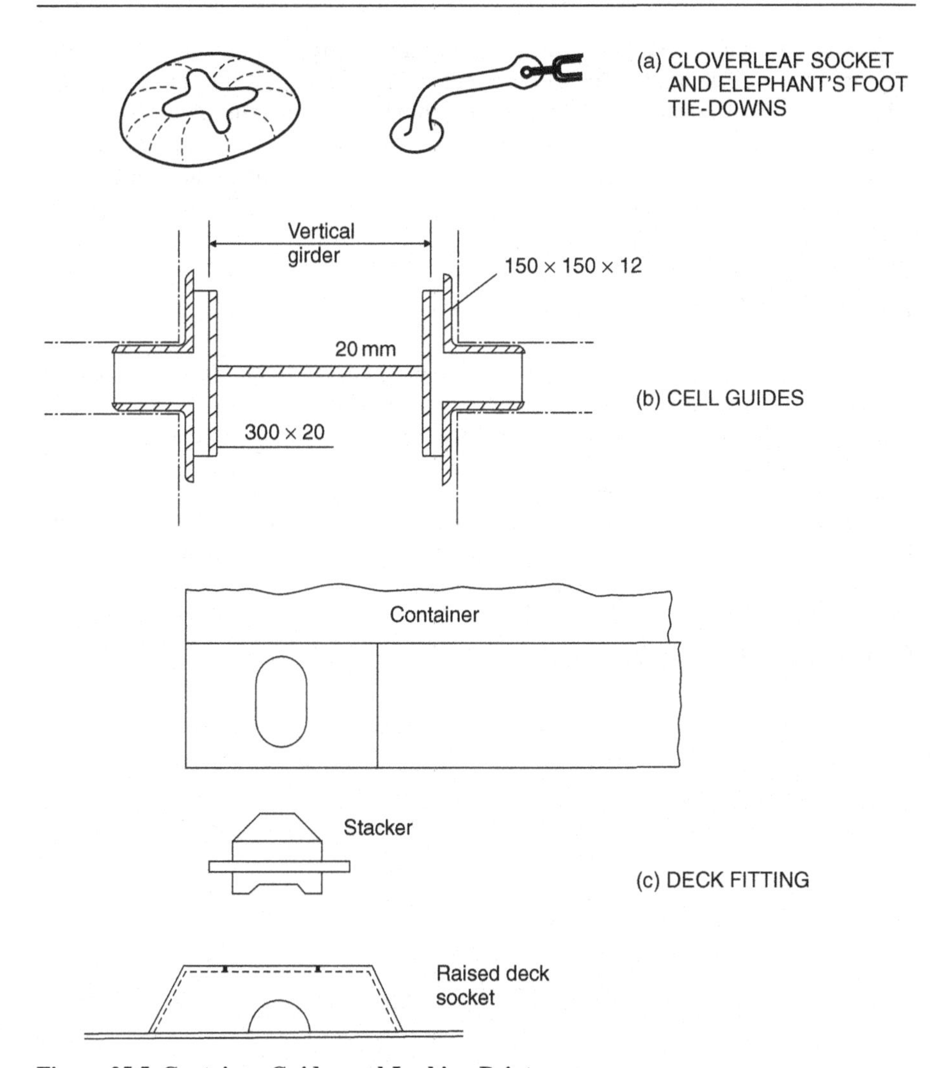

Figure 25.5 Container Guides and Lashing Points.

jam, if it is too large when one container lands on the one below the corner posts and castings that accept a maximum eccentricity may not mate. Lloyd's stipulate a maximum clearance of 25 mm in the transverse direction and 40 mm in the longitudinal direction. The tolerances are such that the cell guides have to be fitted to an accuracy exceeding normal shipyard practice with the use of jigs to ensure the dimensions are maintained following welding. Lloyd's require that the cell guide not deviate from its intended line by more than 4 mm in the transverse direction and 5 mm in the longitudinal direction. Lead-in devices are fitted at the top of the guides.

Above-deck cell guides may also be provided, there being several patented arrangements such as the MacGregor-Navire International AB 'Stackcell' system.

These are not widely used, however, and many ships carrying containers above deck rely on various deck and hatch sockets with locking and nonlocking stackers mating with the standard container corners, plus lashings to secure the containers. With locking stackers fewer lashings are required, therefore the more expensive twistlock is often favored. Deck sockets like the container corner fitting contain the standard ISO hole into which the stackers fit (see Figure 25.5).

Further reading

Anderson, Alexandersson, and Johansson, *Cargo Securing and Cargo Shift on Passenger/ Ro-Ro Vessels*. RINA Conference on RO RO Safety, June 1996.

Code of Safe Practice for Cargo Stowage and Securing. IMO publication (IMO-292E), 2003 edition, 2003.

House DJ: *Cargo Work for Maritime Operations*, ed 7. Elsevier, 2003.

Knot JR: *Lashing and Securing of Deck Cargoes*, ed 3, London, 2002, The Nautical Institute.

RINA Conference, *Safety of Ro-Ro Passenger Vessels*. Conference Proceedings – Royal Institution of Naval Architects Publications, 1996.

Some useful websites

www.tts-se.com For details of ramps, doors, lifts, car decks, side loading systems, etc.

26 Pumping and piping arrangements

Chapter Outline

Bilge and ballast pumping and piping 315
Bilge suctions 316
Bilge pumps and pipe systems 316
Scuppers 318
General service pipes and pumping 318
Air and sounding pipes 319
Sea inlets 319
Cargo pumping and piping arrangements in tankers 320
Single product/crude oil carrier 320
Multi-product tankers 321
Cargo pumps 321
Cargo tank ventilators 321
Cargo tank protection 321
Inert gas system 324
Cargo tank purging and gas freeing 324
Cargo tank washing 324
Further reading 325
Some useful websites 325

In the construction of a merchant ship the shipbuilder is concerned with the installation of the statutory bilge drainage, ballasting, general services, and where required the liquid cargo loading and discharging, pumping and piping arrangements. Arrangements may also be fitted in tankers for tank washing and for introducing inert gas into the tanks.

Bilge and ballast pumping and piping

All cargo ships are provided with pumping and piping arrangements so that any watertight compartment or watertight section of a compartment can be pumped out when the vessel has a list of up to 5° and is on an even keel. In the case of passenger ships, each compartment or section of a compartment may be pumped out following a casualty under all practical conditions whether the ship is listed or not.

Ship Construction. DOI: 10.1016/B978-0-08-097239-8.00026-X

The arrangements in the machinery space are such that this space may be pumped out through two suctions under the above conditions. One suction is from the main bilge line and the other from an independent power-driven pump. An emergency bilge suction is also provided in machinery spaces, and may be connected to the main circulating water pump (for condenser) in steam ships, or the main cooling water pump in motor ships.

Bilge suctions

As the vessel is to be pumped out when listed it is necessary to fit port and starboard suctions in other than very narrow spaces. Generally, vessels are designed to have a moderate trim by the stern in service, and the suctions will therefore be placed in the after part of the compartment. However, where a ship has a single hold that exceeds 33.5 m in length suctions are also arranged in the forward half length of the hold. On many vessels a sloping margin plate is fitted and a natural bilge is formed with the suctions conveniently located within this recess. Adequate drainage to the bilge is provided where a ceiling covers this space. If, however, the tank top extends to the ship sides, bilge wells having a capacity of at least 0.17 m^3 may be arranged in the wings of the compartment. In a passenger ship these bilge wells must not extend to within 460 mm of the bottom shell so as to retain a reasonable margin of safety where the inner bottom height is effectively reduced. The shaft tunnel of the ship is drained by means of a well located at the after end, and the bilge suction is taken from the main bilge line (see Figure 26.1).

At the open ends of bilge suctions in holds and other compartments, outside the machinery space and shaft tunnel, a strum box is provided. The strum box is a perforated plate box welded to the mouth of the bilge line (Figure 26.1), which prevents debris being taken up by the bilge pump suction. Perforations in the strum box do not exceed 10 mm in diameter, and their total cross-sectional area is at least twice that required for the bore of the bilge pipe. Strums are arranged at a reasonable height above the bottom of the bilge or drain well to allow a clear flow of water and to permit easy cleaning. In the machinery space and shaft tunnel the pipe from the bilges is led to the mud box, which is accessible for regular cleaning. Each mud box contains a mesh to collect sludge and foreign objects entering the end of the pipe.

Bilge pumps and pipe systems

Cargo ships have at least two power-driven bilge pumping units in the machinery space connected to the main bilge line, and passenger ships have at least three.

In passenger ships the power-driven bilge pumps are, where practicable, placed in separate watertight compartments, so that all three are not easily flooded by the same damage. Where the passenger ship has a length in excess of 91.5 m it is a requirement that at least one of these pumps will always be serviceable in reasonable damage situations. A submersible pump may be supplied with its source of power above the bulkhead deck. Alternatively, the pumps are so distributed throughout the length of the ship that it is inconceivable that one might not be able to work in the event of reasonable damage.

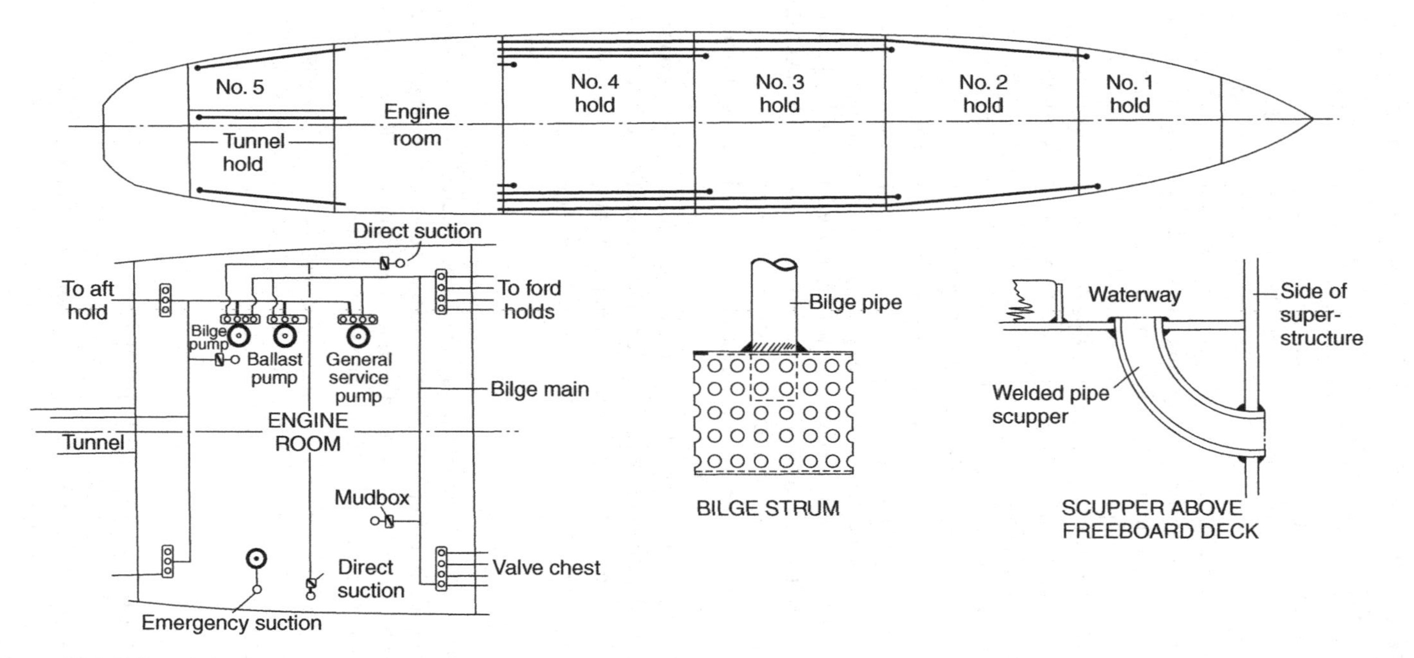

Figure 26.1 Bilge piping arrangement.

Suction connections are led to each hold or compartment from the main bilge line. Valves are introduced to prevent one watertight compartment from being placed in direct communication with another, and to prevent dry cargo spaces and machinery spaces being placed in direct communication with tanks or pumps having sea inlets. These screw-down nonreturn valves are often provided in a bilge valve distribution chest, or may be fitted directly in the connections to the bilge main. The bilge pipes that are used to drain cargo and machinery spaces are kept separate from the sea inlet pipes and ballast pipes, which are used for filling or emptying tanks where the water and oil are carried. Often a separate 'dirty ballast' system is arranged to overcome this problem.

If possible, the bilge pipes are kept out of the double-bottom tanks, and in way of a deep tank are led through a pipe tunnel. If the peaks are used as tanks then a power pump suction is led to each peak. Only two pipes are permitted to pass through the collision bulkhead below the bulkhead deck and a screw-down valve operated from above the bulkhead deck is provided for each pipe in a chest on the forward side of the bulkhead. An indicator is provided to show at the valve operating position whether it is open or closed.

Bilge mains in passenger ships are kept within 20% of the ship's beam of the side shell, and any piping outside this region or in a duct keel is fitted with a nonreturn valve. These requirements are intended to prevent any compartment from becoming flooded when the ship is grounded or otherwise damaged and a bilge pipe is severed. Many passenger ships are provided with divided deep tanks or side tanks that permit cross-flooding arrangements limiting the list after a casualty. This cross-flooding is generally controlled by valves operated from above the bulkhead deck, but self-acting arrangements can also be adopted.

Bilge and ballast piping may be of cast or wrought iron, steel, copper, or other approved materials. Lead or other heat-sensitive materials are not permitted. The piping is fitted in lengths that are adequately supported and have flanged connections, provision being made for expansion in each range of pipes.

Scuppers

Scuppers are fitted at the ship's side to drain the decks. Figure 26.1 shows a scupper fitted above the freeboard deck. Below the freeboard deck and within the intact superstructures and houses on the freeboard deck fitted with weathertight doors, these scuppers may be led to the bilges. Alternatively, they may be led overboard provided that:

1. The freeboard is such that the deck edge is not immersed when the ship heels to 5°; and
2. The scuppers are fitted with means of preventing water from passing inboard (usually a screw-down nonreturn valve capable of being operated from above the freeboard deck).

General service pipes and pumping

Pumps and substantial piping systems are provided in ships to supply the essential services of hot and cold fresh water for personal use, and salt water for sanitary and firefighting purposes.

Many large passenger ships are provided with a large low-pressure distilling plant for producing fresh water during the voyage, as the capacities required would otherwise need considerable tank space. This space is better utilized to carry oil fuel, improving the ship's range. Independent tanks supplying the fresh water required for drinking and culinary purposes, and fresh washing water etc., may be taken from the double-bottom tanks, the pumps for each supply also being independent. Hot fresh water is supplied initially from the cold fresh water system, through a nonreturn valve into a storage-type hot water heater fitted with heater coils, and then the heated water is pumped to the outlet.

The sanitary system supplies sea water for flushing water closets etc., and may be provided with a hydropneumatic pump in cargo vessels, but on larger passenger ships where the demand is heavier, a continuously operating power pump is required.

It is a statutory requirement that a fire main and deck wash system should be supplied. This has hose outlets on the various decks, and is supplied by power-driven pumps in the machinery spaces. Provision may be made for washing down the anchor chain from a connection to the fire main.

Air and sounding pipes

Air pipes are provided for all tanks to prevent air being trapped under pressure in the tank when it is filled, or a vacuum being created when it is emptied. The air pipes may be fitted at the opposite end of the tank to the filling pipe and/or at the highest point of the tank. Each air pipe from a double-bottom tank, deep tanks that extend to the ship's side, or any tank that may be run up from the sea, is led up above the bulkhead deck. From oil fuel and cargo oil tanks, cofferdams, and all tanks that can be pumped up, the air pipes are led to an open deck, in a position where no danger will result from leaking oil or vapors. The heights above decks and closing arrangements are covered by the Load Line Conditions of Assignment (see Chapter 31).

Sounding pipes are provided to all tanks, and compartments not readily accessible and are located so that soundings are taken in the vicinity of the suctions, i.e. at the lowest point of the tank. Each sounding pipe is made as straight as possible and is led above the bulkhead deck, except in some machinery spaces where this might not be practicable. A minimum bore of 32 mm is in general required for sounding pipes, but where they pass through refrigeration spaces, to allow for icing, a minimum bore of 65 mm is required where the temperature is at 0 °C or less. Underneath the sounding pipe a striking plate is provided where the sounding rod drops in the bilge well etc. Sometimes a slotted sounding pipe is fitted to indicate the depth of liquid present, and the closed end must be substantial to allow for the sounding rod striking it regularly. Various patent tank sounding devices are available and can be fitted in lieu of sounding pipes, as long as they satisfy the requirements of the classification society.

Sea inlets

Where the piping system requires water to be drawn from the sea, for example fire and washdeck, ballast and machinery cooling systems, the inlet valve is fitted to

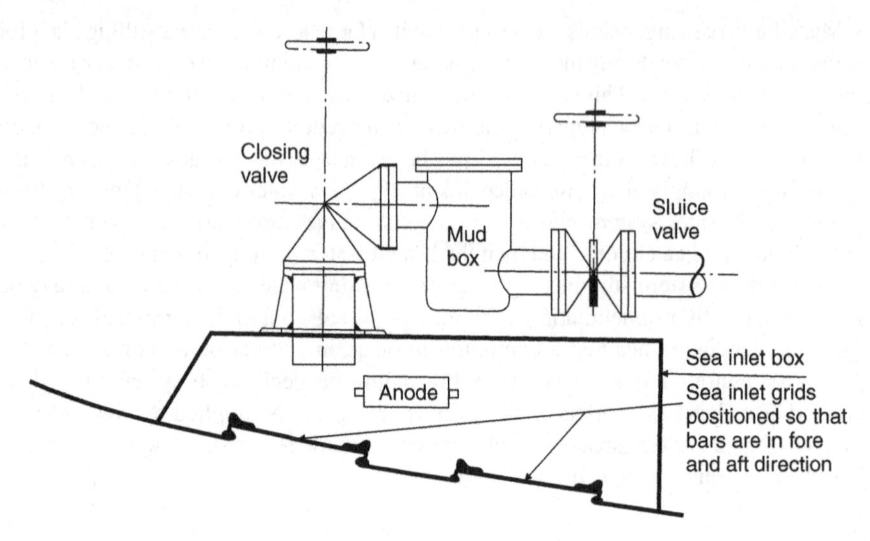

Figure 26.2 Sea inlet.

a substantial box within the line of the shell plate containing the sea inlet opening (see Figure 26.2). This opening is to have rounded corners and be kept clear of the bilge strake if possible. The sea inlet box should have the same thickness as the adjacent shell but is not to be less than 12.5 mm thick and need not exceed 25 mm. Sea inlets in tanker pump rooms within 40% of the ship's midship length are required to have compensation, generally in the form of a heavier insert plate in the shell. A grill may be fitted over the opening and a sacrificial anode will normally be fitted because the valve metal and steelbox set up a galvanic cell (see Chapter 27).

Cargo pumping and piping arrangements in tankers

Cargo pumps are provided in tankers to load and discharge cargo, and also to ballast some of the tanks, which becomes necessary when making voyages in the unloaded condition. Many modern tankers have clean ballast capacity and these tanks are served by a separate pumping system.

The particular cargo pumping system adopted depends very much on the range of cargo carried. A fairly straightforward system is available for the larger bulk oil carrier, carrying a single product. Where smaller tankers carry a number of oil products at one time, which must be kept separate, the pumping system is more complex.

Single product/crude oil carrier

Where a single oil product is carried, and where larger tankers are designed solely to carry crude oil, a single pump room is fitted aft, adjacent to the machinery spaces. The piping system is of the 'direct line' type, three or four lines being provided, each with suctions from a group of tanks (see Figure 26.3). Each pump discharge is led up to the deck mains, which run forward to the transverse loading and discharging connections.

A few large tankers have a discharge system that relies on hydraulically controlled sluice valves in the tank bulkheads. These permit a flow of the oil to a common suction in the after tank space. Many large tankers partially adopt this system, sluice valves being provided in the longitudinal bulkheads, and the oil is allowed to find its way from the wing tanks to the center tanks. Suctions from the main cargo lines are located in the center tanks. Such an arrangement is shown in Figure 26.3, which also indicates the separate stripping system, and clean ballast lines.

Multi-product tankers

Where a number of oil products are carried, the more complex pumping arrangements require two and in some cases three pump rooms to be fitted. One may be fitted aft adjacent to the machinery space, a second amidships, and where a third pump room is provided this is forward. On many older tankers the piping was often arranged on the 'ring main' system to provide flexibility of pumping conditions (see Figure 26.4). To obtain the optimum number of different pumping combinations in modern multi-product carriers the tanks may be fitted with individual suction lines.

Cargo pumps

Initially, on modern tankers the main cargo pumps were of the centrifugal type, either geared turbine or motor driven, and had a very high pumping capacity, those on the large tankers being capable of discharging say 3500 m^3/hour. Because of their high capacities the centrifugal cargo pumps are unsuitable for emptying tanks completely, and for this purpose reciprocating stripping pumps with capacities of, say, 350 m^3/hour are provided with a separate stripping line. More recent developments have seen the use of individual hydraulically driven submerged cargo pumps in the cargo tanks with a single discharge line and the conventional pump room dispensed with. Also, cargo tanks are being fitted with submerged cargo pumps driven by explosion-proof electric motors. Tanker cargo discharge systems are now often fully computerized.

Cargo tank ventilators

The cargo tank ventilators are to be entirely separate from air pipes from other compartments of the tanker and positioned so that flammable vapor emissions cannot be admitted to other spaces or areas containing any source of ignition.

Cargo tank protection

Oil tankers of less than 20,000 tonnes deadweight are required to be fitted with a fixed deck foam system capable of delivering foam to the entire cargo tanks deck area and into any tank the deck of which has been ruptured.

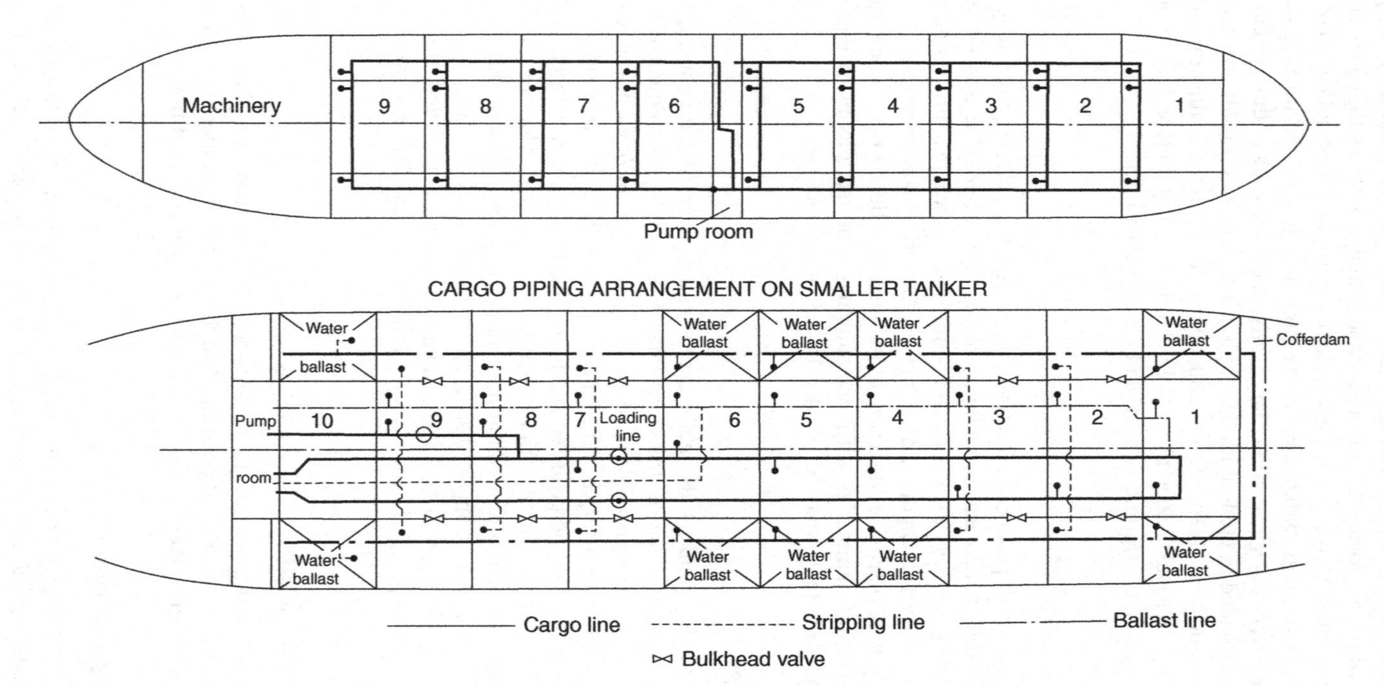

Figure 26.3 Diagrammatic representation of direct line cargo piping arrangement.

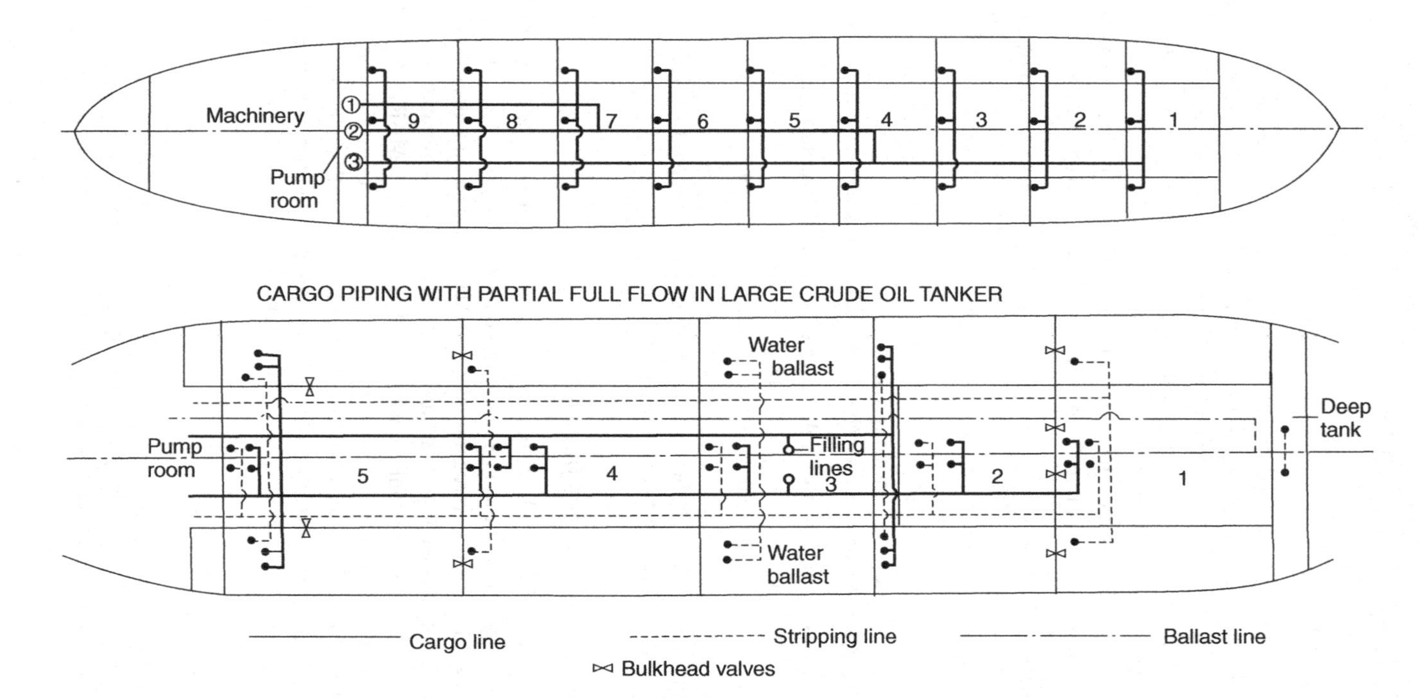

Figure 26.4 Diagrammatic representation of ring main cargo piping arrangement.

Oil tankers of 20,000 tonnes deadweight or more should have a fixed deck foam system and a fixed inert gas system.

Inert gas system

The inert gas system is to be so designed and operated as to render and maintain the atmosphere of the cargo tanks nonflammable, other than when the tanks are gas free. Hydrocarbon gas normally encountered in oil tanks cannot burn in an atmosphere containing less than 11% of oxygen by volume, thus if the oxygen content in a cargo tank is kept below, say, 8% by volume fire or explosion in the vapor space should not occur. Inert gas introduced into the tank will reduce the air (oxygen) content.

On an oil tanker, inert gas may be produced by one of two processes:

1. Ships with main or auxiliary boilers normally use the flue gas, which contains typically only 2–4% by volume of oxygen. This is scrubbed with sea water to cool it and to remove sulfur dioxide and particulates, and it is then blown into the tanks through a fixed pipe distribution system.
2. On diesel-engined ships the engine exhaust gas will contain too high an oxygen level for use as an inert gas. An inert gas generating plant may then be used to produce gas by burning diesel or light fuel oil. The gas is scrubbed and used in the same way as boiler flue gas.

Nonreturn barriers in the form of a deck water seal, and nonreturn valve are maintained between the machinery space and deck distribution system to ensure no petroleum gas or liquid petroleum passes back through the system to the machinery space.

The double-hull and double-bottom spaces of tankers required to have an inert gas system are to have connections for the supply of inert gas.

Cargo tank purging and gas freeing

When the tanker is fitted with an inert gas system, the discharged cargo tanks are purged until the hydrocarbon vapors in the cargo tanks have been reduced to less than 2% by volume. Gas freeing may then take place at the cargo tank deck level.

For tankers not provided with an inert gas system the flammable vapor is discharged initially through the tank venting outlets specified by MARPOL. When the flammable vapor concentration at the outlet has been reduced to 30% of the lower flammable limit, gas freeing can be continued at cargo tank deck level.

A range of portable gas-freeing fans with associated equipment is readily available, with typical air movement ranging from 8000 to 15,000 cubic meters per hour.

Cargo tank washing

After discharge of cargoes, oil tanker cargo tanks can be cleaned by either hot or cold water, fresh or sea water, or by crude oil washing. Water cleaning machines can be fixed in a tank or may be portable, the portable machine being connected by hose to a deck water main before being introduced through a tank-cleaning hatch into the tank. Crude oil carriers can have fixed washing equipment in the tanks connected to the cargo pumps via cross-connections at the ship's manifold to the cleaning main and can use crude oil instead of water as the washing medium. This is usually done while the tanker is discharging cargo and it enables redissolving of oil fractions adhering to the tank surfaces to take place so that these residues can be discharged with the cargo. There may then be no need to water wash tanks for the removal of residues unless clean water ballast is to be carried in the tank.

Further reading

Code for the Construction and Equipment of Ships Carrying Dangerous Chemicals in Bulk (BCH Code). IMO publication (IMO-772E).

International Code for the Construction and Equipment of Ships Carrying Dangerous Chemicals in Bulk (IBC Code). IMO publication (IMO-100E).

MARPOL 73/78 Consolidated Edition. See Annex 1, Chapter I, Regulation 18: Pumping, piping and discharge arrangements for oil tankers, IMO publication (IMO-1B 520E), 2002

MARPOL 2005, Amendments 2005. IMO publication (IMO-1525E).

SOLAS Consolidated Edition. See Chapter II-2, Part D: Fire safety measures for tankers, IMO publication (IMO-1D 110E), 2004

Some useful websites

www.imo.org/about/conventions/listofconventions/pages/international-convention-for-the-prevention-of-pollution-from-ships-(marpol).aspx

www.victorpyrate.com See 'Tank washing machines' and 'Gas freeing fans'.

27 Corrosion control and antifouling systems

Chapter Outline

Nature and forms of corrosion 327
Atmospheric corrosion 328
Corrosion due to immersion 328
Electrochemical nature of corrosion 328
Bimetallic (galvanic) corrosion 330
Stress corrosion 331
Corrosion/erosion 331
Corrosion allowance 333
Corrosion control 333
Cathodic protection 333
Sacrificial anode systems 333
Impressed current systems 334
Protective coatings (paints) 334
Corrosion protection by means of paints 337
Antifouling systems 337
Impressed current antifouling systems 337
Antifouling paints 338
Painting ships 339
Surface preparation 339
Temporary paint protection during building 341
Paint systems on ships 341
Below the waterline 342
Waterline or boot topping region 342
Superstructures 342
Cargo and ballast tanks 342
Further reading 343
Some useful websites 343

Nature and forms of corrosion

There is a natural tendency for nearly all metals to react with their environment. The result of this reaction is the creation of a corrosion product that is generally a substance of very similar chemical composition to the original mineral from which the metal was produced.

Ship Construction. DOI: 10.1016/B978-0-08-097239-8.00027-1

Atmospheric corrosion

Protection against atmospheric corrosion is important during the construction of a ship, both on the building berth and in the shops. Serious rusting may occur where the relative humidity is above about 70%; the atmosphere in many shipyards is unfortunately sufficiently humid to permit atmospheric corrosion throughout most of the year. However, even in humid atmospheres the rate of rusting is determined mainly by the pollution of the air through smoke and/or sea salts.

Corrosion due to immersion

When a ship is in service the bottom area is completely immersed and the waterline or boot topping region may be intermittently immersed in sea water. Under normal operating conditions a great deal of care is required to prevent excessive corrosion of these portions of the hull. A steel hull in this environment can provide ideal conditions for the formation of electrochemical corrosion cells.

Electrochemical nature of corrosion

Any metal, in tending to revert to its original mineral state, releases energy. At ordinary temperatures in aqueous solutions the transformation of a metal atom into a mineral molecule occurs by the metal passing into solution. During this process the atom loses one or more electrons and becomes an ion, i.e. an electrically charged atom, with the production of an electric current (the released energy). This reaction may only occur if an electron acceptor is present in the aqueous solution. Thus, any corrosion reaction is always accompanied by a flow of electricity from one metallic area to another through a solution in which the conduction of an electric current occurs by the passage of ions. Such a solution is referred to as an electrolyte solution, and because of its high salt content sea water is a good electrolyte solution.

A simple corrosion cell is formed by two different metals in an electrolyte solution (a galvanic cell), as illustrated in Figure 27.1. It is not essential to have two different metals, as we shall see later. As illustrated, a pure iron plate and a similar pure copper plate are immersed in a sodium chloride solution that is in contact with oxygen at the surface. Without any connection the corrosion reaction on each plate would be small. Once the two plates are connected externally to form an electrical path, then the corrosion rate of the iron will increase considerably and the corrosion on the copper will cease. The iron electrode, by means of which the electrons leave the cell and by way of which the conventional current enters the cell, is the anode. This is the electrode at which the oxidation or corrosion normally takes place. The copper electrode, by means of which the electrons enter the cell and by way of which the conventional current leaves the cell, is the cathode, at which no corrosion occurs. A passage of current through the electrolyte solution is by means of a flow of negative ions to the anode and a flow of positive ions to the cathode.

Electrochemical corrosion in aqueous solutions will result from any anodic and cathodic areas coupled in the solution, whether they are metals of different potential

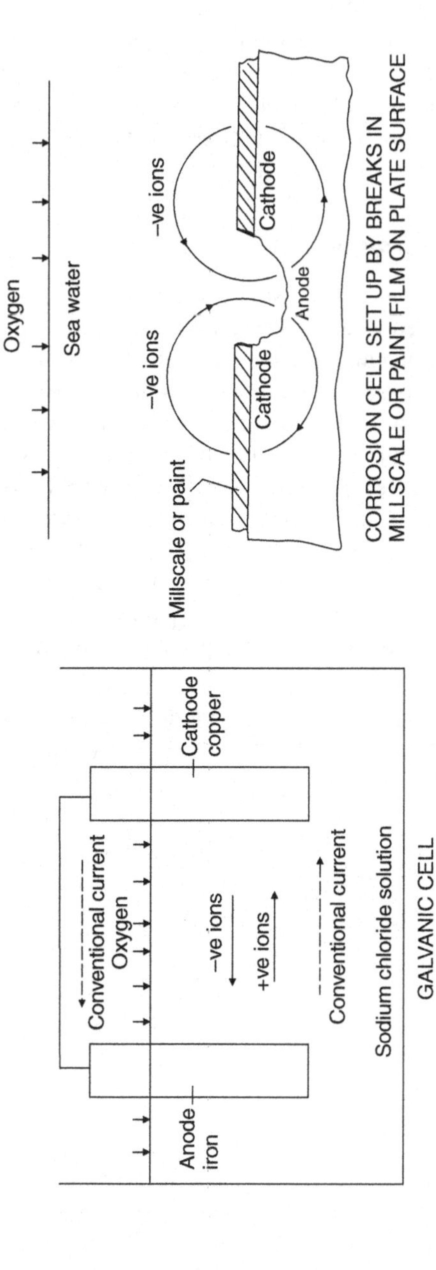

Figure 27.1 Corrosion cell.

in the environment or they possess different potentials as the result of physical differences on the metal surface. The latter is typified by steel plate carrying broken millscale in sea water (Figure 27.1) or corrosion currents flowing between areas of well-painted plate and areas of defective paintwork.

In atmospheric corrosion and corrosion involving immersion, both oxygen and an electrolyte play an important part. Plates freely exposed to the atmosphere will receive plenty of oxygen but little moisture, and the moisture present therefore becomes the controlling factor. Under conditions of total immersion it is the presence of oxygen that becomes the controlling factor.

Bimetallic (galvanic) corrosion

Although it is true to say that all corrosion is basically galvanic, the term 'galvanic corrosion' is usually applied when two different metals form a corrosion cell.

Many ship corrosion problems are associated with the coupling of metallic parts of different potential, which consequently form corrosion cells under service conditions. The corrosion rates of metals and alloys in sea water have been extensively investigated and as a result galvanic series of metals and alloys in sea water have been obtained. A typical galvanic series in sea water is shown in Table 27.1.

The positions of the metals in the table apply only in a sea water environment, and where metals are grouped together they have no strong tendency to form couples with each other. Some metals appear twice because they are capable of having both a passive and an active state. A metal is said to be passive when the surface is exposed to an electrolyte solution and a reaction is expected but the metal shows no sign of

Table 27.1 Galvanic series of metals and alloys in sea water

Noble (cathodic or protected) end
Platinum, gold
Silver
Titanium
Stainless steels, passive
Nickel, passive
High-duty bronzes
Copper
Nickel, active
Millscale
Naval brass
Lead, tin
Stainless steels, active
Iron, steel, cast iron
Aluminum alloys
Aluminum
Zinc
Magnesium
Ignoble (anodic or corroding) end

corrosion. It is generally agreed that passivation results from the formation of a current barrier on the metal surface, usually in the form of an oxide film. This thin protective film forms, and a change in the overall potential of the metal occurs when a critical current density is exceeded at the anodes of the local corrosion cells on the metal surface.

Among the more common bimetallic corrosion cell problems in ship hulls are those formed by the mild steel hull with the bronze or nickel alloy propeller. Also, above-the-waterline problems exist with the attachment of bronze and aluminum alloy fittings. Where aluminum superstructures are introduced, the attachment to the steel hull and the fitting of steel equipment to the superstructure require special attention. This latter problem is overcome by insulating the two metals and preventing the ingress of water, as illustrated in Figure 27.2. A further development is the use of explosion-bonded aluminum/steel transition joints, also illustrated. These joints are free of any crevices, the exposed aluminum-to-steel interface being readily protected by paint.

Stress corrosion

Corrosion and subsequent failure associated with varying forms of applied stress is not uncommon in marine structures.

Internal stresses produced by non-uniform cold working are often more dangerous than applied stresses. For example, localized corrosion is often evident at cold flanged brackets.

Corrosion/erosion

Erosion is essentially a mechanical action but it is associated with electrochemical corrosion in producing two forms of metal deterioration. Firstly, in what is known as 'impingement attack', the action is mainly electrochemical but it is initiated by erosion. Air bubbles entrained in the flow of water and striking a metal surface may erode away any protective film that may be present locally. The eroded surface becomes anodic to the surrounding surface and corrosion occurs. This type of attack can occur in most places where there is water flow, but particularly where features give rise to turbulent flow. Sea water discharges from the hull are a particular case, the effects being worse if warm water is discharged.

Cavitation damage is also associated with a rapidly flowing liquid environment. At certain regions in the flow (often associated with a velocity increase resulting from a contraction of the flow stream) the local pressures drop below that of the absolute vapor pressure. Vapor cavities, i.e. areas of partial vacuum, are formed locally, but when the pressure increases clear of this region the vapor cavities collapse or 'implode'. This collapse occurs with the release of considerable energy, and if it occurs adjacent to a metal surface damage results. The damage shows itself as pitting, which is thought to be predominantly due to the effects of the mechanical damage. However, it is also considered that electrochemical action may play some part in the damage after the initial erosion.

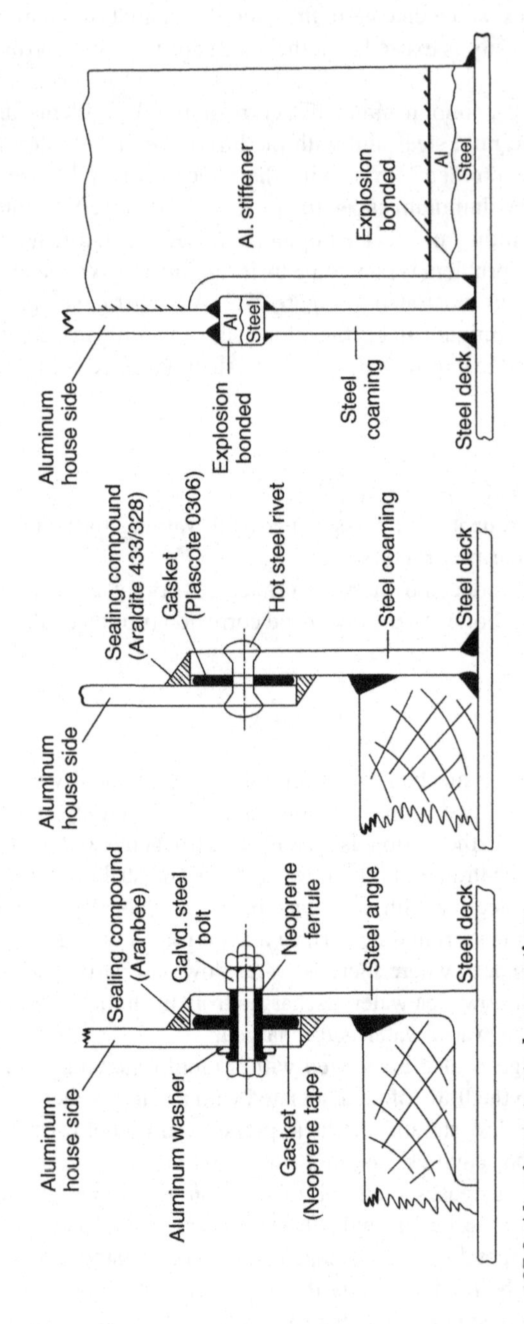

Figure 27.2 Aluminum-to-steel connections.

Corrosion allowance

Plate and section scantlings specified for ships in the rules of classification societies include corrosion additions to the thickness, generally based on a 25-year service life. The corrosion allowance is based on the concept that corrosion occurs on the exposed surface of the material at a constant rate, no matter how much material lies behind it. That is, if a plate is 8 or 80 mm thick, corrosion will take place at the same rate, not at a faster rate in the thicker plate. Some shipowners may specify thicker material than is required by the classification society, where there are plans for long-term ownership.

Corrosion control

The control of corrosion may be broadly considered in two forms, cathodic protection and the application of protective coatings, i.e. paints.

Cathodic protection

Only where metals are immersed in an electrolyte can the possible onset of corrosion be prevented by cathodic protection. The fundamental principle of cathodic protection is that the anodic corrosion reactions are suppressed by the application of an opposing current. This superimposed direct electric current enters the metal at every point, lowering the potential of the anode metal of the local corrosion cells so that they become cathodes.

There are two main types of cathodic protection installation, sacrificial anode systems and impressed current systems.

Sacrificial anode systems

Sacrificial anodes are metals or alloys attached to the hull that have a more anodic, i.e. less noble, potential than steel when immersed in sea water. These anodes supply the cathodic protection current, but will be consumed in doing so and therefore require replacement for the protection to be maintained.

This system has been used for many years, the fitting of zinc plates in way of bronze propellers and other immersed fittings being common practice. Initially, results with zinc anodes were not always very effective owing to the use of unsuitable zinc alloys. Modern anodes are based on alloys of zinc, aluminum, or magnesium, which have undergone many tests to examine their suitability; high-purity zinc anodes are also used. The cost, with various other practical considerations, may decide which type is to be fitted.

Sacrificial anodes may be fitted within the hull, and are often fitted in ballast tanks. However, magnesium anodes are not used in the cargo-ballast tanks of oil carriers owing to the 'spark hazard'. Should any part of the anode fall and strike the tank structure when gaseous conditions exist, an explosion could result. Aluminum anode systems may be employed in tankers provided they are only fitted in locations where the potential energy is less than 28 kgm.

Impressed current systems

These systems are applicable to the protection of the immersed external hull only. The principle of the systems is that a voltage difference is maintained between the hull and fitted anodes, which will protect the hull against corrosion, but not overprotect it, thus wasting current. For normal operating conditions the potential difference is maintained by means of an externally mounted silver/silver chloride reference cell detecting the voltage difference between itself and the hull. An amplifier controller is used to amplify the micro-range reference cell current, and it compares this with the preset protective potential value that is to be maintained. Using the amplified DC signal from the controller, a saturable reactor controls a larger current from the ship's electrical system that is supplied to the hull anodes. An AC current from the electrical system would be rectified before distribution to the anodes. Figure 27.3 shows such a system.

Originally consumable anodes were employed, but in recent systems nonconsumable relatively noble metals are used; these include lead/silver and platinum/palladium alloys, and platinized titanium anodes are also used.

A similar impressed current system employs a consumable anode in the form of an aluminum wire up to 45 meters long that is trailed behind the ship whilst at sea. No protection is provided in port.

Although the initial cost is high, these systems are claimed to be more flexible, to have a longer life, to reduce significantly hull maintenance, and to weigh less than the sacrificial anode systems.

Care is required in their use in port alongside ships or other unprotected steel structures.

Protective coatings (paints)

Paints intended to protect against corrosion consist of pigment dispersed in a liquid referred to as the 'vehicle'. When spread out thinly the vehicle changes in time to an adherent dry film. The drying may take place through one of the following processes:

1. When the vehicle consists of solid resinous material dissolved in a volatile solvent, the latter evaporates after application of the paint, leaving a dry film.
2. A liquid like linseed oil as a constituent of the vehicle may produce a dry paint film by reacting chemically with the surrounding air.
3. A chemical reaction may occur between the constituents of the vehicle after application, to produce a dry paint film. The reactive ingredients may be separated in two containers ('two-pack paints') and mixed before application. Alternatively, ingredients that only react at higher temperatures may be selected, or the reactants may be diluted with a solvent so that the reaction occurs only slowly in the can.

Corrosion-inhibiting paints for application to steel have the following vehicle types:

1. *Bitumen or pitch.* Simple solutions of bitumen or pitch are available in solvent naphtha or white spirit. The bitumen or pitch may also be blended by heat with other materials to form a vehicle.
2. *Oil based.* These consist mainly of vegetable drying oils, such as linseed oil and tung oil. To accelerate the drying by the natural reaction with oxygen, driers are added.

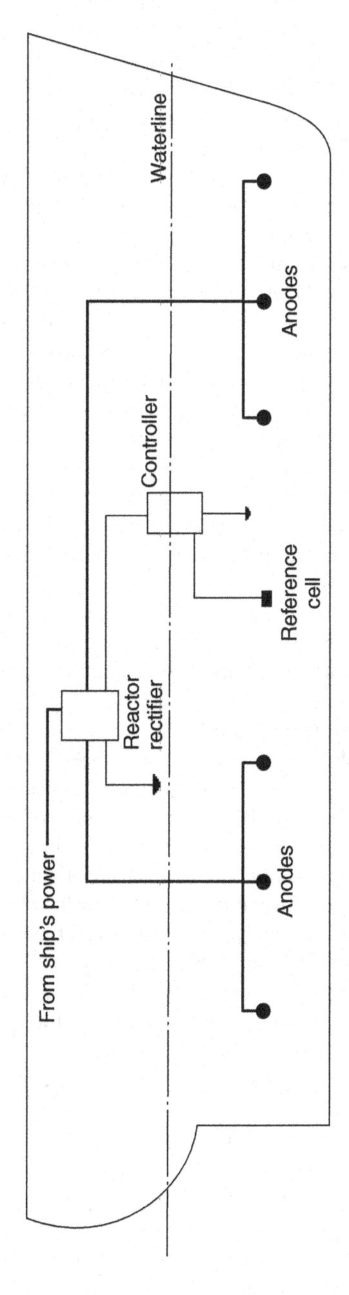

Figure 27.3 Impressed current cathode protection system.

3. *Oleo-resinous.* The vehicle incorporates natural or artificial resins into drying oils. Some of these resins may react with the oil to give a faster-drying vehicle. Other resins do not react with the oil but heat is applied to dissolve the resin and cause the oil to body.
4. *Alkyd resin.* These vehicles provide a further improvement in the drying time and film-forming properties of drying oils. The name alkyd arises from the ingredients, alcohols and acids. Alkyds need not be made from oil, as an oil-fatty acid or an oil-free acid may be used.

   (**Note:** Vehicle types 2 and 4 are not suitable for underwater service, and only certain kinds of type 3 are suitable for such service.)
5. *Chemical resistant.* Vehicles of this type show extremely good resistance to severe conditions of exposure. As any number of important vehicle types come under this general heading, these are dealt with individually.
   a. *Epoxy resins.* Chemicals that may be produced from petroleum and natural gas are the source of epoxy resins. These paints have very good adhesion, apart from their excellent chemical resistance. They may also have good flexibility and toughness where co-reacting resins are introduced. Epoxy resins are expensive owing to the removal of unwanted side products during their manufacture, and the gloss finish may tend to 'chalk', making it unsuitable for many external decorative finishes. These paints often consist of a 'two-pack' formulation, a solution of epoxy resin together with a solution of cold curing agent, such as an amine or a polyamide resin, being mixed prior to application. The mixed paint has a relatively slow curing rate at temperatures below 10 °C. Epoxy resin paints should not be confused with epoxy-ester paints, which are unsuitable for underwater use. Epoxy-ester paints can be considered as alkyd equivalents, as they are usually made with epoxy resins and oil-fatty acids.
   b. *Coal tar/epoxy resin.* This vehicle type is similar to the epoxy resin vehicle except that, as a two-pack product, a grade of coal tar pitch is blended with the resin. A formulation of this type combines to some extent the chemical resistance of the epoxy resin with the impermeability of coal tar.
   c. *Chlorinated rubber and isomerized rubber.* The vehicle in this case consists of a solution of plasticized chlorinated rubber, or isomerized rubber. Isomerized rubber is produced chemically from natural rubber, and it has the same chemical composition but a different molecular structure. Both these derivatives of natural rubber have a wide range of solubility in organic solvents, and so allow a vehicle of higher solid content. On drying, the film thickness is greater than would be obtained if natural rubber were used. High build coatings of this type are available, thickening or thixotropic agents being added to produce a paint that can be applied in much thicker coats. Coats of this type are particularly resistant to attack from acids and alkalis.
   d. *Polyurethane resins.* A reaction between isocyanates and hydroxyl-containing compounds produces 'urethane' and this reaction has been adapted to produce polymeric compounds from which paint film, fibers, and adhesives may be obtained. Paint films so produced have received considerable attention in recent years, and since there is a variety of isocyanate reactions, both one- and two-pack polyurethane paints are available. These paints have many good properties: toughness, hardness, gloss, abrasion resistance, as well as chemical and weather resistance. Polyurethanes are not used underwater on steel ships, only on superstructures etc., but they are very popular on yachts, where their good gloss is appreciated.
   e. *Vinyl resins.* Vinyl resins are obtained by the polymerization of organic compounds containing the vinyl group. The solids content of these paints is low; therefore, the dry film is thin and more coats are required than for most paints. As vinyl resin paints have poor adhesion to bare steel surfaces they are generally applied over a pretreatment

primer. Vinyl paint systems are among the most effective for the underwater protection of steel.

6. *Zinc-rich paints.* Paints containing metallic zinc as a pigment in sufficient quantity to ensure electrical conductivity through the dry paint film to the steel are capable of protecting the steel cathodically. The pigment content of the dry paint film should be greater than 90%, the vehicle being an epoxy resin, chlorinated rubber, or similar medium.

Corrosion protection by means of paints

It is often assumed that all paint coatings prevent attack on the metal covered simply by excluding the corrosive agency, whether air or water. This is often the main and sometimes only form of protection; however, there are many paints that afford protection even though they present a porous surface or contain various discontinuities.

For example, certain pigments in paints confer protection on steel even where it is exposed at a discontinuity. If the reactions at the anode and cathode of the corrosion cell that form positive and negative ions respectively are inhibited, protection is afforded. Good examples of pigments of this type are red lead and zinc chromate, red lead being an anodic inhibitor and zinc chromate a cathodic inhibitor. A second mode of protection occurs at gaps where the paint is richly pigmented with a metal anodic to the basis metal. Zinc dust is a commercially available pigment that fulfills this requirement for coating steel in a salt water environment. The zinc dust is the sacrificial anode with respect to the steel.

Antifouling systems

The immersed hull and fittings of a ship at sea, particularly in coastal waters, are subject to algae, barnacle, mussel, and other shellfish growth that can impair its hydrodynamic performance and adversely affect the service of the immersed fittings.

Fittings such as cooling water intake systems are often protected by impressed current antifouling systems, and immersed hulls today are finished with very effective self-polishing antifouling paints.

Impressed current antifouling systems

The functional principle of these systems is the establishment of an artificially triggered voltage difference between copper anodes and the integrated steel plate cathodes. This causes a minor electrical current to flow from the copper anodes, so that they are dissolved to a certain degree. A control unit makes sure that the anodes add the required minimum amount of copper particles to the sea water, thus ensuring the formation of copper oxide that creates ambient conditions precluding local fouling. A control unit can be connected to the management system of the vessel. Using information from the management system, the impressed current antifouling system can determine the amount of copper that needs to be dissolved to give optimum performance with minimum wastage of the anodes.

Antifouling paints

Antifouling paints consist of a vehicle with pigments that give body and color together with materials toxic to marine vegetable and animal growth. Copper is the best known toxin used in traditional antifouling paints.

To prolong the useful life of the paint the toxic compounds must dissolve slowly in sea water. Once the release rate falls below a level necessary to prevent settlement of marine organisms, the antifouling composition is no longer effective. On merchant ships the effective period for traditional compositions was about 12 months. Demands in particular from large tanker owners wishing to reduce very high docking costs led to specially developed antifouling compositions with an effective life up to 24 months in the early 1970s. Subsequent developments of constant emission organic toxin antifoulings having a leaching rate independent of exposure time saw the paint technologists by chance discover coatings that also tended to become smoother in service. These so-called self-polishing antifoulings, with a lifetime that is proportional to applied thickness and therefore theoretically unlimited, smooth rather than roughen with time and result in reduced friction drag. Though more expensive than their traditional counterparts, given the claim that each 10-micron (10^{-3} mm) increase in hull roughness can result in a 1% increase in fuel consumption, their self-polishing characteristic as well as their longer effective life, up to 5 years' protection between dry dockings, made them attractive to the shipowner.

The benefits of the first widely used self-polishing copolymer (SPC) antifouling paints could be traced to the properties of their prime ingredients, the tributyltin compounds or TBTs. TBTs were extremely active against a wide range of fouling organisms and they were able to be chemically bonded to the acrylic backbone of the paint system. When immersed in sea water a specific chemical reaction took place that cleaved the TBT from the paint backbone, resulting in both controlled release of the TBT and controlled disappearance or polishing of the paint film. Unfortunately, it was found that the small concentrations of TBTs released, particularly in enclosed coastal waters, had a harmful effect on certain marine organisms. This led to the banning of TBT antifouling paints for pleasure boats and smaller commercial ships in many developed countries and the introduction of regulations limiting the release rate of TBT for antifouling paints on larger ships. The International Convention on the Control of Harmful Antifouling on Ships, 2001 subsequently required that:

1. Ships shall not apply or reapply organotin compounds that act as biocides in antifouling systems on or after 1 January 2003; and
2. No ship shall have organotin compounds that act as biocides in antifouling systems (except floating platforms, FSUs, and FPSOs built before 2003 and not docked since before 2003).

The final phase-out of TBT paints was in 2008. In theory, ships could retain the paint if it was sealed under another coating but in practice TBTs are no longer at sea.

(**Note:** Organotin means an organic compound with one or more tin atoms in its molecules used as a pesticide, hitherto considered to decompose safely, now found to be toxic in the food chain. A biocide is a chemical capable of killing living organisms.)

Antifouling paints subsequently applied have generally focused on either the use of copper-based self-polishing antifouling products, which operate in a similar manner to the banned TBT products, or the use of the so-called low-surface-energy coatings. The latter coatings do not polish or contain booster biocides, instead they offer a very smooth, low-surface-energy surface to which it is difficult for fouling to adhere. When the vessel is at rest some fouling may occur but once it is underway and reaches a critical speed the fouling is released.

Painting ships

To obtain the optimum performance from paints it is important that the metal surfaces are properly prepared before application of paints and subsequently maintained as such throughout the fabrication and erection process. Paints tailored for the service conditions of the structure to which they apply, and recommended as such by the manufacturer, only should be applied. It is common, especially for ships undergoing a regular dry-docking, for the coating process to be specified and managed by the coating supplier.

Surface preparation

Good surface preparation is essential to successful painting, the primary cause of many paint failures being the inadequacy of the initial material preparation.

It is particularly important before painting new steel that any millscale should be removed. Millscale is a thin layer of iron oxides that forms on the steel surface during hot rolling of the plates and sections. Not only does the non-uniform millscale set up corrosion cells, as illustrated previously, but it may also come away from the surface removing any paint film applied over it. Any corrosion from storage of the steel outdoors in a stockyard must also be removed.

It is also important to create a good surface for coating. Swedish standards are often quoted, SA 3 being white metal and SA 2.5 being near white metal. SA 2.5 is common as a reasonably achieveable standard for shipyards. The surface is clean and the blasting process also provides a 'key' with very small indentations to which the coating adheres well.

The most common methods employed to prepare steel surfaces for painting are:

- Blast cleaning
- Pickling
- Flame cleaning
- Preparation by hand
- High-pressure water blasting.

Blast cleaning is the most efficient method for preparing the surface and is in common use in all large shipyards. The shipyard preparation process has been outlined in Chapter 13. Smaller shipyards, where the quantity of steel processed does not justify the capital investment in the preparation equipment, usually buy steel already prepared from a steel mill or stockist.

Following the blast cleaning it is desirable to brush the surface and apply a coat of priming paint as soon as possible, since the metal is liable to rust rapidly.

There are two main types of blasting equipment available, an impeller wheel plant where the abrasive is thrown at high velocity against the metal surface, and a nozzle type where a jet of abrasive impinges on the metal surface. The latter type should preferably be fitted with vacuum recovery equipment, rather than allow the spent abrasive and dust to be discharged to atmosphere, as is often the case in ship repair work. Impeller wheel plants that are self-contained and collect the dust and recirculate the clean abrasive are generally fitted within the shipbuilding shops.

Steel shot is the preferred blast medium for preparation of steel plates and profiles for new construction. As the process proceeds the shot is recovered and reused. The shot breaks down during the blasting and a mix of new and broken shot provides good surface preparation. Dust and other debris is filtered out of the recovery system for disposal.

Cast iron and steel grit may be used for the abrasive, also copper slag, but nonmetallic abrasives are also available. Grit has been used in ship repair but is becoming much less popular because of environmental issues associated with disposal of the waste material. The use of sand is prohibited in most countries because the fine dust produced may cause silicosis.

Pickling involves the immersion of the metal in an acid solution, usually hydrochloric or sulfuric acid, in order to remove the millscale and rust from the surface. After immersion in these acids the metal will require a thorough hot water rinse. It is preferable that the treatment is followed by application of a priming coat. Pickling is no longer found in most of the ship construction industry.

Using an *oxyacetylene flame* the millscale and rust may be removed from a steel surface. The process does not entirely remove the millscale and rust, but it can be quite useful for cleaning plates under inclement weather conditions, the flame drying out the plate. Flame, or another heating process, is used as part of an automated preparation line in a shipyard, to preheat the steel, which assists the cleaning and drying of the primer paint coating.

Hand cleaning by various forms of wire brush is often not very satisfactory, and would only be used where the millscale has been loosened by weathering, i.e. exposure to atmosphere over a long period. For ships in service hand cleaning may be used for very small corroded areas.

High-pressure water blasting is superseding grit blasting for ship repair. It is capable of restoring a steel surface to SA 2.5, provided any corrosion has not been allowed to develop. Disposal of the waste material, in this case mainly water, is a main reason for the adoption of water blasting.

Blast cleaning using shot or grit is preferred for best results and economy in shipbuilding, it is essential prior to application of high-performance paint systems used today. Pickling, which also gives good results, can be expensive and less applicable to production schemes; flame cleaning is much less effective; and hand cleaning gives the worst results. Water blasting is mainly a ship repair process.

Temporary paint protection during building

After the steel is blast cleaned it may be several months before it is built into the ship and finally painted. It is desirable to protect the material against rusting in this period as the final paint will offer the best protection when applied over perfectly clean steel.

The formulation of a prefabrication primer for immediate application after blasting must meet a number of requirements. It should dry rapidly to permit handling of the plates within a few minutes and working the plates within a day or so. It should be nontoxic, and it should not produce harmful porosity in welds nor give off obnoxious fumes during welding or cutting. For some high-speed welding processes it is necessary to remove the primer in way of the welds to avoid porosity. After cutting and welding, areas of damaged primer are usually strip coated by hand to cover the affected steel and retain the primer protection against corrosion.

The primer must also be compatible with any subsequent paint finishes to be applied. Satisfactory formulations are available, for example a primer consisting of zinc dust in an epoxy resin.

Some shipyards where work is entirely undercover and where the steel is used quickly may dispense with primer.

Given the sophistication and cost of modern, long-life coatings, their application is critical to performance. It is common for shipyards to take complete units and blocks, once all the hot work is completed, for final coating in a controlled environment. So-called 'paint cells' are temperature- and humidity-controlled buildings. The units are driven into the cells using self-elevating transporters. The units are then blasted, if no primer has been used or the primer is degraded and unsuitable for overcoating. Once cleaned the coatings can be applied in near ideal conditions. Hull and internal coatings are applied.

Where the units are to be joined in the building dock, unpainted areas are left. These are completed after the hull welding, using localized protection from unsuitable weather conditions.

Paint systems on ships

The paint system applied to any part of a ship will be dictated by the environment to which that part of the structure is exposed. Traditionally, the painting of the external ship structure was divided into three regions:

1. Below the waterline, where the plates are continually immersed in sea water.
2. The waterline or boot topping region, where immersion is intermittent and a lot of abrasion occurs.
3. The topsides and superstructure exposed to an atmosphere laden with salt spray, and subject to damage through cargo handling.

However, now that tougher paints are used for the ship's bottom the distinction between regions need not be so well defined, one scheme covering the bottom and waterline regions.

Internally, by far the greatest problem is the provision of coatings for various liquid cargo and salt water ballast tanks.

Below the waterline

The ship's bottom has priming coats of corrosion-inhibiting paint applied that are followed by an antifouling paint. Paints used for steels immersed in sea water are required to resist alkaline conditions. The reason for this is that an iron alloy immersed in a sodium chloride solution having the necessary supply of dissolved oxygen gives rise to corrosion cells with caustic soda produced at the cathodes. Further, the paint should have a good electrical resistance so that the flow of corrosion currents between the steel and sea water is limited. These requirements make the standard nonmarine structural steel primer red lead in linseed oil unsuitable for ship use below the waterline. Suitable corrosion-inhibiting paints for ships' bottoms are pitch or bitumen types, chlorinated rubber, coal tar/epoxy resin, or vinyl resin paints. The antifouling paints may be applied after the corrosion-inhibiting coatings and should not come into direct contact with the steel hull, since the toxic compounds present may cause corrosion.

Waterline or boot topping region

Generally, modern practice requires a complete paint system for the hull above the waterline. This may be based on vinyl and alkyd resins or on polyurethane resin paints.

Superstructures

Red lead- or zinc chromate-based primers are commonly used. White finishing paints are then used extensively for superstructures. These are usually oleo-resinous or alkyd paints that may be based on 'nonyellowing' oils, linseed oil-based paints that yellow on exposure being avoided on modern ships.

Where aluminum alloy superstructures are fitted, under no circumstance should lead-based paints be applied; zinc chromate paints are generally supplied for application to aluminum.

Cargo and ballast tanks

Severe corrosion may occur in a ship's cargo tanks as the combined result of carrying liquid cargoes and sea water ballast, with warm or cold sea water cleaning between voyages. This is particularly true of oil tankers. Tankers carrying 'white oil' cargoes suffer more general corrosion than those carrying crude oils, which deposit a film on the tank surface, providing some protection against corrosion. The latter type may, however, experience severe local pitting corrosion due to the non-uniformity of the deposited film, and subsequent corrosion of any bare plate when sea water ballast is carried. Epoxy resin paints are used extensively within these tanks, and vinyl resins and zinc-rich coatings may also be used.

Further reading

Antifouling Systems 2005 *Antifouling Systems—International Convention on the Control of Harmful Antifouling Systems on Ships*. IMO publication, 2005 edition.

Paint terminology explained, *The Naval Architect*, June 2003.

Paints and coatings technology, *The Naval Architect*, June 2005.

Some useful websites

www.imo.org See under 'Marine environment—Antifouling systems'.

www.jotun.com See under 'Technical papers—Marine coatings' and 'Products—Cathodic protection'.

www.hempel.com See under 'Fouling control', 'Antifouling coatings', and 'Fouling release coatings'.

www.sigmacoatings.com/marine See under 'Technical data—Systems sheets' (these provide up-to-date examples of typical paint systems for new ships).

www.international-marine.com

www.ameroncoatings.com

www.wheelabratorgroup.com Has information on shipyard steel preparation.

www.hammelmann.de Shows water-blasting capabilities.

www.iacs.org.uk See 'Guidelines and Recommendations': Recommendation 87 'Guidelines for Coating Maintenance and Repairs for Ballast Tanks and Combined Cargo/Ballast Tanks on Oil Tankers'.

28 Heating, ventilation, air-conditioning, refrigeration, and insulation

Chapter Outline

Ventilation 345
Refrigeration 346
Stowage of refrigerated cargo 346
Refrigeration systems 346
Insulation 349
Refrigerated container ships 349
Further reading 351
Some useful websites 351

Ships can expect to operate worldwide, and therefore experience extremes of temperature and humidity. It is important that adequate ventilating and air-conditioning systems are installed in ships to provide reasonable comfort for the crew and passengers, and to maintain cargo at the correct temperature and humidity. Insulation is provided to maintain refrigerated cargo and domestic store rooms, and to a lesser extent to maintain air-conditioning requirements, and to overcome acoustic problems in accommodation and control rooms.

Heating, ventilation, and air-conditioning (HVAC) is usually provided by a specialist contractor, often as a package including design and installation. CAD systems are available that manage the allocation of space for equipment and trunking, alongside pipe and other ship systems.

Ventilation

Mechanical ventilation is provided in the machinery spaces of most ships, but some may have natural exhaust provisions. Stores and working spaces may have both natural supply and exhaust ventilation, but where fumes are present mechanical exhaust provisions will be provided. The mechanical supply is by means of light steel sheet trunking, with louvres at each outlet. Natural and fan exhaust outlets are appropriately placed at the exterior sides or deckhead of the space being ventilated. Fans may be of the quiet running centrifugal type with a separately mounted motor.

Ship Construction. DOI: 10.1016/B978-0-08-097239-8.00028-3

Air-conditioning is a common feature in crew accommodation, within machinery control rooms, and in the accommodation and public spaces of passenger ships. Room temperatures are controlled by a thermostat; heated or chilled air may then be supplied as required and humidity control is also provided. Trunking and louvres are similar to those for mechanical ventilation, smaller bore trunking being possible if a high-velocity system is introduced. Local air-conditioning units are available and may serve an individual passenger suite if desired, with their own control.

The holds of most dry cargo ships are ventilated by a mechanical supply and natural exhaust system. Here the object is to reduce the hold temperatures if necessary and prevent large amounts of condensation accumulating on the hull and cargo. Often, the cargo hold fans, which are of the axial type, are located in houses on the weather deck that enclose openings to the holds. In older ships they may have been positioned in derrick posts where these posts are used to ventilate the tweens and holds (see Figure 28.1). Dry cargo ships may also be fitted with dehumidification facilities, controls being provided so that each hold can be supplied with dry air or outside air.

If dry air is desired when the weather dew point approaches or is above the temperature in the hold, the air supply or recirculating air may be drawn through a conditioning plant, where it comes into contact with a moisture-absorbing solution. A dry air fan then passes this dehumidified air to the cargo hold ventilation system.

Refrigeration

Many perishable cargoes are carried in refrigerated compartments on dry cargo ships, and there are still a number of existing ships carrying refrigerated cargo in bulk, but a large proportion of such cargoes are now carried by container. A number of hybrid ships have been built that accommodate containers on deck and in some holds, and also have insulated cargo hold space. The midship section of a ship specifically designed for carrying only refrigerated cargo and its line of insulation is shown in Figure 28.2.

Stowage of refrigerated cargo

Chilled meat cargo is hung from the strengthened deck stiffening members, and the tween deck height is arranged to provide space below the hung carcasses for the circulation of air. Frozen meat is stacked in the holds of the ship. Fruits and vegetables are stowed in a manner that permits an adequate flow of air to be maintained around the crates, etc.

As a rule the refrigerated rooms in general cargo ships are made rectangular to keep down insulation costs.

Refrigeration systems

Brine, made by dissolving calcium chloride in fresh water, will have a freezing point well below the desired temperatures of the refrigerated compartments. Cold brine

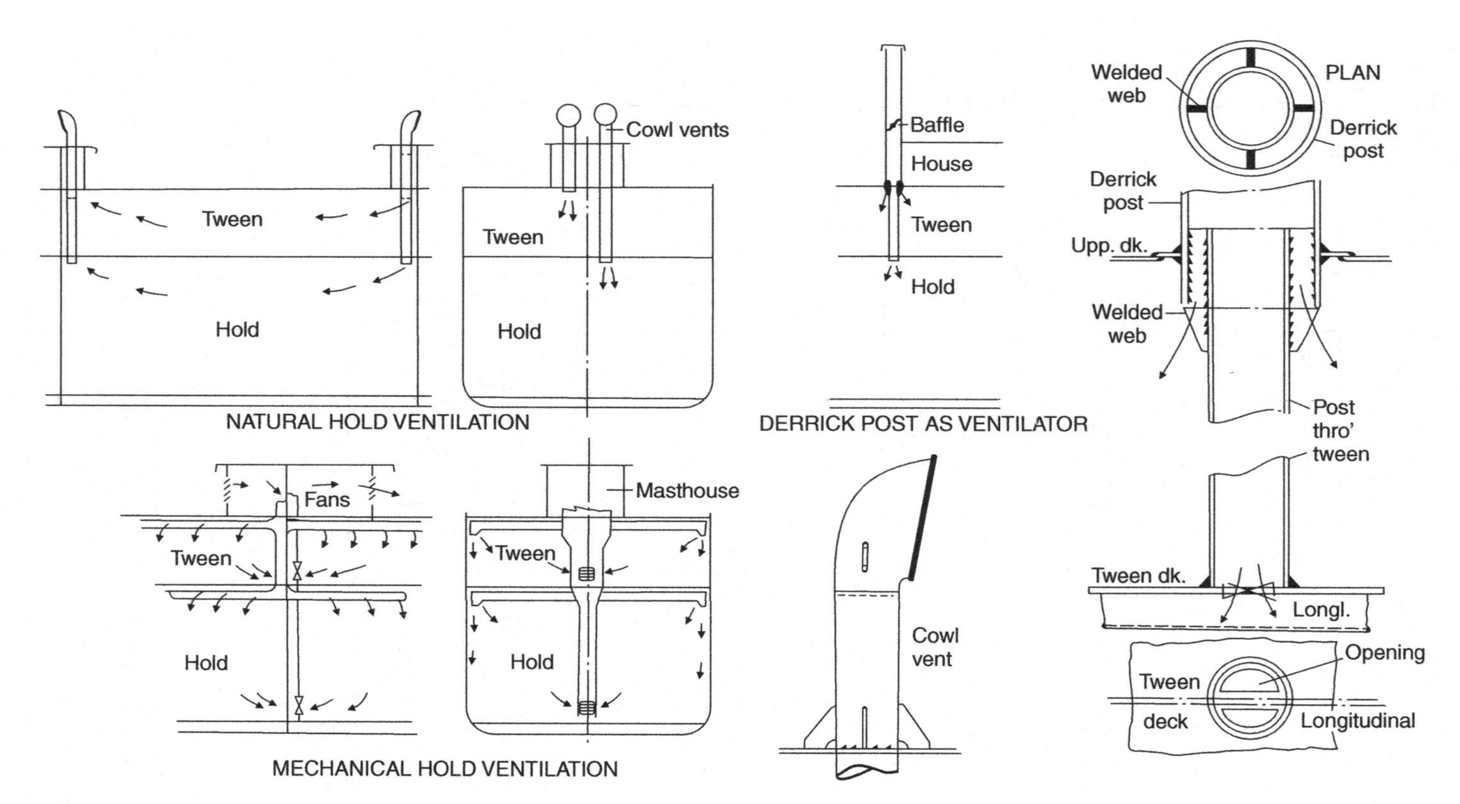

Figure 28.1 Hold ventilation.

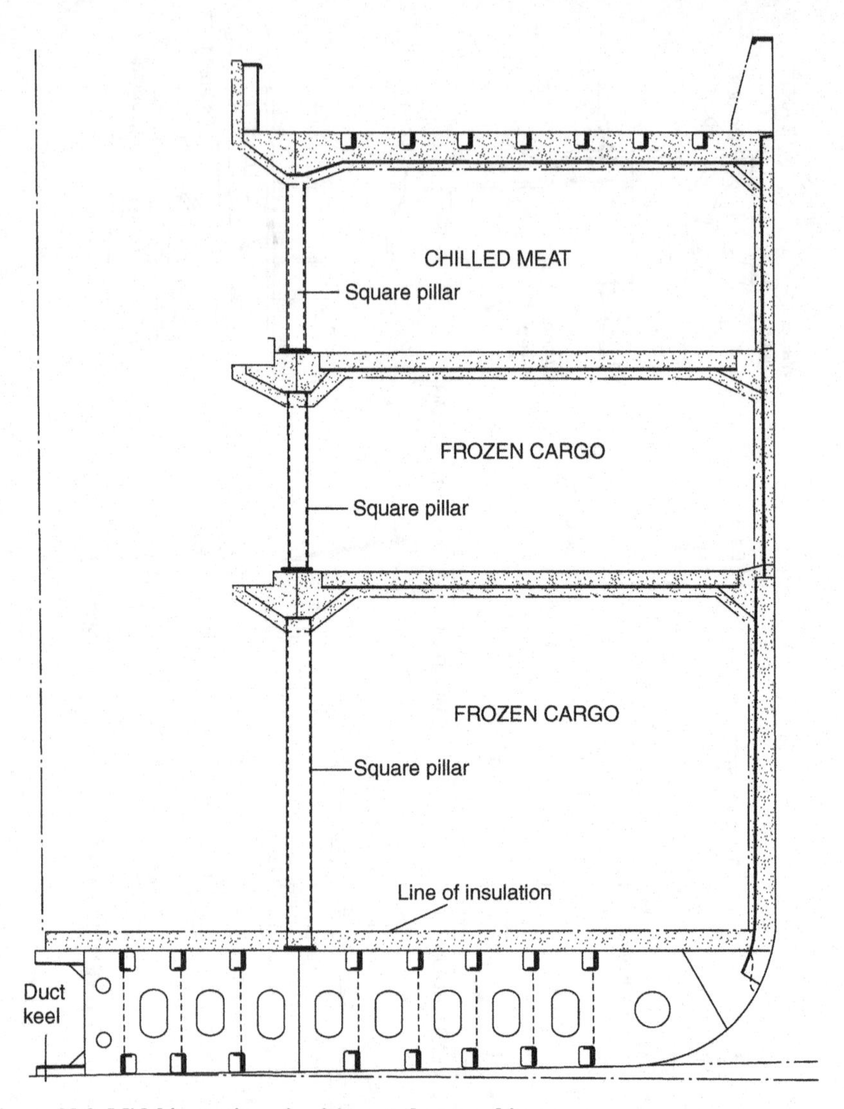

Figure 28.2 Midship section of refrigerated cargo ship.

may be pumped at controlled rates to give the correct working temperature, and it is led from the evaporator of the refrigerating machine to pipes at the top of the cold compartment. The brine absorbs heat from the compartments and returns to the evaporator, where it is again cooled and recirculated.

Air must be continually circulated where fruit is carried to disperse any pockets of carbon dioxide gas given off by the ripening fruit. The brine is then led into grid boxes, and air drawn from the bottom of the compartments by fans is blown over the brine grids into the compartments via trunking arranged along the ceiling.

Insulation

As the steel hull structure is an excellent conductor of heat, some form of insulation must be provided at the boundaries of the refrigerated compartments if the desired temperatures are to be maintained economically.

Cork, glass fiber, and various foam plastics in sheet or granulated form may be used for insulating purposes, also air spaces that are less efficient. Glass fiber is often used in modern ships as it has a number of advantages over the other materials; for example, it is extremely light, vermin proof and fire resistant, and it will not absorb moisture. Mineral wool slabs are also used and such products have the benefit of fire protection as well as insulation.

On the decks and particularly at the tank top, the insulation must often be load-bearing material, and cork might be preferred, but fiberglass can be supported by tongue and grooved board linings and wood bearers. The thickness of the insulation depends on the type of material used and the temperature to be maintained in the compartment. However, the depth of stiffening members often determines the final depth. Insulating material is retained at the sides by galvanized sheet steel or aluminum alloy sheet screwed to wood grounds on the frames or other stiffening members (see Figure 28.3).

Insulation on the boundaries of oil tanks, e.g. on the tank top above an oil fuel double-bottom tank, has an air space of at least 50 mm between the insulation and steel. If a coating of approved oil-resisting composition with a thickness of about 5 mm is applied the air gap may be dispensed with.

Suitable insulated doors are provided to cold rooms in general cargo ships, and in refrigerated cargo ships the hold and tween hatches may be insulated. Patent steel covers or pontoon covers may be filled with a suitable insulating material to prevent heat losses.

A particular problem in insulated spaces is drainage, as ordinary scuppers would nullify the effects of the insulation. To overcome this problem brine traps are provided in drains from the tween deck chambers and insulated holds. The brine in the trap forms an effective seal against ingress of warm air, and it will not freeze, preventing the drain from removing water from the compartment (Figure 28.3).

Refrigerated container ships

Many of the container ships operating on trade routes where refrigerated cargoes were carried in conventional refrigerated cargo liners ('reefer ships') have provision for carrying refrigerated containers and have in many cases replaced the latter.

The ISO containers (usually 20-foot size, since with most refrigerated cargoes 40-foot size would be too heavy) are insulated, and below decks the end of each hold may be fitted with brine coolers that serve each stack of containers. Air from the brine coolers is ducted to and from each insulated container. Connection of each container to the cold air ducts is by means of an automatic coupling, which is remotely controlled and can be engaged when the container is correctly positioned in the cell guides.

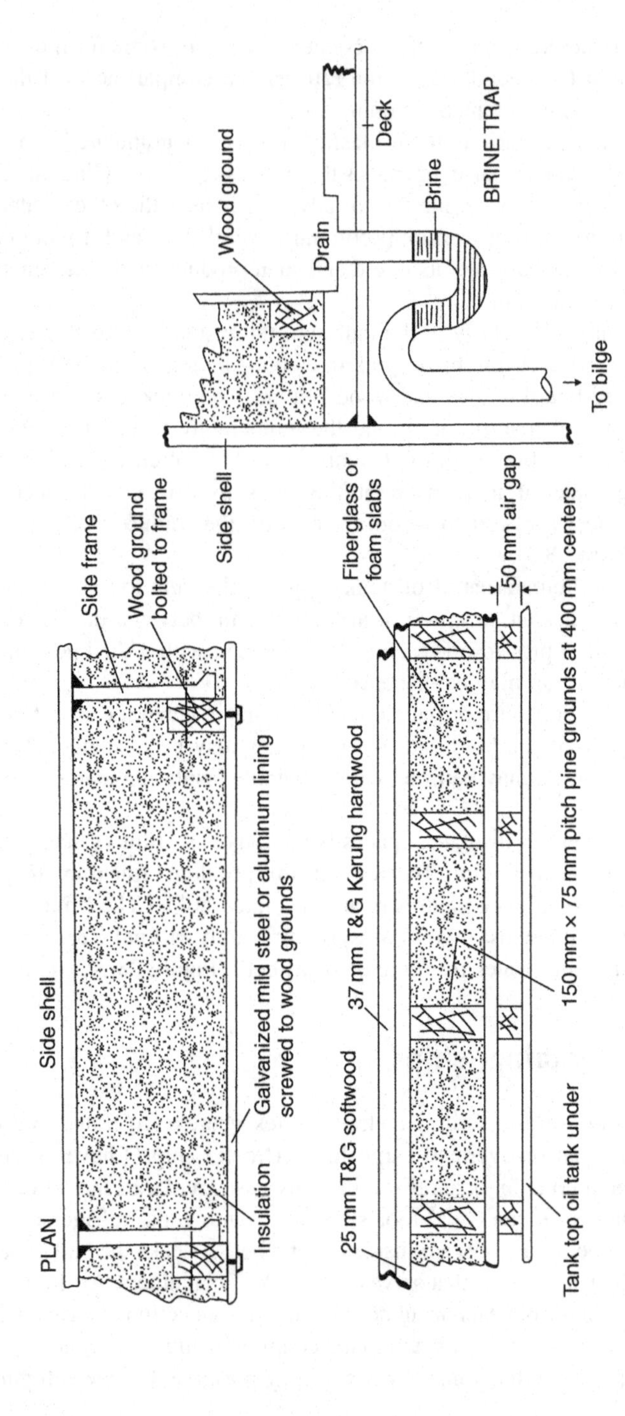

Figure 28.3 Insulation details.

The below-decks system described with fully insulated containers means that heavy insulation of the hold space is unnecessary. On the ship's sides, bulkheads, and deckhead about 50 mm of foam insulation with a fire-retardant coating may be fitted and the tank top covered with 75 mm of cork and bitumastic. If provision is only made for the ship to carry a part load of under-deck refrigerated containers, these are generally arranged in the after holds adjacent to the machinery space.

On-deck refrigerated containers are generally serviced by clip-on or integrated air-cooled electric motor drive cooling units. The units are plugged into the ship's electrical system by way of suitable deck sockets. Similar water-cooled units have been used for below-deck containers on short-haul voyages.

In recent years there has been a drive to install systems that enable use of standard containers with integrated air-cooled motor drive cooling units in the holds as well as on deck. It is reasonably argued that this would be in keeping with the concept of total container flexibility across all forms of transport.

Further reading

Environmental aspects of shipboard HVAC, *The Naval Architect*, (July/August 2002).

First Agrexco hybrid reefer ship nearing completion at ENVC, *The Naval Architect*, (May 2003).

Hold carriage of air-cooled refrigerated containers, *The Naval Architect*, (September 2003).

Some useful websites

www.rockwool.co.uk

Part Seven

International Regulations

29 International Maritime Organization

Chapter Outline

Organization of the IMO 355
Work of the IMO 355
Relationship with national authorities 356
Relationship with classification societies 357
Further reading 357
Some useful websites 357

The International Maritime Organization (IMO) is a specialized agency of the United Nations. It has as its most important objectives the improvement of maritime safety and the prevention of marine pollution. The functions of the IMO, only as it affects ship construction, are dealt with in this book.

Organization of the IMO

The Assembly, which is the supreme governing body of the IMO and consists of representatives of all member states (170 in October 2011) meets every two years and determines policy, the work program, votes the budget, approves all recommendations made by the IMO, and elects members of the Council. The Council consists of an agreed number (40 in 2011) of representatives of member states elected for a term of two years. It normally meets twice a year and is the IMO's governing body between Assembly sessions. The Maritime Safety Committee deals with the technical work of the IMO and in order to facilitate this work various subcommittees are set up to deal with specific subjects such as fire protection, ship design and equipment, etc. The Marine Environmental Protection Committee is responsible for coordinating IMO activities in the prevention of pollution from ships. The latter two committees meet once, or sometimes twice, a year and all member states may participate in their activities and those of their subcommittees.

Work of the IMO

The IMO is responsible for convening and preparing international conferences on subjects within its sphere of action, for the purpose of concluding international

Ship Construction. DOI: 10.1016/B978-0-08-097239-8.00029-5

conventions or agreements. Conventions do not come into force until stipulated numbers of member countries have ratified, i.e. adopted, them. Provided it is approved by an agreed majority of the members party to a convention, a technical amendment to the convention proposed by a party to the convention may be adopted at meetings of the Maritime Safety Committee and Marine Environmental Protection Committee.

Of particular relevance, and dealt with in the following chapters, are the following conventions:

- International Convention on Tonnage Measurement, 1969
- International Convention on Load Lines of Ships, 1966
- International Convention for the Safety of Life at Sea (SOLAS), 1974.

The latter, and its subsequent amendments and protocols, includes requirements in respect of fire protection in ships, which are dealt with in Chapter 32.

The International Convention for the Prevention of Pollution from Ships (MARPOL) 1973 and its Protocol of 1978 also prescribe ship construction requirements, particularly in respect of tankers (see Chapter 22).

A significant recent development in the work of the IMO was the approval of IMO Resolution MSC.287(87), which adopted the International Goal-Based Ship Construction Standards for Bulk Carriers and Oil Tankers to take effect on 1 January 2012. That is, the IMO have set the goals that have to be achieved in respect of ship safety and marine pollution prevention, and national regulators and/or classification societies will have to develop technical rules and standards that can achieve those goals. The intention is that the IMO should play a larger role in determining the fundamental standards to which new ships are built. It is not intended that the IMO would take over the detailed rule making of the classification societies, but the IMO would state what goals have to be achieved, leaving the classification societies, designers, and builders to determine how the required goals can be best met.

Relationship with national authorities

Member countries have their own governmental agency concerned with maritime safety, which drafts and enforces the shipping legislation of that country. The conventions and amendments are ratified by the member country when they are incorporated in that country's national legislation relating to ships registered in that country and which make international voyages. A national authority also has responsibility for ensuring that ships that are not registered in that country, but visiting its ports, are complying with the provisions of the conventions in force and to which they are party. The conventions require ships that are trading internationally to have current convention certificates issued by, or on behalf of, the governmental agency of the country in which they are registered. In the case of SOLAS (and under the harmonized system of survey and certification), these consist of:

1. For passenger ships a certificate called a Passenger Ship Safety Certificate valid for not more than one year.

2. For cargo ships of 500 gross tonnage or more, the following certificates:
   a. A Cargo Ship Safety Construction Certificate valid for not more than five years.
   b. A Cargo Ship Safety Equipment Certificate valid for not more than five years.
3. For cargo ships of 300 gross tonnage or more, a Cargo Ship Safety Radio Certificate valid for not more than five years.

Under the 1988 SOLAS Protocol, which came into force in February 2000, a cargo ship may be issued with a single Cargo Ship Safety Certificate rather than separate construction, equipment, and radio certificates.

Oil tankers of 150 gross tonnage or more and other ships of 400 gross tonnage or more complying with MARPOL Annex I are required to have a current International Oil Pollution Prevention Certificate valid for not more than five years. Vessels registered with a state that has signed up to MARPOL Annex IV and/or Annex VI are required to have a current International Sewage Pollution Prevention Certificate and/or an International Air Pollution Prevention Certificate respectively.

Both passenger and cargo ships would have an International Load Line Convention Certificate valid for not more than five years. In addition, each vessel would have an International Tonnage Certificate.

Relationship with classification societies

The major classification societies represented by the IACS (see Chapter 4) attend the IMO meetings on a consultative basis. Many of the member countries of the IMO have authorized different classification societies to issue one or more of the convention certificates on their behalf. This is particularly true in respect of assignment and issuing of Load Line Certificates, where load line surveys are often undertaken in a foreign port. The initials of the assigning classification society, rather than the governmental authority, are commonly observed on a ship's Plimsoll mark. Smaller countries with limited maritime technical expertise to service their governmental authority, particularly if they have a large register of ships trading internationally, may rely entirely on the classification societies to survey and issue their convention certificates.

Further reading

SOLAS Consolidated Edition, IMO publication (IMO-110E), 2009.
MARPOL 73/78 Consolidated Edition, IMO publication (IMO-1B 520E), 2006.

Some useful websites

IMO website at www.imo.org
IACS website at www.iacs.org.uk
Lloyd's Register website at www.lr.org

30 Tonnage

Chapter Outline
International Convention on Tonnage Measurement of Ships 1969 359
Tonnages 359
Gross tonnage 359
Net tonnage 360
Measurement 360
Panama and Suez Canal tonnages 361
Compensated gross tonnage (CGT) 361
Further reading 362

Gross tonnage is a measure of the internal capacity of the ship and net tonnage is intended to give an idea of the earning or useful capacity of the ship. Various port dues and other charges may be assessed on the gross and net tonnages. It will be noted from the previous chapter that the gross tonnage of a ship may determine its requirement to comply with international standards.

International Convention on Tonnage Measurement of Ships 1969

An International Conference on Tonnage Measurement was convened by the IMO in 1969 with the intention of producing a universally acceptable system of tonnage measurement. The International Convention on Tonnage Measurement of Ships 1969 was prepared at this conference and this convention came into force on 8 July 1982. All ships constructed on or after that date were measured for tonnage in accordance with the 1969 Convention. Ships built prior to that date were, if the owner so desired, permitted to retain their existing tonnages for a period of 12 years from that date, i.e. all ships were required to be measured in accordance with the 1969 Convention by 18 July 1994.

Tonnages

Gross tonnage

The gross tonnage (GT) is determined by the following formula:

$$\mathrm{GT} = K_1 V$$

Ship Construction. DOI: 10.1016/B978-0-08-097239-8.00030-1

where

$$K_1 = 0.2 + 0.02 \log_{10} V$$

$V =$ total volume of all enclosed spaces in cubic meters.

Net tonnage

The net tonnage (NT) is determined by the following formulae:

1. For passenger ships (i.e. ships carrying 13 passengers or more)

$$\mathrm{NT} = K_2 V_\mathrm{c} = \left(\frac{4d}{3D}\right)^2 + K_3\left(N_1 + \frac{N_2}{10}\right)$$

2. For other ships

$$\mathrm{NT} = K_2 V_\mathrm{c} = \left(\frac{4d}{3D}\right)^2$$

where

$V_\mathrm{c} =$ total volume of cargo spaces in cubic meters

$d =$ molded draft amidships in meters (summer load line draft or deepest subdivision load line for passenger ships)

$D =$ molded depth in meters amidships

$$K_2 = 0.2 + 0.02 \log_{10} V_c$$

$$K_3 = 1.25 \times \frac{\mathrm{GT} + 10{,}000}{10{,}000}$$

$N_1 =$ number of passengers in cabins with not more than eight berths

$N_2 =$ number of other passengers

$N_1 + N_2 =$ total number of passengers the ship is permitted to carry.

The factor $(4d/3D)^2$ is not taken to be greater than unity. The term $K_2V_\mathrm{c}(4d/3D)^2$ is not to be taken as less than 0.25GT, and NT is not to be taken as less than 0.30GT.

It will be noted that vessels with high freeboards, i.e. low draft to depth (d/D) ratios will have low net tonnages. Squaring this ratio can result in excessively low net tonnages, hence the limiting value of 0.30GT.

Measurement

Measurement for tonnage and issue of an International Tonnage Certificate is the responsibility of the appropriate maritime authority in the country of registration of the ship. Most maritime authorities have authorized various classification societies and perhaps other bodies to act on their behalf to measure ships and issue the International Tonnage Certificate.

The volumes to be included in the calculation of gross and net tonnages are measured irrespective of the fitting of insulation or other linings to the inner side shell or structural boundary plating in metal, i.e. the molded dimensions are used. It is possible for the surveyor to compute tonnages directly form the molded lines of the ship or computer-stored offsets. This is in contrast to the previous regulations, where the surveyor was required to physically measure spaces to the inside of frames and linings.

Panama and Suez Canal tonnages

Tolls for transiting the Suez Canal are based on a net tonnage of the ship determined in accordance with the regulations of the Suez Canal authority. Whilst the IMO recommended acceptance of the use of the Convention rule for the Suez and Panama Canals in 1981, the Suez Canal Authority has maintained its own system.

The Panama Canal Authority introduced a net tonnage measurement system based on the 1969 Tonnage Convention in 1994. It has, however, subsequently changed its charging regime from a flat rate for all ships to one based on ship size and type, and has separate locomotive usage rates. This is considered a more equitable system, since charges are now applied according to the transit needs of each ship. Further, since 2005 a new measurement and pricing system is being phased in for container ships and other vessels with on-deck container-carrying capacity. For container vessels the Authority have adopted the TEU as the measurement unit. For other vessels with on-deck container-carrying capacity the Authority will continue to apply the universal tonnage requirement to the underdeck spaces and will charge a per-TEU fee for the actual number of containers carried on deck.

Both the Suez and Panama Canal Authorities have recognized a number of classification societies to undertake measurement and issue of tonnage certificates on their behalf.

Compensated gross tonnage (CGT)

For many years the gross tonnage was used as a measure for comparing the output of various shipbuilding countries. Gross tonnage was never intended to be used for statistical purposes and, although it may have served this purpose, it can give very misleading impressions. For example, the building of a 65,000 gross tonnage passenger liner may involve considerably more man-hours and capital than the construction of a 150,000 gross tonnage oil tanker.

A system of compensated gross tonnage was introduced and adopted as a means of overcoming this problem. Factors were developed for each ship type and the gross tonnage of the vessel then multiplied by the appropriate factor to give the compensated gross tonnage. Compensated gross tonnage factors were developed by the

Table 30.1 *A* and *B* factors

Ship type	**A**	**B**
Oil tankers (double hull)	48	0.57
Chemical tankers	84	0.55
Bulk carriers	29	0.61
Combined carriers	33	0.62
General cargo ships	27	0.64
Reefers	27	0.68
Full container	19	0.68
Ro-ro vessels	32	0.63
Car carriers	15	0.70
LPG carriers	62	0.57
LNG carriers	32	0.68
Ferries	20	0.71
Passenger ships	49	0.67
Fishing vessels	24	0.71
NCCV	46	0.62

OECD in cooperation with the Association of West European Shipbuilders and the Shipbuilders Association of Japan in 1968.

In 2007 a revised system for CGT was developed. This replaced fixed factors for ship type and size by a formula, $\text{CGT} = A \times \text{GT}B$, where *A* is a factor to account for the type of ship, *B* is a factor to account for ship size, and GT is the gross tonnage of the vessel. The *A* and *B* factors are shown in Table 30.1.

Further reading

International Conference on Tonnage Measurement of Ships, 1969, IMO publication (IMO-713E), 1969.

31 Load Line Rules

Chapter Outline

Freeboard computation 363
Minimum freeboards 366
Timber freeboards 366
Conditions of assignment of freeboard 366
Special conditions of assignment for Type 'A' ships 370
Further reading 370

Reference to 'rules' in this chapter means the requirements of the International Convention on Load Lines of Ships 1966.

Freeboard computation

Basic freeboards are given in the rules that are dependent on the length and type of vessel.

Ships are divided into types 'A' and 'B'. Type 'A' ships are those that are designed to carry only liquid cargoes in bulk, and in which the cargo tanks have only small access openings closed by watertight gasketed covers of steel or equivalent material. These vessels benefit from the minimum assignable freeboard. All ships that do not come within the provisions regarding Type 'A' ships are considered as Type 'B' ships.

As a considerable variety of ships will come within the Type 'B' category, a reduction or increase from the basic table Type 'B' freeboard is made in the following cases:

1. Vessels having hatchways fitted with portable beams and covers on exposed freeboard or raised quarter decks, and within 25% of the ship's length from the FP on exposed superstructure decks, are to have the basic freeboard increased.
2. Vessels having steel weathertight covers fitted with gaskets and clamping devices, improved measures for the protection of the crew, better freeing arrangements, and satisfactory subdivision characteristics may obtain a reduction in the basic freeboard given for a Type 'B' ship. This reduction may be increased up to the total difference between the values for Type 'A' and Type 'B' basic freeboards. The Type 'B' ship, which is effectively adopting a Type 'A' basic freeboard, is referred to as a Type 'B-100' and its final calculated freeboard will be almost the same as that for a Type 'A' ship. Other Type 'B' vessels that comply with not such severe subdivision requirements can be assigned a basic freeboard reduced by up to 60% of the difference between 'B' and 'A' basic values.

Ship Construction. DOI: 10.1016/B978-0-08-097239-8.00031-3

Obtaining the maximum possible draft can be important in many Type 'B' vessels, and careful consideration at the initial design stage with regard to subdivision requirements can result in the ship being able to load to deeper drafts. This is particularly the case with bulk carriers, since these vessels can often be designed to obtain the 'B-60' freeboards, and where this is impossible some reduction in freeboard may still be possible. The Convention allows freeboards between that assigned to a Type 'B' and a Type 'B-60' where it can be established that a one-compartment standard of subdivision can be obtained at the draft of a Type 'B' vessel, but not at the draft of a Type 'B-60'.

Ore carriers of normal layout arranged with two longitudinal bulkheads and having the side compartments as water ballast tanks are particularly suited to the assignment of Type 'B-100' freeboards, where the bulkhead positions are carefully arranged. In the case of Type 'A', Type 'B-100', and Type 'B-60' vessels over 225 m, the machinery space is also to be treated as a floodable compartment. The full subdivision requirements are given in Table 31.1.

Damage is assumed as being for the full depth of the ship, with a penetration of one-fifth the beam clear of main transverse bulkheads. After flooding the final waterline is to be below the lower edge of any opening through which progressive flooding may take place. The maximum angle of heel is to be 15°, and the metacentric height in the flooded condition should be positive.

Having decided the type of ship, the computation of freeboard is comparatively simple, a number of corrections being applied to the rule on basic freeboard given for Type 'A' and

Table 31.1 Subdivision requirements

Type	Length	Subdivision requirements
A	Less than 150 m	None
A	Greater than 150 m Less than 225 m	To withstand the flooding of any compartment Within the cargo tank length, which is designed to be empty when the ship is loaded to the summer waterline at an assumed permeability of 0.95
A	Greater than 225 m	As above, but the machinery space is also to be treated as a floodable compartment with an assumed permeability of 0.85
B+	—	None
B	—	None
B-60	100–225 m	To withstand the flooding of any single damaged compartment within the cargo hold length at an assumed permeability of 0.95
B-60	Greater than 225 m	As above, but the machinery space is also to be treated as a floodable compartment at an assumed permeability of 0.85
B-100	100–225 m	To withstand the flooding of any two adjacent fore and aft compartments within the cargo hold length at an assumed permeability of 0.95
B-100	Greater than 225 m	As above, but the machinery space, taken alone, is also to be treated as a floodable compartment at an assumed permeability of 0.85

Type 'B' ships against length of ship. The length (L) is defined as 96% of the total length on the waterline at 85% of the least molded depth, or as the length measured from the fore side of the stem to the axis of the rudder stock on the waterline, if that is greater.

The corrections to the basic freeboard are as follows:

a. *Flush deck correction.* The basic freeboard for a Type 'B' ship of not more than 100 m in length having superstructures with an effective length of up to 35% of the freeboard length (L) is increased by:

$$7.5(100 - L)\left(0.35 - \frac{E}{7}\right) \text{ mm}$$

where E = effective length of superstructure in meters.

b. *Block coefficient correction.* Where the block coefficient C_b exceeds 0.68, the basic freeboard (as modified above, if applicable) is multiplied by the ratio:

$$\frac{C_b + 0.68}{1.36}$$

C_b is defined in the rules as $\nabla/(L \cdot\cdot B \cdot\cdot d)$, where ∇ is the molded displacement at a draft d, which is 85% of the least molded depth.

c. *Depth correction.* The depth (D) for freeboard is given in the rules. Where D exceeds $L/15$ the freeboard is increased by $(D - (L/15))R$ mm, where R is $L/0.48$ at lengths less than 120 m and 250 at lengths of 120 m and above. Where D is less than $L/15$ no reduction is made, except in the case of a ship with an enclosed superstructure covering at least $0.6L$ amidships. This deduction, where allowed, is at the rate described above.

d. *Superstructure correction.* Where the effective length of superstructure is $1.0L$, the freeboard may be reduced by 350 mm at 24 m length of ship, 860 mm at 85 m length, and 1070 mm at 122 m length and above. Deductions at intermediate lengths are obtained by linear interpolation. Where the total effective length of superstructures and trunks is less than $1.0L$ the deduction is a percentage of the above. These percentages are given in tabular form in the rules, and the associated notes give corrections for size of forecastles with Type 'B' ships.

e. *Sheer correction.* The area under the actual sheer curve is compared with the area under a standard parabolic sheer curve, the aft ordinate of which (S_A) is given by $25(L/3 + 10)$ mm and the forward ordinate (S_F) by $2S_A$ mm. Where a poop or forecastle is of greater than standard height, an addition to the sheer of the freeboard deck may be made.

The correction for deficiency or excess of sheer is the difference between the actual sheer and the standard sheer multiplied by $(0.75 - S/2L)$, where S is the total mean enclosed length of superstructure. Where the sheer is less than standard the correction is added to the freeboard. If the sheer is in excess of standard a deduction may be made from the freeboard if the superstructure covers $0.1L$ abaft and $0.1L$ forward of amidships. No deduction for excess sheer may be made if no superstructure covers amidships. Where superstructure covers less than $0.1L$ abaft and forward of amidships the deduction is obtained by linear interpolation. The maximum deduction for excess sheer is limited to 125 mm per 100 m of length.

If the above corrections are made to the basic freeboard, the final calculated freeboard will correspond to the maximum geometric summer draft for the vessel.

The final freeboard may, however, be increased if the bow height is insufficient, or the owners request the assignment of freeboards corresponding to a draft that is less than the maximum possible. Bow height is defined as the vertical distance at the forward perpendicular between the waterline corresponding to the assigned summer freeboard and the top of the exposed deck at the side. This height should not be less than the values quoted in the Convention.

Minimum freeboards

The minimum freeboard in the summer zone is the freeboard described above; however, it may not be less than 50 mm. In the tropical and winter zones the minimum freeboard is obtained by deducting and adding respectively 1/48th of the summer molded draft. The tropical freeboard is, however, also limited to a minimum of 50 mm. The freeboard for a ship of not more than 100 m length in the winter North Atlantic zone is the winter freeboard plus 50 mm. For other ships, the winter North Atlantic freeboard is the winter freeboard. The minimum freeboard in fresh water is obtained by deducting from the summer or tropical freeboard the quantity:

$$\frac{\text{Displacement in SW}}{4 \times \text{TPC}} \text{ mm}$$

where TPC is the tonnes per cm immersion at the waterline, and displacement is in tonnes.

Timber freeboards

If a vessel is carrying timber on the exposed decks it is considered that the deck cargo affords additional buoyancy and a greater degree of protection against the sea. Ships thus arranged are granted a smaller freeboard than would be assigned to a Type 'B' vessel, provided they comply with the additional conditions of assignment for timber-carrying vessels. No reduction of freeboard may be made in ships that already have Type 'A' or reduced Type 'B' freeboards. The freeboards are computed as described above, but have a different superstructure correction, this being modified by the use of different percentage deductions given in the rules for timber freeboards. Winter timber freeboard is obtained by adding to the summer timber freeboard 1/36th of the molded summer timber draft. Winter North Atlantic timber freeboard is the same as for normal freeboards, and the tropical timber freeboard is obtained by deducting from the summer timber freeboard 1/48th of the molded summer timber draft. The fresh water timber freeboard is determined as for normal freeboards.

Conditions of assignment of freeboard

1. The construction of the ship must be such that her general structural strength will be sufficient for the freeboards to be assigned. The design and construction of the

ship must be such that her stability in all probable loading conditions is sufficient for the freeboards assigned. Stability criteria are given in the Convention.

2. *Superstructure end bulkheads.* To be of efficient construction to the satisfaction of the Administration. The heights of the sills of openings at the ends of enclosed superstructures should be at least 380 mm above the deck.
3. *Hatchways closed by portable covers with tarpaulins.* The coamings should be of substantial construction with a height above deck of at least 600 mm on exposed freeboard and RQD and on exposed superstructure decks within 25% of the ship's length from FP (Position 1) and at least 450 mm on exposed superstructure decks outside 25% of the ship's length from FP (Position 2).

The width of bearing surface for the covers should be at least 65 mm. Where covers are of wood the thickness should be at least 60 mm with a span of not more than 1.5 m. For mild steel portable covers, the strength is calculated with assumed loads. The assumed loads on hatchways in Position 1 may be not less than 1 tonne/square meter for ships 24 meters in length and should not be less than 0.75 tonnes/square meter for hatchways in Position 2. Where the ship's length is 100 m or greater the assumed loads on hatchways at Positions 1 and 2 are 1.75 and 1.30 tonnes/square meter respectively. At intermediate lengths the loads are obtained by interpolation.

The product of the maximum stress thus calculated and the factor 4.25 should not exceed the minimum ultimate strength of the material. The deflection is limited to 0.0028 times the span under these loads.

For portable beams of mild steel the assumed loads above are adopted, and the product of the maximum stress thus calculated and the factor 5 should not exceed the minimum ultimate strength of the material. Deflections are not to exceed 0.0022 times the span under these loads.

Mild steel pontoon covers used in place of portable beams are to have their strength calculated with the assumed loads above. The product of the maximum stress so calculated and the factor 5 should not exceed the minimum ultimate strength of the material, and the deflection is limited to 0.0022 times the span. Mild steel plating forming the tops of the covers should have a thickness not less than 1% of the stiffener spacing or 6 mm if that is greater. Covers of other material should be of equivalent strength. Carriers and sockets are to be of substantial construction and where rolling beams are fitted it should be ensured that beams remain in position when the hatchway is closed.

Cleats are set to fit the taper of wedges. They are at least 65 mm wide, spaced not more than 600 mm center to center, and not more than 150 mm from the hatch covers. Battens and wedges should be efficient and in good condition. Wedges should be of tough wood or equivalent material, with a taper of not more than 1 in 6, and should be not less than 13 mm thick at the toes.

At least two tarpaulins in good condition should be provided for each hatchway, and should be of approved material, strength, and waterproof. Steel bars or equivalent are to be provided to secure each section of the hatchway covers after the tarpaulins are battened down, and covers of more than 1.5 m in length should be secured by at least two such securing appliances.

4. *Hatchways closed by weathertight steel covers.* Coaming heights are as for those hatchways with portable beam covers. This height may be reduced or omitted altogether on condition that the Administration is satisfied that the safety of the ship is not thereby impaired. Mild steel covers should have their strength calculated assuming the loads given previously. The product of the maximum stress thus calculated and the factor of 4.25 should not exceed the minimum ultimate strength of the material, and deflections are limited to not more than 0.0028 times the span under these loads. Mild steel plating forming the tops of the covers should not be less in thickness than 1% of the spacing of stiffeners or 6 mm if that is greater. The strength and stiffness of covers made of other materials is to be of equivalent strength.

Means of securing weathertightness should be to the satisfaction of the Administration, the tightness being maintained in any sea condition.

5. *Machinery space openings.* These are to be properly framed and efficiently enclosed by steel casings of ample strength. Where casings are not protected by other structures their strength is to be specially considered. Steel doors to be fitted for access should have the sills at least 600 mm above the deck in Position 1, and at least 380 mm above the deck in Position 2. Fiddley, funnel, or machinery space ventilator coamings on exposed decks are to be as high above deck as is reasonable.
6. *Other openings in freeboard and superstructure decks.* Manholes and flush scuttles in Positions 1 or 2 or within superstructures other than enclosed superstructures should be closed by substantial weathertight covers. Openings other than those considered are to be protected by an enclosed superstructure or deckhouse, or companionway of equivalent strength. Doors for access should be of steel, and the sills should have the same heights as above.
7. *Ventilators.*Should have steel coamings and where they exceed 900 mm in height they should be specially supported. In Position 1 ventilator coamings should be of height 900 mm above deck, and in Position 2 760 mm above deck. Vent openings should be provided with efficient weathertight closing appliances except in the case of coamings exceeding 4.5 m in height in Position 1 and 2.3 m in height in Position 2, above deck.
8. *Air pipes.* Exposed parts of pipe should be of substantial construction. The height from the deck should be at least 760 mm on the freeboard deck and 450 mm on superstructure decks. A lower height may be approved if these heights interfere with working arrangements. Permanently attached means of closing the pipe openings should be provided.
9. *Cargo ports and other similar side openings.* Below the freeboard deck to be fitted with watertight doors to ensure the ship's structural integrity. Unless permitted by the Administration the lower edge of such openings should not be below a line drawn parallel to the freeboard deck at the side, which has at its lowest point the upper edge of the uppermost load line.
10. *Scuppers, inlets, and discharges.* Discharges led through the shell either from spaces below the freeboard deck or from within superstructures and deckhouses

on the freeboard deck fitted with weathertight doors should be fitted with efficient and accessible means for preventing water from passing inboard. Normally this should be an automatic nonreturn valve with means of closing provided above the freeboard deck. Where the vertical distance from the summer waterline to the inboard end of the discharge pipe exceeds $0.02L$ the discharge may have two automatic nonreturn valves without positive means of closing, provided the inboard valve is always accessible. Where the distance exceeds $0.02L$ a single automatic nonreturn valve without positive means of closing may be accepted. In manned machinery spaces, main and auxiliary sea inlets and discharges in connection with the operation of machinery may be controlled locally. Scuppers and discharge pipes originating at any level and penetrating the shell either more than 450 mm below the freeboard deck or less than 600 mm above the summer waterline should be fitted with an automatic nonreturn valve. Scuppers leading from superstructures or deckhouses not fitted with weathertight doors should be led overboard.

11. *Side scuttles.* Below the freeboard deck or within the enclosed superstructures, side scuttles should be fitted with efficient hinged, watertight, inside deadlights. No side scuttle should be fitted with its sill below a line drawn parallel to the freeboard deck at the side and having its lowest point 2.5% of the ship's breadth above the summer waterline or 500 mm, whichever is the greater distance.
12. *Freeing ports.* The minimum freeing port area (A) on each side of the ship where sheer in way of the well is standard or greater than standard, is given, in square meters, by $A = 0.7 + 0.035\ l$ where l is the length of bulwark in the well and is less than 20 m, and $A = 0.07\ l$ where l is greater than 20 m.

In no case need l be greater than $0.7L$. If the bulwark is greater than 1.2 m in height, A is increased by 0.004 square meters per meter of length of well for each 0.1 m difference in height. If the bulwark is less than 0.9 m in height, A is reduced by 0.004 square meters per meter of length of well for each 0.1 m difference in height. Where there is no sheer, A is increased by 50% and with less than standard sheer the percentage increase is obtained by interpolation.

The lower edges of freeing ports should be as near the deck as practicable. Two-thirds of the freeing port area is required to be provided in the half of the well nearest the lowest point of the sheer curve, where the deck has sheer. Openings in the bulwarks are protected by bars spaced approximately 230 mm apart. If shutters are fitted, these should be prevented from jamming.

13. *Protection of crew.* Efficient guard-rails or bulwarks of minimum height 1 meter are to be fitted on all exposed parts of freeboard and superstructure decks. A lower rail may be permitted by the Administration. The maximum vertical spacing between deck and lower rail is 230 mm, and between other rails is 380 mm.

Satisfactory means should be provided for protection of crew in getting to and from their quarters and other parts used in the working of the ship.

Special conditions of assignment for Type 'A' ships

1. *Machinery casings.* To be protected by an enclosed poop or bridge of standard height, or deckhouse of equivalent strength and height. The casing may be exposed if there are no doors fitted giving access from the freeboard deck, or if a weathertight door is fitted and leads to a passageway separated from the stairway to the engine room by a second weathertight door of equivalent material.
2. *Gangway and access.* An efficiently constructed fore and aft gangway should be fitted at the level of the superstructure deck between poop and midship bridge or deckhouse, or equivalent means such as passages below deck. If houses are all aft, satisfactory arrangements should be made to allow crew to reach all parts of the ship for working purposes.
3. *Hatchways.* All exposed hatchways on freeboard and forecastle decks or on top of expansion trunks are to be provided with efficient watertight covers of steel or equivalent material.
4. *Freeing arrangements.* Should have open rails fitted for at least half the length of the exposed parts of the weather deck, with the upper edge of the sheer strake being kept as low as possible. Where superstructures are connected by trunks, open rails should be fitted for the whole length of the exposed parts of the freeboard deck in way of the trunk.

Further reading

International Conference on Load Lines, 1966. IMO publication (IMO 701E), 2005 edition.

32 Structural fire protection

Chapter Outline

Requirements 371
'A', 'B', and 'C' class divisions 372
Openings in fire protection divisions 373
Protection of special category spaces 375
Fire protection arrangements in high-speed craft 375
Further reading 375
Some useful websites 376

Of the requirements of the International Conventions for the Safety of Life at Sea, those having a particular influence on ship construction are the requirements relating to structural fire protection. Varying requirements for vessels engaged in international voyages are given for passenger ships carrying more than 36 passengers, passenger ships carrying not more than 36 passengers, cargo ships, and tankers.

Requirements

Ships carrying more than 36 passengers are required to have accommodation spaces and main divisional bulkheads and decks that are generally of incombustible material in association with either an automatic fire detection and alarm system or an automatic sprinkler and alarm system. The hull, superstructure, and deckhouses are subdivided by 'A' class divisions into main vertical zones, the length of which on any one deck should not exceed 40 m. Main horizontal zones of 'A' class divisions are fitted to provide a barrier between sprinklered and nonsprinklered zones of the ship. Bulkheads within the main vertical zones are required to be 'A', 'B', or 'C' class divisions depending on the fire risk of the adjoining spaces and whether adjoining spaces are within sprinkler or nonsprinkler zones.

Passenger vessels carrying not more than 36 passengers are required to have the hull, superstructure, and deckhouses subdivided into main vertical zones by 'A' class divisions. The accommodation and service spaces are to be protected either by all enclosure bulkheads within the space being of at least 'B' class divisions or only the corridor bulkheads being of at least 'B' class divisions, where an approved automatic fire detection and alarm system is installed.

Ship Construction. DOI: 10.1016/B978-0-08-097239-8.00032-5

Cargo ships exceeding 500 gross tonnage are generally to be constructed of steel or equivalent material and to be fitted with one of the following methods of fire protection in accommodation and service spaces:

- 'Method Ic'. All internal divisional bulkheads constructed of noncombustible 'B' or 'C' class divisions and no installation of an automatic sprinkler, fire detection, and alarm system in the accommodation and service spaces, except smoke detection and manually operated alarm points, which are to be installed in all corridors, stairways, and escape routes.
- 'Method IIc'. An approved automatic sprinkler, fire detection, and fire alarm system is installed in all spaces in which a fire might be expected to originate, and in general there is no restriction on the type of divisions used for internal bulkheads.
- 'Method IIIc'. A fixed fire detection and fire alarm system is installed in all spaces in which a fire might be expected to originate, and in general there is no restriction on the type of divisions used for internal bulkheads except that in no case must the area of any accommodation space bounded by an 'A' or 'B' class division exceed 50 square meters.

Crowns of casings of main machinery spaces are to be of steel construction and insulated. Bulkheads and decks separating adjacent spaces are required to have appropriate A, B, or C ratings depending on the fire risk of adjoining spaces. Cargo spaces of ships of 2000 gross tonnage or more are to be protected by a fixed gas fire-extinguishing system or its equivalent unless they carry bulk or other cargoes considered by the authorities to be a low fire risk. Cargo ships carrying dangerous goods are subject to special fire protection precautions.

In the construction of tankers, particular attention is paid to the exterior boundaries of superstructures and deckhouses that face the cargo oil tanks. Accommodation boundaries facing the cargo area are insulated to A60 standard, no doors are allowed in such boundaries giving access to the accommodation, and any windows are to be of non-opening type and fitted with steel covers if in the first tier on the main deck. Bulkheads and decks separating adjacent spaces of varying fire risk are required to have appropriate A, B, and C ratings within the accommodation space. For new tankers of 20,000 tons deadweight and upwards the cargo tanks' deck area and cargo tanks are protected by a fixed deck foam system and a fixed inert gas system (see Chapter 26). Tankers of less than 2000 tons deadweight are provided with a fixed deck foam system in way of the cargo tanks.

'A', 'B', and 'C' class divisions

'A' class divisions are constructed of steel or equivalent material and should be capable of preventing the passage of smoke and flame to the end of a one-hour standard fire test. A plain stiffened steel bulkhead or deck has what is known as an A-O rating. By adding insulation in the form of approved incombustible materials to the steel an increased time is taken for the average temperature of the unexposed side to rise to 139 °C above the original temperature or not more than 180 °C at any one point above the original temperature during the standard fire test. The 'A' class division rating is related to this time as follows:

Class	Time (min)
A-60	60
A-30	30
A-15	15
A-0	0

Figure 32.1 shows typical steel divisions with typical proprietary non-asbestos fiber-reinforced silicate board insulation. 'B' class divisions are those that are constructed as to be capable of preventing the passage of flame to the end of half an hour of the standard fire test. Various patent board materials are commonly used where 'B' class divisions are required and there are two ratings B-0 and B-15. These relate to the insulation value such that the average temperature of the unexposed side does not rise more than 139 °C above the original temperature and at any one point more than 225 °C above the original temperature when the material is subjected to the standard fire test within the following times:

Class	Time (min)
B-15	15
B-0	0

'C class divisions are constructed of approved incombustible materials but do not need to meet any specified requirements relative to passage of smoke and flame nor temperature rise.

The standard fire test referred to is a test in which a specimen of the division with a surface area of not less than 4.65 square meters and height or length of 2.44 m is exposed in a test furnace to a series of time–temperature relationships, defined by a smooth curve drawn through the following points:

At end of first 5 minutes	538 °C
At end of first 10 minutes	704 °C
At end of first 30 minutes	843 °C
At end of first 60 minutes	927 °C

Some typical examples of fire divisions are given in Table 32.1 for a passenger ship carrying more than 36 passengers.

Openings in fire protection divisions

Generally, openings in fire divisions are to be fitted with permanently attached means of closing that have the same fire-resisting rating as the division. Suitable

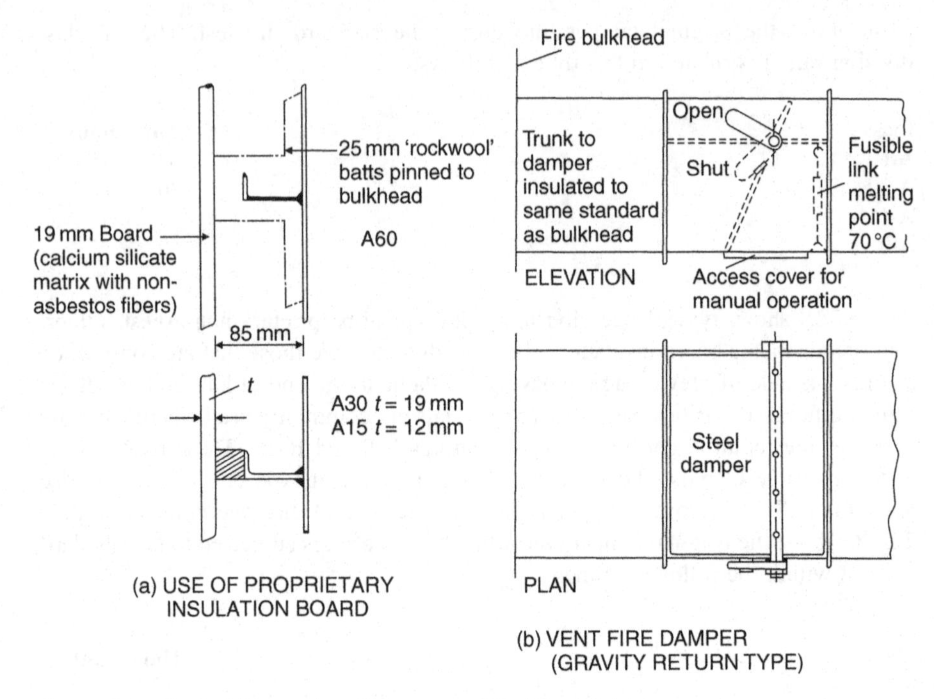

Figure 32.1 Typical Examples of Passenger Ship Fire Divisions.

Table 32.1 Typical examples of fire divisions for a passenger ship carrying more than 36 passengers

Bulkhead	Adjacent compartments	Class
Main fire zone	Galley/passageway	A-60
Main fire zone	Wheelhouse/passageway	A-30
Within fire zone	Fan room/stairway	A-15
Within fire zone	Cabin/passageway (nonsprinklered zone)	B-15
Within fire zone	Cabin/passageway (sprinklered zone)	B-0

arrangements are made to ensure that the fire resistance of a division is not impaired where it is pierced for the passage of pipes, vent trunks, electrical cables, etc.

Greatest care is necessary in the case of openings in the main fire zone divisions. Door openings in the main fire zone bulkheads and stairway enclosures are fitted with fire doors of equivalent fire integrity and are self-closing against an inclination of 3.5° opposing closure. Such doors are capable of closure from a control station either simultaneously or in groups, and also individually from a position adjacent to the door. Vent trunking runs are ideally contained within one fire zone, but where they must pass through a main fire zone bulkhead or deck a fail-safe automatic-closing fire damper is fitted within the trunk adjacent to the bulkhead or deck. This usually takes the form of a steel flap in the trunk, which is held open by a weighted hinge secured by an external fusible link. The flap must also be capable of being released manually and there is some form of indication as to whether the flap is open or closed (see Figure 32.1).

Protection of special category spaces

A special category space is an enclosed space above or below the bulkhead deck used for the carriage of motor vehicles with fuel for their own propulsion in their own tanks and to which passengers have access. Obvious examples are the garage spaces in ro-ro passenger ferries and vehicle decks in ro-ro cargo ships. Such spaces cannot have the normal main vertical fire zoning without interfering with the working of the ship.

Equivalent protection is provided in such spaces by ensuring that the horizontal and vertical boundaries of the space are treated as main fire zone divisions and an efficient fixed fire-extinguishing system is fitted within the space. This takes the form of a fixed-pressure water-spraying system, generally in association with an automatic fire detection system. Special scupper arrangements are provided to clear the deck of the water deposited by the system in the event of a fire to avoid a drastic reduction in stability.

Fire protection arrangements in high-speed craft

The IMO High-Speed Craft Code (HSC Code) recognizes the use of lightweight construction materials such as aluminum alloy and fiber-reinforced plastics that have a lesser fire rating than steel. Consequently, that Code has applied a new approach to fire insulation to that taken by SOLAS for conventional steel ships. The HSC Code has introduced the concept of a 'fire-restricting material', which may be a combustible material or combination of combustible materials provided it can comply with a prescribed fire test limiting heat release, smoke production, and spread of flame. Also introduced is the concept of a 'fire-resisting division' to prevent flame spread from one compartment to another within a prescribed time, which is related to a passenger evacuation time for the craft. A fire-resisting division can be constructed from noncombustible material, fire-resisting material, or a combination of both.

Further reading

International Code of Safety for High Speed Craft (HSC Code). IMO publication (IMO-1185E), 2000 edition.

SOLAS Consolidated Edition. Chapter II-2, Part B: Fire safety measures for passenger ships; Part C: Fire safety measures for cargo ships, IMO publication (IMO 110E), 2009.
Testing for fire safety in modern marine designs, *The Naval Architect*, October 2004.

Some useful websites

www.imo.org/home See 'Safety', then 'Fire protection, fire detection and fire extinction' for history of SOLAS fire protection requirements, summary of requirements, fire test requirements, relevant publications, etc.

Subject Index

Note: Page references followed by "f" indicate figure, and by "t" indicate table.

A

'A' brackets, 258–260, 261f
  construction of, 260
'A' class divisions, 372–373
A60 standard, 372
Aframax, 13
Aft end structure, 249–252, 258, 262
Aft peak, 251f, 259f
Aft peak bulkhead, 208–209, 213, 218
After end structure, 273
After perpendicular (AP), 11–12
Air pipes, 319, 368–369
Air-conditioning, 31, 126, 345–346
Alkyd resin, 334–337
'Alternative tonnages', 20
Aluminum
  alloy, 53–55, 56f, 65
  alloy sandwich panels, 57
  alloy tests, 65
  alloying elements, 55
  extrusions, 54, 97
  fiber-reinforced composites (FRCs), 58–59
  fire protection, 57
  high speed ferries, 54
  numeric designation, 55
  production, 54–57
  riveting, 55–57
  superstructure, 53–54, 57
Aluminum-to-steel connections, 332f
Amidships, 11–12
Anchor stowage, 248
Annealing, 48
Annual surveys, 41
Antifouling paints, 338–339
Antifouling systems, 337–339
A-O rating, 372–373
Assembly, 149–151, 153f
Assembly plate parts listing, 133f
Assembly plate parts nesting, 133f
Association of West European Shipbuilders, 361–362
Association of Western European Shipyards (AWES), 7–8
Atmospheric corrosion, 328, 330
Automatic arc welding, 88f
Automatic welding with cored wires, 86
Auxiliary steering gear, 256–258
Awning deck, 19–20, 19f

B

'B' class divisions, 373
'B-60' freeboards, 364
Backstep weld method, 107, 108f
Balanced rudders, 254
Ballast
  capacity, 24, 29, 177, 184, 214–216, 320
  dirty system, 316
  pumping and piping arrangements, 315–318
Bar keel, 175, 176f
Barge-carrying ships, 21
Base line, 12–13
Bending stress, 68–73
Beam
  extreme, 12–13
  knees, 138–140
  molded, 12–13
Bending moments, 20–21, 29, 31, 256, 297
  in seaway, 68
  wave, 70f
Bending stresses, 68–73
  ship as beam, 71

Bending stresses (*Continued*)
  strength deck, 71–73
Bilge, 126
  blocks, 162, 163f
  keel, 197–198, 198f
  piping arrangement, 315–318, 317f
  pumping, 315–318
  scuppers, 318
  suctions, 316
  wells, 180, 319
Bimetallic corrosion, 330–331, 330t
Bitumen, 334–337
Blast cleaning, 339–340
Block
  assembly, 152
  building, 162, 163f
Block coefficient correction, 365
Boiler bearers, 187
Boot topping region, 342
Bottom girders, 269
Bottom structure
  double, 177–184, 179f, 181f, 182f
  keels, 175–177, 176f
  machinery seats, 184–187
  single, 177, 178f
Bow
  doors, 307–308
  ramps, 308
  steering arrangements, 248
  thrust units, 248
Bracket floors, 180–183
Brackets, tank side, 192f, 193, 194f
Breadth, 5, 12–13
    see also Beam
'Breast hooks', 243, 245f
Bridge structures, 238
Brine, 346–348
  coolers, 349
  traps, 348–349, 350f
Brittle fracture, 75–76
Brittleness, 61
Buckling, 76–78
Building berths, 122–123
Building docks, 122–123, 163f, 170–171
Building hall, 154–155, 171
Building slipway, 162
Bulbous bows, 243–244, 245f
Buckling, 76–78
Bulk carriers, 13, 23–26, 25f
  bottom structure of, 184, 185f
  bulkhead stool, 213f
  single shell side block unit, 149f
  single side skin midship section, 200f, 201f
Bulk carrier single shell side block unit, 149f
Bulkhead stool, 210, 213f
Bulkheads, 207–213, 271
  watertight, 211f, 212f
Bulwarks, 235–237, 236f
  construction, 235–237
Buoyancy, 67–68
Butterfly' rig, 297–299
Butt welds, 190, 197–198
Butts, 189
    see also Butt welds

C

'C' class divisions, 373
Cabin module, 154f
Camber, 12–13
Canals, 5
Capesize bulk carriers, 293–294
Capesize ships, 13
Car carriers, 26, 27f
Cargo
  access, 307–308
  and ballast tanks, 342
  pumps, 321
  restraint, 312–314
  tank washing, 324–325
Cargo handling equipment, 23
Cargo lifting
    see also Shipboard cranes
Cargo ports, 368–369
Cargo ships
  dry, 17–23
  watertight bulkheads spacing, 208–209
Cargo tank
  protection, 321–324
  purging and gas freeing, 324
  ventilators, 321
  washing, 324–325
Cathodic protection, 333–334
Cavitation, 331
Centre line girder, 250f
Chain locker, 244–245
  construction of, 244–245, 246f
Chain pipes, 244–245, 246f

Charpy impact test, 63–65, 64f
Chemical additions (steel), 47
Chemical tankers, 275f, 276
Chlorinated rubber, 334–337
Classification societies, 6
  damage repairs, 43
  hull planned maintenance scheme, 43
  Lloyd's Register, 38–39
  periodical surveys, 41–43
  rules and regulations, 38
  ship operating in ice, 40
  structural design programs, 40–41
  tests for hull materials, 63–65
  weld tests, 114
Clean water ballast, 29, 272
Coal tar, 334–337
Coastal tanker, 266f
Code for the Construction and Equipment of Ships Carrying Dangerous Chemicals in Bulk (BCH Code), 276
Cold frame bending, 144
Collision bulkhead, 208–209, 213
Community of European Shipyards Associations (CESA) contract form, 7–8
Compensated gross tonnage (CGT), 361–362
Computer-aided design (CAD), 130–134
Computer-aided manufacturing (CAM), 130–134
Consumables, 114
Container
  guides, 313f
  lashing, 312
  ships, 15, 21, 22f, 202f
  stackers, 313f
Contracts, 7–8
Controllable pitch propellers, 260–262
Conveyor, 136–138
Corrosion, 331
  electrochemical nature of, 328–330
  due to immersion, 328
  nature and forms, 327–333
Corrosion allowance, 333
Corrosion cell, 329f
Corrosion control, 333–337
  cathodic protection, 333–334
  paints, 337
  protective coating, 334–337
Corrosion-inhibiting paints, 334–337
Corrosion-resistant steels, 50
Corrugated bulkheads, 210
Cranes, 120
Crew protection, 369
Cross ties, 164
Crude oil carrier, 320–321
Cruise ships, 31, 262
Cruiser stern, 250f
Curved panel, 151
Cutting, 98–101
  gas, 98
  gouging, 100
  laser, 100–101
  metal, 99f
  plasma-arc, 98–100
  water jet, 101
Cutting machines, plate profile, 138–140

D

Damage repairs, 43
Deadrise, 12–13
Deadweight, 12–13
Deck, 226–229
  beams, 105, 229
  cranes, 294
  construction, 230f
  girders, 220, 223, 226, 229, 269
  loads, 226, 229
  longitudinals, 269
  plating, 226–228
  sheathing, 228f
  stiffening, 229
  supports, 227f
  transverses, 269
Deck girders, 269
Deckhouses, *see* Superstructures and deckhouses
Deep tanks, 214–216
  construction, 216, 217f
  testing, 216
Depth correction, 365
Derrick rigs, 297–305, 299f
  forces in, 300–303, 302f
  initial tests and re-tests of, 304–305
Design
  concept, 3–4
  contract, 3–4
  one-off, 6–7

Design (*Continued*)
preliminary, 3–5
preparation of, 3–4
spiral, 3, 4f
Det Norske Veritas (DNV), 51
Dimensions, 5, 11–13, 14f
Direct line cargo piping arrangements, 322f
Displacement, 5
Docking surveys, 41–42
Docks
building, 122–123, 170–171
floating, 171–172
Doors
watertight, 21–23, 41, 213–214, 215f, 368–369
weathertight, 214, 240, 308, 318, 368–369
Double-bottom compartments, testing, 184
Double-hull oil tanker, 268f
Double-hull tanker, 28f
erection sequence for, 156f
Draft, 5, 14f
extreme, 12–13
molded, 12–13
Drag chains, 168, 170
Drilling machines, 140
Dry cargo bulk carrier, 156f
Duct keels, 177
Ductility, 61
Dye penetrant testing, 112

E

Economic criteria, 3–4
Egg box structure, 149–151
construction, 105
Elastic limit, 62
Elastomers, 51
Electric arc welding, 84–93, 85f
Electric furnaces, 46
Electric podded propulsors, 262–264
advantages of, 262
Electrochemical corrosion, 328–331
Electro-gas welding, 94
Electro-slag welding, 94, 95f
Elephant's foot type cargo lashing, 312, 313f
Elevators, 311f
End launches
arresting arrangements, 168–170, 169f
building slipway, 162
launching sequence, 168
launching ways and cradle, 164–165
lubricant, 165
releasing arrangements, 165–168
Engine seats, 186f
Entrance, 12–13
Epoxy resins, 334–337
Erection welding sequences, 110
Erosion, 331
Expansion trunks, 28f, 265, 370
Extreme beam, 12–13
Extreme depth, 12–13
Extreme dimensions, 12–13
Extreme draft, 12–13
Extrusion, 54–55

F

Fairing ship lines, 126–131, 132f
Fatigue failures, 76
Feeder ships, 15
Fiber-reinforced composites (FRCs), 58–59
Fillet welds, 105, 107
Fire doors, 374–375
Fire protection, 57
in cargo ships, 372
in high-speed craft, 375
Fire resistance, 373–375
Fire zone bulkheads, 374–375
Fitting out basin, 121, 123f
Fitting out berth, 121, 148–149
Flame planers, 140
Flare, 12–13
Flat panel, 151
Flat plate keel, 175, 176f, 189
Floating docks, 171–172
Floating production, storage, and offloading vessels (FPSOs), 274
Floating storage units (FSUs), 274
Floors, 180
Flush deck correction, 365
Flux-cored wires (FCAW), 86
Folding covers, 232–235
Fore end construction, 242f
Fore end structure, 243f, 272–273
deep tank, 272–273
forepeak, 273
ice strengthening, 273

Forecastle, 237, 241, 244–245
Forty-foot equivalent unit (FEU), 13
Forward perpendicular (FP), 11–12
Frame bending, 143–146, 145f
cold frame bending, 144
robotics, 144–146
section profilers, 144
Framing, 191–193
longitudinal, 193
transverse, 192–193, 192f
Freeboard, 12–13
Freeboard computation, 363–366
minimum freeboard, 366
timber freeboards, 366
Freeboard corrections, 365–366
Freeing arrangements, 370
Freeing ports, 235, 368–369
Friction stir welding, 96–97, 97f
Fully pressurized tanks, 282
Fully refrigerated tanks, 283

G

Galvanic corrosion, 330–331, 330t
Galvanic series, 330, 330t
Gangway and access, 370
Gap presses, 141f
Garboard strake, 176f, 178f
Gas
cutting, 98
shield arc welding, 89–93
welding, 82–84, 83f
Gas carriers
general arrangement, 287–288
Lloyd's classification, 288–289
Gas-freeing fans, 324
Gas-shielded arc welding process, 89–93
GAZ Transport membrane system, 283
Gearless carriers, *see* Capesize bulk carriers; Panamax carriers
General arrangement, 6
General cargo ship
erection sequence for, 155f
masts on, 294–295
midship section, 199f
General service pipes, 318–319
General service pumping, 318–319
Glass fiber-reinforced plastic (GRP), 58
Goal-based standards, 38
Gouging cutting, 100
Grade E steel, 268t
Green material, 148, 158
Greenfield, 120–121
Gross tonnage (GT), 359–360
Ground ways, 162, 165
Guillotines, 140
Gunwale, rounded, 190–191, 191t, 268t

H

Half beams, 229
Half block model, 128
Half breadth, 12–13
Half siding of keel, 12–13
Hand cleaning, 340
Handymax, 13
Handysize, 13
Hardness, 61
Hatch, 229–235
coamings, 232
covers, 232–235, 233f, 234f
opening, 231f
Hatchways, 271, 370
Hawse pipes, 247f, 248
Heating, *see* Heating, ventilation, and air-conditioning (HVAC)
Heating, ventilation, and air-conditioning (HVAC), 345
Heat-line bending, 142–143
Heat treatment of steels, 48
Heavy lifting, 300
see also Patent Stülken derricks
High speed craft, 204f
Higher tensile steels, 267–268
High-pressure water blasting, 340
High-speed craft, 31–33
fire protection in, 375
types, 32f
Hogging, 68
Hold ventilation, 347f
Horizontal girders, 216
Hovercraft, 18f, 31–33
Hull form, 6, 11–12, 23
Hull planned maintenance scheme, 43

I

Ice
classes, 40
strengthening, 195–197

IMO High-Speed Craft Code (HSC Code), 375
IMO International Gas Carrier Code, 280–282
independent tanks, 281
integral tanks, 280–281
internal insulation tanks, 281
membrane tanks, 281
secondary barrier protection, 282, 282t
semi-membrane tanks, 281
IMO length, 11–12
Impact tests, 63–65
'Impingement attack', 331
Impressed current antifouling systems, 337
Impressed current systems, 334, 335f
Independent tanks, 281
Independent Type A tanks, 283–286
Independent Type B tanks, 286
Inert gas system, 324
Inner bottom plating, 180
Insulation, 349, 350f, 372
A60 standard, 372
Integral tanks, 280–281
Intermediate surveys, 41
Internal deck access ramps, 309f
Internal insulation tanks, 281
International Air Pollution Prevention Certificate, 357
International Association of Classification Societies (IACS), 37–38, 357
International Certificate of Fitness for the Carriage of Dangerous Chemicals in Bulk, 276
International Code for the Construction and Equipment of Ships Carrying Dangerous Chemicals in Bulk (IBC Code), 276
cargo tank types, definition, 276
International Conference on Tonnage Measurement was convened by the IMO in 1969, 359
International Convention for the Prevention of Pollution from Ships (MARPOL), 324, 356
convention, 267, 271
tankers, 15, 29–30
International Convention for the Safety of Life at Sea (SOLAS), 1974, 356, 375
International Convention on Load Lines of Ships, 1966, 356
International Convention on the Control of Harmful Antifouling on Ships, 2001, 338
International Convention on Tonnage Measurement, 1969, 356
International Conventions for the Safety of Life at Sea, 371
International Labour Organization (ILO) Convention, 305
International Load Line Convention Certificate, 357
International Maritime Organization (IMO)
oil tanker categories, 15
organization of, 355
relationship with classification societies, 357
relationship with national authorities, 356–357
work of, 355–356
International Sewage Pollution Prevention Certificate, 357
International Tonnage Certificate, 357, 360–361
Inverse curve, 144
In-water surveys, 42
ISO hole, 313–314
Isomerized rubber, 334–337

J

'Jack-knifing', 303
Joining ship sections afloat, 158–159

K

Keel blocks, 162
Keel rake, 12–13
Keels, 175–177, 176f
Killed steels, 47, 86
Kort nozzle, 262
Kvaerner-Moss spherical tank, 287f

L

Laser cutting, 100–101
Laser welding, 95–96
Lashing points, 313f
Launching, 161–172, 166f
arresting arrangements, 168–170, 169f

building docks, 170–171
cradle, 164–165, 169f
declivity, 162, 164, 170
drag chains, 168, 169f, 170
end, 162–170
floating docks, 171–172
lubricant, 165
marine railways, 172
releasing arrangements, 165–168, 167f
sequence, 168
ship lifts, 171
side, 170
slewing arrangements, 167f
triggers, 165–168
ways and cradle, 164–165
Length
overall (LOA), 11–12
between perpendiculars (LBP), 11–12
Lighter aboard ship (LASH), 21
Lines plan, 126–128, 127f
Liquefied gas carriers, 284f
Liquefied natural gas (LNG), 280
Liquefied natural gas ships, 283–286
independent Type A tanks, 283–286
independent Type B tanks, 286
membrane tanks, 286
semi-membrane Type B tanks, 286
Liquefied petroleum gas (LPG), 279–280
Liquefied petroleum gas ships, 282–283
fully pressurized tanks, 282
fully refrigerated tanks, 283
semi-pressurized (or semi-refrigerated) tanks, 282–283
Liquid methane carrier, 285f
Lloyd's classification, 288–289
Lloyd's length, 11–12
Lloyd's Register (LR), 38–39, 49–50, 55, 76, 78, 110, 208, 229–232
classification symbols, 39
Load Line Convention, 357
Load line rules
freeboard assignment conditions, 366–370
freeboard computation, 363–366
Local stresses, 73–75
Loftwork, 128–130
scale lofting, 10:1, 130
Longitudinal bottom framing, 182f, 183
Longitudinal deck framing, 226, 269
Longitudinal shear forces, 68
Longitudinal side framing, 193
Lubricant, 165

M

MacGregor-Navire International AB 'Rotoloader', 311
MacGregor-Navire International AB 'Stackcell' system, 313–314
Machinery
casings, 370
positions, 20–21, 364, 366–369
seats, 184–187
space openings, 368–369
Magnetic particle testing, 112
Mangles, 137
Manual welding electrodes, 86, 87f
Manufacture of steels, 46–47
Marine Environmental Protection Committee (MEPC), 355–356
Marine pollution prevention, 355–356
Marine railways, 172
Maritime Administration (MARAD), USA, 7
Maritime Safety Committee (MSC), 355–356
Mast construction, 297
Masts, 294–297
Matrix assemblies, *see* Egg box structure
Mechanical planers, 140
Mechanical ventilation, 345
Mega-blocks, 158
Membrane systems, 288f
Membrane tanks, 281, 286
Merchant Shipping Act (1894), 30
Metal cutting, 99f
Metal inert gas (MIG) welding, 89–93, 91f
Mid-deck tanker, 29–30
Midship section
bulk carrier, 200f, 201f
cargo ship, 192, 199f
container ship, 202f
refrigerated cargo ship, 348f
ro-ro ship, 203f
coastal tankers, 266f
Mild steel, 267
Millscale, 137, 329f
Minor assembly, 151

Modern rudders, 254
Mold, 58–59
Mold loft, 130
Molded beam, 12–13
Molded depth, 12–13
Molded dimensions, 12–13
Molded draft, 12–13
Monitoring Ship Stresses at Sea, 78
Multi-product tankers, 321

N

Nesting plate, 133f
Net tonnage (NT), 360
Neutral axis, 69–71
New Panamax ships, 15
Nondestructive testing, 112–114
Normalizing, 48
Norwegian Shipbuilders Association and Norwegian Shipowners Association, 7
Notation, 39–40
Notch
  ductility, 75–76
  tough steel, 75–76
Numerical flame cutting control system, 139f

O

OECD, 361–362
Oil Pollution Act, 1990 (OPA 90), 29
Oil tankers, 13, 15, 26–30, 28f
  longitudinal framing of, 270f
Oiltight hatchways, 271
Oleo-resinous, 334–337
Open floors, 175, 177
Open hearth process, 46
Open shelter deck, 20
Open water stern, 23
Outfit modules, 152–154
Oxyacetylene flame, 340
Oxyacetylene, 82–84, 98
Oxygen process, 47

P

Paint systems on ships, 341–342
  below waterline, 342
  superstructures, 342
  waterline or boot topping region, 342
Painting ships, 339–342
  cargo and ballast tanks, 342
  paint systems on ships, 341–342
  surface preparation, 339–340
  temporary protection during building, 341
Paints, 334–337
  corrosion protection by, 337
    see also Protective coatings
Panama Canal Authority, 361
Panama Canal limits, 15
Panama Canal tonnages, 361
Panamax, 13, 15
Panamax carriers, 293–294
Panamax ships, 15
Panel assemblies, 122f, 133, 151–152
Panting, 73–74
  additional stiffening for, 195
  arrangements forward, 196f
Parallel middle body, 12–13
Passenger ship fire divisions, 374f, 374t
Passenger Ship Safety Certificate, 356–357
Passenger ships, 30–33
  superstructures, 238–240
  watertight bulkhead spacing, 209
Patent Stülken derricks, 300
Payment schedule, 8
Periodical surveys, 41–43
Photogrammetry, 158
Perpendicular after, 11–12, 14f, 126
Perpendicular forward, 11–12, 126
Pickling, 340
Pig iron, 45–47
Piggyback covers, 232–235
Pillars, 220–223, 221f
  construction, 220–223
  small, 223
  small solid, 222f
  spacing of hold, 220
Pipe module, 150f
Piping arrangements, 320–325
Planing machines, 140
Plasma welding, 93, 93f
Plasma-arc cutting, 98–100
Plate
  butts, 108–110, 189
  edge preparation, 106f
  handling in machine shops, 138
  preparation, 135–138
  profilers, 144
  rolls, 141
  seams, 108–110, 189

Plate preparation, 135–138
heating, 137
part preparation, 138–143
plate leveling rolls (mangles), 137
priming paint, 137
shot-blasting, 137
stockyard, 136
Polyurethane resins, 334–337
Pontoon covers, 232–235
Poop structure, 238
Poppets, 162, 164–165, 166f
Portable decks, 311–312
Post-Panamax ships, 15
Pounding, 74
Pounding region, additional stiffening in, 183–184
Powder cutting, 98
Prefabrication, 147–148
Pre-MARPOL tankers, 15
Presses, 140–141
Primer, 342
Priming paint, 137
Proof stress, 63
Propeller post, 252–254
Propellers, 260–262
controllable pitch propellers, 260–262
nozzle, 263f
shrouded propellers, 262
Propulsive performance, 5
Protective coatings, 334–337
Pumping and piping arrangements
in cargo ships, 315
in tankers, 320–325
Purchase, of new vessel, 6–7

Q

Quarter access ramps, 310f
Queen Mary 2, 249, 264
Quenching, 48

R

Rabbet, 128, 252–254
Racking, 73
Radiographic testing, 110, 112, 114
Raised quarter deck, 17–19
Rake
of keel, 12–13
of stem, 12–13
Ramps, 308–309
'Reefer ships', 349
Refrigerated cargo ship, 348f
Refrigerated cargo stowage, 346
Refrigerated container ships, 349–351
Refrigeration, 346–348
Register length, 11–12
Rimmed steels, 47, 86
Ring main cargo piping arrangements, 323f
Rise of floor, 12–13, 164
Riverside layout, traditional, 119f
Riveting aluminum, 55–57
Robotics, 107, 144–146
Roll-on roll-off (ro-ro) vessels, 21–23, 22f, 203f, 307
stern ramps in, 308
Rudders, 254–256, 255f
bearing, 256, 257f
construction, 254
pintles, 254–256
stock, 256
trunk, 256
Run, 12–13

S

Sacrificial anode systems, 333
Safe working load (SWL), 294
Safety convention, 355–357
Sagging, 68
Sampson posts, 294–297
Scale lofting, 10:1, 130
Scantling, 19–20, 229
Scissor lift, 312
Scrieve board, 229
Scuppers, 318
inlets, and discharges, 368–369
Sea inlets, 319–320, 320f
Seams, 189
Secondary barrier protection, 282, 282t
Section machining, 123f, 138–143
Section preparation, 135–138
Section profilers, 144
Segregated ballast tanks (SBTs), 267
Self-polishing copolymer (SPC) antifouling paints, 338
Semi-membrane tanks, 281
Semi-membrane Type B tanks, 286
Semi-pressurized tanks, 282–283
Semi-refrigerated tanks, 282–283
Service speed, 6

Shaft bossing, 258–260, 261f
  construction of, 260
Shaft tunnel, 218, 219f
  construction, 218
Sheaves, 300–303
Sheer, 12–13
Sheer correction, 365
Sheerstrake, 190
Shell
  butts, 108–110, 189
  expansion, 128, 129f
  forming, 142f
  plating, 189–204, 190f
  bilge keel, 197–198, 198f
  bottom, 189–190
  framing, 191–193
  grades of steel for, 191
  local strengthening of, 195–197
  seams, 108–110, 189
  side, 190–191
  tank side brackets, 193, 194f
Shell plating, three-dimensional representations, 128
Shelter deck, 19–20
Ship
  arresting arrangements, 168–170, 169f
  as beam, 71
  building process, 122f
  classification, operating in ice, 40
  design, 3–5
  drawing office, 125–128
  lifts, 171
  product model, 131–134, 132f
  releasing arrangements, 165–168, 167f
  spiral, 4f
  stresses, 78
  structure assembly, 147–159
  types, 18f
Ship drawing office, 126–128
  lines plan, 126–128, 127f
  loftwork followed, 128–130
  shell expansion, 128, 129f
  shell plating, three-dimensional representations, 128
Ship lifts, 171
Ship openings, 368–369
Ship product model, 132f
Ship types, 364t
Shipboard cranes, 293–294
Shipbuilders Association of Japan (JSA), 7, 361–362
Shipbuilding process, 122f, 135
Shipyard layout, 119–123, 123f
  modern large, 121f
Shipyard planning, decisions involved, 121
Shipyard replanning, decisions involved, 121
Shot-blasting, 137, 340
Shrouded propellers, 262
Side
  doors, 309–311
  girders, 183–184
  launching, 170
  loaders, 309–311
  scuttles, 368–369
Single bottom structure, 177
  construction, 166f
Single product carrier, 320–321
Single pull covers, 232–235
Single-hull bulk carrier, block erection for, 157f
Slag-shielded processes, 84–89
Slewing ramp, 309, 310f
Sliding ways, 164–165, 170
Slipways, 119–120, 162
Small waterplane area, twin hull craft (SWATH), 18f, 31–33, 32f
Solid plate floors, 180–184, 269
Sounding pipes, 319
Spacing of watertight bulkheads:
  cargo ships, 208–209
  passenger ships, 209
Spar deck, 19–20, 19f
Special category space, protection of, 375
Special surveys, 42–43
Special trade passenger (STP) ships, 31
'Spectacle frame', 260
Spurling pipes, 244–245
Stability, 5
Standard fire test, 373
Steel
  castings, 52
  chemical addition to, 47
  construction, 372
  corrosion-resistant, 50
  forgings, 52
  grade, 46

heat treatment of, 48
high tensile, 50
Lloyd's requirement for mild, 191t
manufacture, 46–47
sandwich panels, 50–51
sections, 48, 49f
shipbuilding, 49
Steering gear, 256–258
Stem rake, 12–13
Stern, 241–243
construction, 252
doors, 307–308, 308f
frame, 252–254, 253f
ramps, 308f, 310f
tube, 258, 259f
Stiffening, 297
in mast and heavy derricks, 298f
Stockyard, 136
Strain, 61–62
Strength deck, 71–73
Stress, 61–62
corrosion, 331
relieving, 48
and strain relationship, 104f
Structural design programs, 40–41
Structural fire protection, 371–375
requirement, 371–372
special category space, 375
see also Fire protection
Stud welding, 88–89, 90f
Subassembly, 151
Submerged arc welding, 86–88
Submerged turret loading (STL) system, 274
Suez Canal Authority, 361
Suez Canal limits, 16
Suez Canal tonnages, 361
Suezmax, 13
Superstructure correction, 365
Superstructures, 273–274
Superstructures and deckhouses, 237–240
bridge structures, 238
effective, 238, 239f
forecastle, 237
passenger ship, 238–240
poop structure, 238
weathertight doors, 240
Surface effect ships (SESs), 31–33
Surface preparation for paint, 339–340
Syncrolift, 171

T

Tack weld, 105
Tank
cleaning, 30, 271
side brackets, 193, 194f
top, 177, 180, 193, 210, 223, 349
Tank spaces, construction in, 268–269
bottom girders, 269
deck girders, 269
floors and transverses, 269
longitudinal framing, 269
Tanker construction materials, 267–268
higher tensile steel, 267–268
mild steel, 267
Tankers
cargo piping arrangements in, 320–325, 322f
cargo pumping arrangements in, 320–325
Tempering, 48
Template drawing, 130
Tensile strength, 61, 63
Tensile test, 63
Testing
deep tanks, 216
derrick rigs, 304–305
double-bottom compartments, 184
material, 61–65
rudder, 254
tanks, 272
watertight bulkheads, 213
Thermit welding, 96
Three island type, 17, 19–20
Timber freeboard, 366
Toilet module, 154f
Tonnage, 12–13, 20, 356–357, 359–360
compensated gross tonnage (CGT), 361–362
gross tonnage, 359–360
measurement for, 360–361
net tonnage, 360
Topside tanks, 218
Torsion, 73, 74f
Toughness, 61
Transom stern, 251f
Transverse framing, 269

Transverse stresses, 73
Transverse webs, 193, 195, 268–269, 268f, 273
Transversely framed double bottom, 180–183
Trial speed, 6
Tributyltin compounds (TBTs), 338
Triggers, 165–168
Tumblehome, 12–13
Tungsten inert gas welding, 89
Turret deck, 24
Tween deck height, 12–13
Twenty-foot equivalent unit (TEU), 13
'Two-pack paints', 334–337
Type 'A' ships, 363–365
  special conditions of assignment, 370
Type 'B' ships, 363–365

U

Ultimate strength, 61, 63
Ultra-large container ships, 15
Ultra-large crude carrier (ULCC), 13
Ultrasonic inspection, 114
Unbalanced rudders, 254
Union purchase rig, 303
  forces in, 304f
Unit:
  assembly, 151–152
  erection, 154–158, 155f, 156f, 157f

V

Vehicle lashing, 312
Ventilation, 345–346
  fire damper, 374f
Ventilators, 368–369
Vertical shear and longitudinal bending, in still water, 67–68, 69f
Vertical stiffeners, 210
Very large crude carrier (VLCC), 13
Vinyl resins, 334–337
Virtual reality, 134
Visual inspection of welds, 112

W

Water blasting, 340
Water jet cutting, 101
Watertight bulkheads:
  construction, 209–210
  corrugated, 212f
  plain, 211f
  spacing, 208–209
  testing, 213
Watertight doors, 213–214, 215f
Wave bending moments, 68, 70f
Wave piercer, 18f, 31–33
Weather deck, 26, 226–229, 235, 346, 370
Weathertight doors, 240
Weathertight vehicle ramp, 307–308
Welding
  automation, 105–107
  classification society tests and, 114
  distortion, 107
  electric arc, 84–93, 85f
  electro-gas, 94
  electro-slag, 94, 95f
  and faults, 110, 111f
  flux and, 82–88, 94
  friction stir, 96–97, 97f
  gas, 82–84, 83f
  and inspection, 113f
  laser, 95–96
  and nondestructive testing, 112–114
  practice, 103–105
  sequences, 107–110, 108f
  and testing, 110
  thermit, 96
Wing in ground effect craft (WIG), 18f
Wire model, 131
Wood ceiling, 177, 180

Y

Yield point, 63

Z

Zinc-rich paints, 334–337

Printed in the United States
By Bookmasters